Wilhelm Reich and the Cold War

The True Story of How a
Communist Spy Team, Government
Hoodlums and Sick Psychiatrists
Destroyed Sexual Science and
Cosmic Life Energy Discoveries

by James Edward Martin

Second Edition, Revised & Updated

Natural Energy Works
Ashland, Oregon, USA
2014

Publication and worldwide distribution rights:

Natural Energy Works
PO Box 1148
Ashland, Oregon 97520
United States of America
http://www.naturalenergyworks.net

E-mail: info@naturalenergyworks.net
Also available through Lightning Source / Ingram Distribution

WILHELM REICH AND THE COLD WAR
Copyright © 2000, 2014 by James E. Martin. All Rights Reserved. Printed in the United States of America. No part of this book may be used or reproduced in any manner whatsoever without written permission, except in the case "fair use" quotations embodied in critical articles or reviews, with proper citation. For information, send inquiries to Natural Energy Works.

ISBN: 978-0-9802316-8-7, 0-9802316-8-X
Second Edition, 2014

Library of Congress Cataloging-in-Publication data
Martin, James E.
Wilhelm Reich and the Cold War
1. Reich, Wilhelm, (1897-1957) — Biography.
2. Cold War History — Russian espionage.

140505

All photo images have been identified as to their sources to the best extent humanly possible. In most cases, unattributed photos come from old archives where photos and newspaper clippings were found, but without identifications. In those cases, when so informed of proper sources we will be happy to either add the credits or remove the item from subsequent printings. The maps on page xii were prepared courtesy of James DeMeo. Cover design by James Martin. Front cover photo of Reich courtesy of the Wilhelm Reich Museum and Trust (www.wilhelmreichtrust.org).

Wilhelm Reich
and the Cold War

Table of Contents   Page

Acknowledgements ix

Preface to the Second Edition xi

Maps of Reich's Travels xii

Introduction 1
    Topic of book; scope of research.

I. Biographical Sketch of Wilhelm Reich (1897-1957) 7
    A brief overview of Reich's life.

II. The Org: a New Beginning for Professional
    Revolutionaries 19
    Providing an historical context for Reich's political psychology and exposure to the Communist Party prior to his emigration to the US; discussion of the emigré contribution to the Allied war effort.

III. Doctor and General Eitingon:
    The "Sword" and the "Shield" 83
    New evidence regarding the relationship between the Soviet Union and Psychoanalysis.

IV. Wilhelm Reich and the Cambridge Five 99
    Introducing Arnold Deutsch, Reich's colleague who secretly joined the NKVD and became the most important Soviet agent ever. Dialogue with Michael Straight concerning his role in Cold War history.

Table of Contents (continued)   Page

Photo Section 169

V. The Secret Life of Mildred Edie Brady 207
    Brady, who launched the FDA's investigation of Reich in 1947, had many connections to Soviet intelligence.

VI. The Einstein Affair 253
    Reich's meeting with Albert Einstein to explore the orgone accumulator.

VII. The Final Years 265
    Reich's cloudbuster and his expedition to Tucson, Arizona, where the government was experimenting with weather modification and top scientists observed UFOs in the desert sky. Trial, prison and death.

VIII. Documentary Appendix 317
    Secret files verbatim

IX. Chronology 339
    Timeline of historical events.

X. Bibliography 395

XI. Index 405

Abbreviations used in this book

AEC	Atomic Energy Commission
Amtorg	American-Soviet Trading Corporation, New York
Cheka	Soviet Security and Intelligence Service (1918-1922)
CIA	Central Intelligence Agency
COI	Coordinator of Information (forerunner of OSS)
CPUSA	Communist Party of the United States
CPGB	Communist Party of Great Britain
DOE	Department of Energy
FBI	Federal Bureau of Investigation
FDA	US Food and Drug Administration
FOIA	Freedom of Information Act
GRU	Soviet military intelligence
IAP	Institute of Atmospheric Physics, Tucson, Arizona
IPA	International Psychoanalytic Association
KGB	Soviet Security and Intelligence Service (1954-1991)
KPD	German Communist Party
KPO	Communist Party - Opposition
KPÖ	Austrian Communist Party
MASCH	Marxist Worker's Evening School
MI5	British security service for counterintelligence
MI6	British Secret Intelligence Service (also known as SIS)
Mosfurco	Moscow Fur Company, Inc.
NKVD	Soviet State Security Police and Foreign Intelligence Service (1922-3, 1934-43)
NSA	National Security Agency
OGPU	Soviet Political Intelligence, a predecessor to the KGB
ONI	Office of Naval Intelligence
Org	Leninist Organization, New Beginning, the O, *Neu Beginnen*
OSS	Office of Strategic Services
SAJ	Socialist Workers' Youth
SAP	Socialist Workers' Party
Sexpol	Sex-politics, Sex-political organization
SIS	British Secret Intelligence Service (see MI6)
SPD	Social-Democrat Party of Germany
Sopade	Social-Democrat Party in exile
SPÖ	Austrian Social-Democrat Party

Acknowledgments

The author thanks Beth Benson, Emlie Mazepa, Larry Mazepa, Dr. Eva Reich, Dr. Lore Reich, Ilse Ollendorff, Kenn Thomas, Phyllis McCullough, John Wilder, Peter Nasselstein, Dave Huisken, Thierrie Cook, Dr. Stefan Müschenich, Jon East, Bernd Laska, Joe Blankenship, Walter Bowart, Harvey "Job" Matusow, Desiree Eden, Lois Wyvell, Sam Doust, Ed Gehrman, Matt Furey, William Sellers, James DeMeo, Bernard Grad, Dr. Stephen Nagy, Myron and Giselle Sharaf, Joan Brady, Jerome Greenfield, Steve Dickison, Marilyn Bacon, and Peter Robbins. The following archival resources provided important assistance to this book: Cornell University, and Michael Whitney Straight, for his permission to inspect his papers there; The Federal Bureau of Investigation, for the Mildred Edie Brady files; The Hoover Institution, for the papers of Karl Frank; Institute of Atmospheric Physics, Tucson, Arizona, for the time-lapse films, and for James E. McDonald's news clip file; Phyllis McCullough, for the papers of Robert C. McCullough; Mudd Library, Princeton, for the papers of Arthur Garfield Hays; Northeast Historical Film Archive, Maine, for the films of Eisenhower in Rangeley; The Orgone Biophysical Research Laboratory, for exhaustive archival support; Rutgers University, New Jersey, for its collection of the Consumers' Research archives; University of Arizona, Tucson, Special Collections, for Lew Douglas' papers; the Western Historical Manuscript Collection, University of Missouri-St. Louis, Thomas Jefferson Library; and the Leo Baeck Institute. Finally, but most importantly, the author gratefully thanks the subscribers to this edition without whom this book would have never been realized.

Preface to the Second Edition

The first edition of this book was published in 2000 in a very limited edition and at that time it was my intention to re-issue the book shortly thereafter as a paperback. These plans were delayed by the simple fact that I was so drained by the research and the subject material that it was impossible to pick it up again for a long time.

With the encouragement of Dr. James DeMeo of Natural Energy Works, this edition fulfills the original plan for an accessible edition. While a number of new books published over the last fifteen years have touched on this subject matter, nothing has been published that contradicts the essential theme of this book: most of Wilhelm Reich's troubles with the US government at the end of his life were instigated by Communists who viewed his work as a serious threat to their efforts to promulgate a Stalinist ideology and to influence world events.

While no contradictory information has been published since the first edition of this book, much confirming evidence has been established. Some of this information has been included in this edition, including the telling "Document 20" in which Comintern agents in Russia compiled a list of German exiles to be arrested or "liquidated" as enemies of Stalinism. Wilhelm Reich's name is included in this 1936 hit-list in connection with a German associate who was arrested and spent five years in the gulags, primarily for his connection to Reich.

Some writers continue to argue that Wilhelm Reich was a victim of "McCarthyism" or the alleged "right-wing, conservative" tenor of America in the 1950s. It is true that a few lone conservative individuals took issue with the content of Reich's writing, but nothing came of these efforts and they had no connection to each other. Meanwhile, the leftists who sprang to Stalin's defense had strong connections to each other through the international Communist web and fit perfectly into Reich's description of a "Red Thread of Conspiracy."

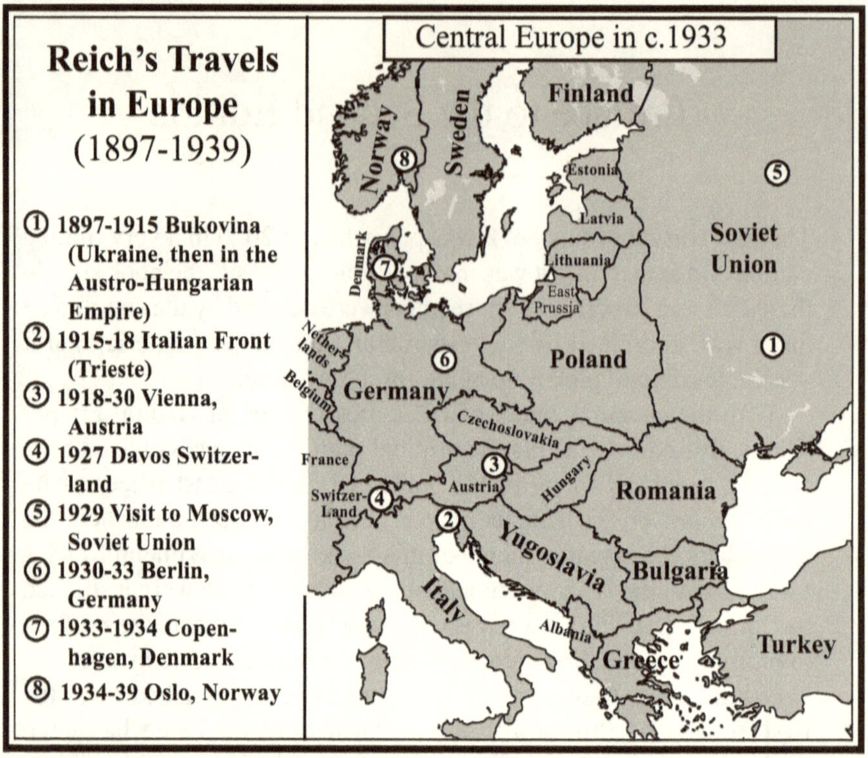
Reich's Travels in Europe
(1897-1939)
① 1897-1915 Bukovina (Ukraine, then in the Austro-Hungarian Empire)
② 1915-18 Italian Front (Trieste)
③ 1918-30 Vienna, Austria
④ 1927 Davos Switzerland
⑤ 1929 Visit to Moscow, Soviet Union
⑥ 1930-33 Berlin, Germany
⑦ 1933-1934 Copenhagen, Denmark
⑧ 1934-39 Oslo, Norway
Central Europe in c.1933
Finland
Norway
Sweden
Estonia
Latvia
Lithuania
East Prussia
Soviet Union
Denmark
Netherlands
Belgium
Germany
Poland
Czechoslovakia
France
Austria
Hungary
Romania
Switzer-Land
Yugoslavia
Bulgaria
Italy
Albania
Greece
Turkey

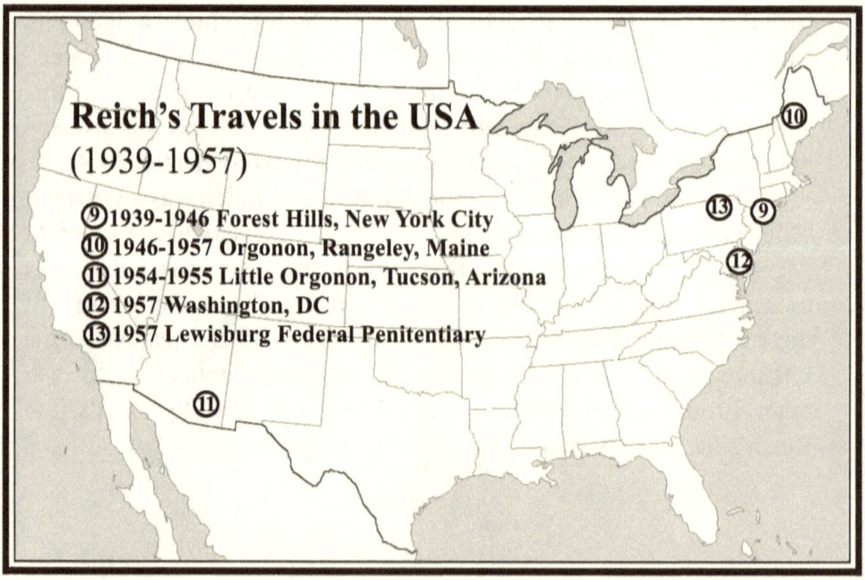
Reich's Travels in the USA
(1939-1957)
⑨ 1939-1946 Forest Hills, New York City
⑩ 1946-1957 Orgonon, Rangeley, Maine
⑪ 1954-1955 Little Orgonon, Tucson, Arizona
⑫ 1957 Washington, DC
⑬ 1957 Lewisburg Federal Penitentiary

Introduction

I. Topic Of Book

In the absence of any scientific refutation of Wilhelm Reich's experimental work, I continue to be outraged at the miscarriage of justice that was perpetrated on him. He died in a federal prison in Pennsylvania in 1957, and his passing was hardly noted at the time. I have never been satisfied with the standard-issue explanations for the chain of events that led to his death, and I continue to find valuable insight in his published books. How did this happen? The government certainly went to a great deal of trouble, and considerable expense, to neutralize a "madman."

There is a short story of Wilhelm Reich's life that is as well-known as it false: the great disciple of Sigmund Freud went crazy, discovered life in a test tube, biological energy radiating in a dark room, and the orgone "box" that supercharged the orgasm. This book is the product of my research into the controversial work of Wilhelm Reich. In order to understand Reich's "American period," I had to investigate his life before he emigrated from a hostile Europe to New York in the US. I began by reading *The Mass Psychology of Fascism*, a book that enlightened me about "political psychology." That's a term that Reich may have coined to describe his observations of the powerful appeal the Nazis commanded among the working people of Germany, and how this appeal could be used to subvert the people's own interests. Reich's writing spoke to me, and there are still quite a few people around the world who feel that way too.

I was shocked to learn that Reich went on to develop what appeared to me at that time as fantasies about energetic auras, rain-making tubes, and ideas of persecution at the hands of a Red Fascist conspiracy by the Communist Party. Yet, as I read Reich's later works, he described his experiments so simply that I was able to confirm the results of some of the basics for myself. I remain impressed with the accuracy of Reich's observations. At that point, I resolved to give Reich the benefit of the doubt about his descriptions and accounts of a political conspiracy against his work. I collected documentation from public sources that confirmed elements of the defense he tried to present in court.

In my research, I asked myself: What was the evidence for a Communist conspiracy against Wilhelm Reich?

I had no idea where all this would lead me, and I was perfectly prepared to discover that Reich was fantasizing about his scientific research as well as his immediate everyday life, if such were the case. I started picking up the loose ends from the fabric of Reich's story and pulled on the "red thread" that Reich so often mentioned as a metaphor for the continuity of his body of work.

I have been interested in history since I was a teen, and I love the way we can have wide-ranging historical discussions in America, on a populist level, under the guise of "conspiracy theory." Above all, I am committed to responsible freedom of expression, and what Reich called "the right to be wrong."

In Reich's case there is an opportunity to examine the records of both sides in the Cold War, since the fall of the Soviet Union has resulted in the release of numerous state secrets. If the Communists covertly harassed Reich, then the evidence should be publicly available now. This book is my attempt to sort through the evidence and present it in an objective way, while maintaining my contact with the tragic problem of secrecy and conspiracy as a social disease.

This book, then, is not a conspiracy theory, but a documentation of conspiracy as a social force. The elements of a conspiracy are simple and omnipresent. Literally a *breathing together*, conspiracy is a well-defined criminal act involving two or more conspirators make a secret agreement to conduct an illegal act, or a legal act in an unlawful manner. Conspiracy is the rule, not the exception.

Most "conspiracy theory" books miss the point, I believe, when they try to identify one ruling group of conspirators responsible for all malfeasance and social control. The evidence everywhere suggests that far more insidious to the social welfare may be concurrent, overlapping, interpenetrated, and cross-conflicted groups with no discernible goal other than *secret power*.

Reich himself had two working hypotheses about conspiracy. One was the idea of an *emotional conspiracy* that worked without secret agreements and depended upon the mass sexual suppression worldwide. The other conspiracy identified by Reich, an actual conspiracy of American agents of the USSR, who were dedicated to global political

Introduction

control and subversion. Both of these viewpoints are valuable. The more concrete description of the conspiracy was impossible to verify or disprove until recently, when the Soviet Union collapsed and for a time, even opened the archives of the most lethal and feared police state apparatus, the KGB.

What has emerged, like a stale miasma from the Pharaoh's tomb, has shocked and outraged the Western public. It will take years to digest the new revelations, especially when it involves collating the information with documents from the CIA and NSA, which have been slow in coming.

II. Scope Of Research

This book encompasses new revelations about Cold War espionage, traces of Reich's influence on the people in positions of social power, and Reich's 1954-1955 expedition to Tucson, alongside contemporary developments in America and Europe.

While I wrote *Wilhelm Reich and the Cold War* for the general reader, this isn't an introductory essay about Reich. I consider this an addition to Reich's biography, with attention to the role of some of his lesser-known colleagues and nemeses who shaped Cold War history. To those readers who are unfamiliar with Reich's books, this investigation might explain how and why the suppression of his work occurred.

The standard biography of Wilhelm Reich, *Fury on Earth* by Myron Sharaf, is a moving book, but it left a gap in the historical record. Not only was the essential story of Reich tragic in and of itself, but I was touched by Myron Sharaf's unsuccessful struggle to come to terms with the memory of his time with Reich. The biography was accurate, in the main, but Sharaf did very little research into the validity of Reich's claim that his work was being destroyed by a conspiracy of Communists, and maintained a disturbing ambiguity about science itself throughout the book.

Another significant area mishandled by Sharaf is Reich's expedition to Tucson, Arizona, in 1954 and 1955, a fascinating interlude between the federal injunction against interstate commerce of orgone accumulators and the contempt trial in 1956. In his last book to be published in his lifetime, *Contact With Space*, Reich described his efforts to break a stubborn drought in the desert. Reich's critics point to this book, with its steadfast descriptions of UFO sightings, atom blasts, high Geiger-counter readings, and weather modification, and roll their eyes.

Until now, however, nobody has ever attempted to confirm the basic details of Reich's account. For example, I was the first person to ever request Mildred Edie Brady's FBI files through the Freedom of Information Act (FOIA) even though Wilhelm Reich's FBI file is one of the most frequently requested files held at the agency.

I've spoken with Reich's daughter, Eva, on numerous occasions, and I came to value her as a friend. She was present throughout many of the events described in *Contact With Space*, and she confirms what she saw. Eva gave me a better understanding of the human scale of the injustice done to Wilhelm Reich, and I admire Eva for seeing things through as best she could. She has never shied away from expressing her criticisms of her father and his work, and she has likewise never recanted her testimony on his behalf. My work has been guided by her spirit of independent-minded inquiry, and her passionate demand for justice, openness, and accountability. I seek an official recognition of the facts regarding the case *Wilhelm Reich versus the USA,* and so did Eva Reich.

To my way of thinking, and I'm not a mental health professional, if Reich was fantasizing to the degree that he hallucinated objects in the sky, and imaginary contacts with Air Force intelligence, then similarly the more mundane details of the book ought to be fantastic and delusional. It should be easy to confirm. I began with an investigation of these details in Reich's account in *Contact With Space*. For example, Reich mentions almost in passing, that the director of the Tucson Savings and Loan, Lewis Douglas, was a highly-placed associate of President Eisenhower. Who was this Douglas? Did he, in fact, correspond with Reich? Did he, as Reich suggests, assist Reich's operation through his bank? An intriguing story revealed itself over time, as I peeled back the onionskin layers in long-forgotten public archives.

III. Attitude Of The Archivists

> *"One day the motives and legalistic maneuvers of the technical winner of today, the drug and cosmetic Hig* [hoodlums in government]*, will emerge from archives and see the clean light of day."*
> – From Wilhelm Reich's letter to Judge Sweeney,
>   May 25, 1956, at his sentencing hearing.

Introduction

I shared some of the material I had found with a number of fellow researchers, and they often encouraged me with new leads and traces of buried evidence. We shared an open, critical and still positive attitude toward Reich's research. In particular, I credit Peter Nasselstein with supplying me with important sources in German and allowing me to include information—heretofore unknown here in the US—in this book. By all rights, he and my other friends should have been credited with coauthoring this book but I bear the responsibility of choosing which sources are included, and possibly mangling them as well. I also relied heavily on David Huisken, Kenn Thomas, and John Wilder, who were available whenever I needed a sounding board and provided many details presented here. All four of these men have provided text for this book.

We suspected all superficial statements about Reich, and gave him credit for his honesty and integrity, at least to the extent that we sought either confirmation or rebuttal of his assertions. I had no premonition as to where these facts might lead, and I never imagined stumbling across so many webs of intrigue. At times, it's felt like sorting shards of pottery at an archaeological dig, at others, a foray into an unfamiliar terrain among speakers of a foreign tongue.

I can't say I have answered each and every remaining question about Reich's work and research, but I hope I've set the story in its historical context. I've learned a great deal from the mistakes of this rogues' gallery, and I sure hope I don't end up like this when today's secrets are revealed a century from now. As Ben Franklin put it: if you don't want to read it in the newspaper tomorrow, don't do it today. There really are no secrets.

Wilhelm Reich and the Cold War

Chapter One:
Wilhelm Reich:
A Biographical Sketch

"My dear little Lorli-girl!

"It was all just a bad dream that the little girl had dreamt!!! Her father hadn't disappeared at all. Instead he had never grown up but remained a young boy who wandered off into the big wide world to ask all the parents who were nasty to their little girls and boys and who scolded them and bawled them out to cease and desist. He thought a lot about the girl back home and sometimes wept a silent tear and looked forward to the day he would return to her."
– from Wilhelm Reich's letter to his daughter, Lore, 14 November 1935.[1]

That little boy who had never grown up was born in an old country called the Austro-Hungarian Empire. Like empires throughout history, this one was born in a war and ended in one (it lasted just over fifty years).

People spoke many different languages in the Empire. By the time Wilhelm Reich was born, the Austro-Hungarian Empire spread across a wide part of southeastern Europe, including Czechoslovakia, Galicia, Slovenia, Croatia, Bosnia-Herzegovina, and occupied wide areas that today belong to Italy, Poland, Ukraine and Rumania.

Reich's father, Leon, was a farmer-rancher who raised beef cattle for sale in the empire's big cities and to feed the German armies. Leon Reich considered himself as a part of the wider Germanic culture. Leon ignored religion, even though his forbears had been noted rabbis; he was a hard-driving cattleman looking forward to the new century.

Cäcilie and Leon Reich's first-born child, their son Wilhelm, was born on March 24, 1897 in the province of Galicia and soon his family moved to the area known as the Bukovina. This was out on the far southeastern frontier of the Empire, an area that now lies straddled across northeastern Rumania and southwestern Ukraine. This part of Eastern Europe had been repeatedly ruined by warfare and invasion over the past thousand years. The Huns and other nomadic tribes overran the Bukovina in the Third Century, AD. It was ceded to Austria by the Turkish Ottoman Empire in 1775. "Bukovina" means beechwood in Rumanian.

"Willi" Reich was born at home and grew up on his family's farm. People in the Bukovina knew Willi's father as a stern and powerful man, a modern-thinker for his time. He didn't permit Yiddish to be spoken in his home, and limited his children's exposure to ethnic clans in the region that included Ukrainians, Germans, Poles, Rumanians, Russians, Jews, Roma, "Huns" and the odd "Turk." He brought in Austrian tutors to educate Willi and his brother, Robert, science, mathematics and literature.

As a boy, Willi Reich loved observing animals on the farm and in nature in the wooded hills of the Bukovina. His tutor encouraged his interest in biology and Willi collected specimens and eventually constructed his own breeding laboratory. Like any child, Willi was keenly interested in the sexual activity of animals. On the farm, it's easy to have a more natural attitude about sex because breeding animals is just part of the job. Willi started having sex at a young age, eleven or twelve, with an older girl who worked on the farm.

Still sexually and emotionally immature, the experience wasn't completely satisfying for him. His curiosity about his mother's growing attraction to his tutor caused Willi to spy on her while she was with her lover and Leon was away on business. Willi would stay up late and listen to them.

Leon Reich was extremely jealous over his wife, and Willi was caught in a bind. Eventually he let slip word of his mother's affair to his father, and Leon flew into a rage. For months, from February through October, 1910, Leon subjected his wife to awful beatings and public denunciations. It was more than Willi's mother, Cäcilie, could bear, and she eventually committed suicide by drinking lye.

That too, was more than Leon Reich could bear. He had bought a life insurance policy and slowly tried to kill himself by intentionally weakening his constitution. He would stand in the cold rain in his shirt-sleeves and wade into the river in his boots while "fishing." Drowning

Biographical Sketch

himself was out of the question, for it would invalidate his insurance, so Leon could only contemplate the currents beneath the surface of the water.

Willi developed a severe case of psoriasis in January, 1913, and the medical treatment he received made it worse: it ended with X-rays and left him with a lifelong skin condition. He later described it as "reacting to the slightest irritation,"[2] and it could be that this condition, caused by a series of "treatments" that had burned off the surface of his skin, granted Reich an extraordinary sensitivity to atmospheric states later in his life. Like sunburned skin that feels the slightest cool breeze, Reich's body sensed the subtlest of stimuli. He would later write of the scientist's most important research instrument, his own organ sensations, and the importance of keeping this "tool clean."

Leon Reich died of tuberculosis in 1914. Willi was left to complete his studies and continue running his family farm at the age of seventeen. Once more, war came to the Bukovina. In his autobiographical writing, he would later remember his surprise when the invading Cossacks "could behave humanly and do me no harm." He noted that his naive belief in propaganda caused him problems throughout life. He "discovered that he had taken it more seriously than the propagandizer."[3] He tried to remain at his family estate, but the whole area became a battle zone between the Austrians and the Russians, and Wilhelm was forced to leave everything he knew and start out on his own. He left his home hidden in a horse cart.

He joined the Austrian Army, served with the cadre unit of the 80th Infantry Regiment in Hungary. Enrolled in officer's school, he later recalled that his drilling and training was "merely a logical continuation of the school situation." In 1916 he was promoted to full corporal and given command of a platoon in an infantry battalion and sent to war. Later he would remember how much the soldiers looked like robots, without wills of their own. He suffered another outbreak of his skin condition and was sent to the rear, then back to the front, to the rear, and then back to the front in April, 1918, for an offensive at the Isonzo River in Italy. Reich described the ghastly trench warfare in *Passion of Youth,* an autobiographical work published after his death. The Austro-Hungarian Empire was broken up after Austria, Germany and its allies were defeated. Even before the war was formally over, Reich was demobilized and went to the venerable capital, Vienna, to pursue a career.

Wilhelm Reich and the Cold War

He enrolled in the University of Vienna in August, 1918, the year of the Great Influenza plague that killed nearly twenty million people world-wide, and briefly studied law before taking up medicine. After his first term he began tutoring other students in the material.

As an indigent provincial, Reich had some dues to pay in those early Vienna days. At the beginning, he only survived with the help of his younger brother, Robert, who had preceded him to Vienna. Reich's predicament: visiting distant relatives for food, living in an unheated closet. Later, his more wealthy girlfriends brought him food. He spent the days in one of the city's cavernous cafes to stay warm, under the watch of a proprietor whose eyes invoked him for another cup of coffee, which had to be refused for lack of funds. He later recalled that he hardly knew he was "Jewish" before he came to Vienna.

Vienna could be a degrading place, full of tenement-style architecture placed next to classical feudal palaces. The old city is still defined by *Der Kreis,* the notable Ring configuration to the city, a series of concentric boulevards emanating from the center district, call to mind the layers of an onion skin. The ornate facades of the buildings masked miserable overcrowding and housing shortage.

Several years ago I had the opportunity to visit Vienna for the first time, and as I walked around the University of Vienna, located near the heart of the urban core and next to a balkanized working class district, it didn't look as though the cold, sterile environment had changed much in centuries. It was a cold and rainy day, and I stepped into a stone-blocked courtyard where I stood and listened to the echoes of a professor's voice droning in a classroom. I reached into my pocket and pulled out an Austrian bank note, and took a look at Sigmund Freud's stern visage engraved upon it.

Reich felt lonely in Vienna, even though he was quickly drawn into a circle of fellow ethnic Jewish intellectual young people. We will meet his friends and acquaintances in the next chapter. By dint of his tutoring other students, he soon became self-supporting and was able to complete his medical degree. Even before graduation, however, he was encouraged by Sigmund Freud to begin seeing psychiatric patients.

During the twenties, Reich lived and worked in Vienna, built his practice as a physician, and contributed to the establishment of psychoanalysis as an accepted academic discipline. He describes his own contributions to the field in each of his books, and few of Reich's critics have denied the influence of his work here. His books remain a

Biographical Sketch

trove of insight for mental health professionals.

What Freud saw in Reich should be evident from a reading of any of Reich's early psychoanalytic papers. These have been republished under the title *Early Writings*. Two of Reich's books from this period have been revised and translated into English, and they are still studied as psychotherapeutic guides: *Die Funktion des Orgasmus*, which was later published in English as *Genitality*, and also *Character Analysis*, a core text in psychoanalysis. They reach out beyond the narrow discipline of analysis and speak to issues that each human can relate to, but they are couched in a technical jargon that requires considerable decoding.

Few natural scientists have offered the range of practical and concrete discoveries on the scale that Reich did: he invented the psychoanalytic technique in distinction from Freud's discovery of psychoanalysis per se. In his twenties, he organized the first technical seminar in Freud's Viennese Psychoanalytic Society. Surprisingly, no such course had been offered, one which dealt specifically with the treatment of patients, and while the popular culture was soon to realize the validity of Freud's analysis, the treatment foundered on a vague tool kit: word association, hypnotic regression, and dream analysis. Within the context of that seminar Reich developed the technique of character analysis, a form of therapy that centered on the patients resistance to analysis, and the emotional defenses that blocked the way to progress.

Reich trained most of the young analysts who came to America in droves after the Anschluss. Even today, only two broad analytic approaches attract patients in any significant number: "body-work" therapies based on Reich's Character Analysis, and Jung's heirs of the Archetype. Had Reich packaged his Character Analysis in the manner of contemporary therapeutics (such as Lowen's Bioenergetics™, Primal Scream™, Radix™ and Rolfing™, to name only a few) he would have been an extremely wealthy man at age thirty.

Reich's personal life was shaped by the social mores of the times, as he was forced into an early marriage to Annie Pink after their sexual relationship became apparent to her relatives. The couple had two daughters, Eva (born in 1924) and Lore (born in 1928). The girls were raised according to "modern" ideas about childcare: scheduled feedings, hired nannies, and private education at local Montessori schools.

The period 1921-1928 was devoted to building his career as a psychoanalyst, establishing himself near the top of the professional hierarchy, endearing himself to Sigmund Freud, and working out the basic lines of his orgasm theory, which set the foundation for all his later work. Reich was the first to place sexual satisfaction at the center of mental health, opposed to the entire profession of psychoanalysts who cautiously authorized sex as a psychological phenomenon. This idea is so commonplace in the US today, that it is part of the folklore. Reich rested his scientific reputation on the primacy of sexual gratification as the keystone for mental health.

Around 1927, Wilhelm Reich joined his cause with that of the Austrian Communist Party. At the same time, he extended the working principles of the open mental health clinics in Vienna, seeking to join the revolutionary movement to well-established sexual hygiene. In organizations that as a group have become known as *Sexpol,* Reich and his comrades distributed sex-education and birth control among the workers of Austria, and after 1930, in Germany. This activity took place in the context of a wider social movement toward planned parenthood, but this movement was largely dominated by eugenic theories of race hygiene, to which Reich and other Sexpol activists were opposed. The Communists were at first glad to have Reich pulling in the crowds of young people who were thirsty for reliable information about sex, but the Party apparatus was repulsed by the basic message of Sexpol: the protection of a freer sexual existence for everyone.

Reich extended his scientific ideas during the years 1928-1936 through the discoveries of the respiratory block and the formation of biophysical character attitudes in muscular armor. He observed the way the organism conducted its "housekeeping" through sexuality, and termed this process *sex-economy,* or the "self-regulation of *primary* natural drives in their distinction from *secondary, perverted* drives."[4] During the early thirties, Reich's marriage to Annie failed and he began a passionate relationship with a dancer from the Berlin State Opera, Elsa Lindenberg. Although they never married, he considered her his second wife. She was a working-class Communist comrade.

By 1933, Reich was expelled from the Communist Party, but still searching for an adequate political response to fascism. He was expelled from the German Psychoanalytic Society that same year under murky circumstances that have never been fully explained by the psychoanalytic establishment. Reich viewed the split as the result of Freud's inability to accept Reich's *orgasm theory* and Reich's insistence that all neurotics suffer some type of genital disturbance. Reich sought a biological basis for the instinctual life of human beings. He spent the years 1934-1939 mostly in the Scandinavian countries of Denmark, Sweden, and Norway, while Europe slid inexorably into the conflagration. Despite the chaos, Reich succeeded in attracting several dozen professionals to his work, and this period is marked by his first experiments in biophysical research, beginning with the bioelectrical experiments that recorded the biological energy associated with pleasure and anxiety in 1935 and 1936. Using sensitive meters to measure the slight variations of the electrical potential at the surface of the skin, Reich demonstrated experimentally the energetic expansion and contraction of the whole body as it experiences different kinds of stimuli, showing that the expansion associated with, for example, a lingering kiss could be registered as a higher charge on the surface of the hands.

Reich's experiments continued with the microscopic work of the *bion experiments* that shed light on the old question of biogenesis. Studying living preparations of amoeba, Reich found that microscopic organisms followed the same lawful processes of expansion and contraction as did human beings. When he went to a the Botanical Institute in Oslo, asking to obtain some amoeba for his experiments, he was told to simply lay out some dead grass and soil in water on a windowsill, and the amoeba would appear, delivered by "air germs," or invisible spores of the protozoa. This sparked Reich's curiosity and he set about observing the process in living preparations under the microscope. At extremely high magnifications—Reich used to say that the natural scientist should spend as much on his microscope as he would on a new car—he observed the decay of soil and grass as a swelling and breakdown of the inorganic material into motile, pulsating vesicles that Reich called *bions*. These orgone energy vesicles, as he would later describe them, developed directly out of the decayed grass infusions, but more startlingly Reich was able to culture bions from inorganic matter such as iron filings and coal dust. The bions in turn spontaneously organized into protozoa. With the discovery of the bions in 1936, Reich hypothesized that cancer cells were caused by the bionous breakdown of animal tissue, and through extensive experiments with laboratory mice, he found that cancer tumors contained a rot bacteria that he named *T-bacilli*, for *Tod* (German for death).

In special bion preparations Reich heated beach sand to

incandescence and froze it, repeatedly, and then sealed it in a nutrient solution. These particular preparations—called "sandpaket" or SAPA bions—radiated visibly in a darkened room. At first, Reich was alarmed by the possibility that the radiation was harmful. He and his coworkers received full-body tans underneath their labcoats. They all felt a tingling sensation and warm skin. He was assured by a physicist that the radiation was non-nuclear. Soon thereafter, in 1939, Reich named this visible energy *orgone radiation*. Reich spent the rest of his life studying this new science called *orgonomy*, the study of life energy.

Just prior to the German invasion of Norway, Reich emigrated to the US with the help on Theodore Wolfe, MD, who would become Reich's American translator. The strain of the highly publicized campaign against Reich's work in Norway was more than Elsa Lindenberg could bear, and the couple separated when Reich moved to the US. In America, Reich became an admirer of Franklin Roosevelt and shook himself free of the European language of revolutionary socialism. He cut most, if not all, of his ties to his former "comrades." Instead of socialist bureaucracy and the politics of pandering, he articulated a vision of "work democracy" in his final days in Norway and developed the idea in the US. Decisions should be made by those who do the work, not by "intellectual dilettantes" bound to ideologies.

The period of 1939-1941 saw Reich getting settled in the new home, establishing a long marital partnership with Ilse Ollendorff (they had a son, Peter, in 1944), and inventing the *orgone accumulator*, which isolated and intensified the orgone energy. In 1941 Reich met with Albert Einstein to discuss orgone energy, and later that year Reich was arrested as an "enemy alien" and detained at Ellis Island for several weeks after the US declared war on Japan & Germany. Einstein was under surveillance by the FBI because of his well-known sympathy for the Soviet Union. Reich paid the price.

From that time until 1947, Reich enjoyed a period of calm, productive work developing his ideas. He experimented with the medical use of the orgone accumulator in cancer diseases, and developed an understanding of the cancer process as a *biopathy*, or what might be called traditionally as a "collapse of the immune system." He founded the Orgone Institute in 1942, and in the same year purchased a farm in Rangeley, Maine, which would become the home of his research facilities called "Orgonon." Reich invented the orgone energy field meter, which registered what has been called the human "aura,"

Biographical Sketch

in 1944. In 1945, Reich followed the bion experiments to their conclusion with Experiment XX, culturing plasmatic "flakes" from a particle-free solution; the preparations were filtered such that they were free of microscopic structures, yet the bions still appeared.

During this time his books, one by one, were translated into English and rewritten for an American audience. By 1946, Reich had breached the public sphere once again, as his books and ideas gained influence, especially among young people and elements of a new "bohemian" culture centered mainly in San Francisco and New York.

Then in the spring of 1947 a carefully planned attack by the career Stalinist propagandist, Mildred Edie Brady, set the Food and Drug Administration against Reich's very being. As a lobbyist with the Consumers Union, Brady had written much of the legislation that empowered that agency, and her opinion carried much weight with the FDA. This agency began an investigation of Reich, and were unsuccessful in obtaining evidence that Reich had been operating a "vice ring;" the FDA also targeted the medical use of the orgone accumulators but "no dissatisfied users were located." After ten years of harassment, coupled with a staggeringly corrupt repetition of Brady's articles in the leading *medical journals* of the day, including the *Bulletin of the Menninger Clinic* and the *Journal of the American Medical Association*, Reich found himself in a legal predicament of Kafkaesque proportions, and in his defense, Reich's canny insistence on setting the terms of the trial resulted in saving his work, even if they killed him.

In spite of the massive harassment, Reich remained productive throughout the Cold War years. He began work on a system of orgonometric equations to describe the laws of orgone energy, and during the period 1947-1951 Reich investigated the anti-nuclear effects of orgone energy. In 1951, he published the results of the *Oranur Experiment* that had alarming implications for the burgeoning "peaceful use of the atom" and the nuclear power industry. In 1954 the FDA applied for an injunction against Wilhelm Reich and Reich's former private attorney, Peter Mills, then the US Attorney in Maine, filed the complaint for an injunction in the federal courts.

Reich's written response, wherein he asserted his right to conduct basic research without interference from federal agencies, was completely ignored and the injunction was granted. The injunction not only ordered Reich to refrain from interstate commerce of the orgone accumulators, but also ordered that Reich and anyone associated with him "be perpetually enjoined from directly or indirectly doing or causing to be done any act whether oral, written, or otherwise [...] with respect to any orgone accumulator device" and so legal history was made. To obey the order, Reich would have had to cease *thinking* about orgone energy.

In spite of all this, Reich obeyed the basic terms of the injunction but continued working and brought his *cloudbuster*, a device designed to bring rain by working with the atmospheric orgone energy potentials to Tucson, Arizona, right under the noses of a well-funded government study on weather modification. From this period, Reich's book *Contact With Space* survives as a testimony to the reach of orgonomy. Around 1953 Reich became interested in UFOs, anti-gravity and space travel. He not only read the best literature on the subject of UFOs, but sighted them many times while operating his cloudbuster in Maine and Arizona.

In America, Reich harbored increasing doubts about the lasting value of individual therapy. He had long placed his emphasis on the social prevention of neurosis, and late in his life he studied newborns, infants and children with the goal of protecting their health from the start. He founded the Orgonomic Infant Research Center for prenatal care and fostering maternal bonding. He had always loved children, and his final testament bequeathed a major portion of his estate to his "little friends" who had taught him so much. (His estate is administered today by the Wilhelm Reich Infant Trust Fund, which operates his home in Maine as a museum and nature retreat.)

Reich's life-long concern with social madness, beginning with his ideas in the 1920s about "political psychology" and continuing up to his court case, which he pursued as an experiment in "social psychiatry," is an important binding tie throughout his work. All of the great works of American fiction touch upon this subject of social insanity, from Mark Twain's Huckleberry Finn down to the present day. Reich called it the "trap." Everyone is aware of the trap at all times, but few can find the exit. As Reich wrote in The Murder of Christ, "You see it, you feel it, you smell it, you eternally long for it, yet you can never, never get through the exit out of the trap. To get out of the trap simply has become an impossibility. It can only be had in dreams and in poems and in great music and paintings, but it is no longer in your motility. The keys to the exit are cemented into your own character armor and into the muscular rigidity of your body and soul." [5]

Biographical Sketch

Reich's final years, 1955-1957, were consumed with his defense against the citation of contempt obtained by the FDA after Reich's assistant, Michael Silvert, MD, had transported orgone accumulators across state lines. Had Silvert not done this, the court injunction obtained by the FDA would have been unenforceable, and Reich would not have been hampered in any way. Reich handled his own legal defense and tried to enter the overwhelming body of evidence that the original injunction had been obtained fraudulently, but the courts would not admit any of this evidence. He received the maximum penalty and was sent to a federal penitentiary in Lewisburg, Pennsylvania where he died of apparent heart failure shortly before he was due to have a parole hearing. His daughter, Eva, said at the time, "he died of a broken heart."

Reich forfeited his life so that the essence of his life's work might be preserved. At the time of his death, the prospects for this preservation might have seemed bleak, but over the forty years since his death, his books have been reissued, numerous groups throughout the world have been established along the lines of his research and ideas. As this book went to press, a new flurry of attacks on Reich's work appeared in mass-media such as *The New York Times, Time* magazine (which listed Reich as one of the century's "crackpots"), the Democrat Party's internet magazine *Salon,* as well as minor "paranormal" periodicals such as *Fortean Times.* The anxiety of these people becomes more palpable. Perhaps, as Reich predicted, we are approaching the "last barrier, in scientific terms, to the final break-down of the human armoring."

Notes

1. *Beyond Psychology,* p. 57.
2. *Passion of Youth,* p. 63.
3. *Passion of Youth,* p. 55.
4. Reich, Wilhelm. "Background and Scientific Development of Wilhelm Reich", from the *Bibliography on Orgonomy,* 1953, p. 5.
5. *The Murder of Christ,* (1955) p. 5.

Wilhelm Reich and the Cold War

Chapter Two:
The Org: a New Beginning for Professional Revolutionaries

"What is the hardest thing of all?
That which seems the easiest
For your eyes to see,
That which lies before your eyes"
     – Goethe

Wilhelm Reich had long-standing relationships with other Austrian and German exiles in America. These relationships were not set in stone, nor were they mutually beneficial. Reich knew the psychoanalysts who relocated to various spots in America: his ex-wife Annie Reich, Edith Jacobson and Sandor Rado settled in New York; Otto Fenichel and Siegfried Bernfeld went to Los Angeles. As documented elsewhere, these analysts worked to destroy Wilhelm Reich's reputation when they came to the US.[1] Reich also had known other German exiles in the literary and political worlds, including Arthur Koestler, who shared Reich's experience within a red cell in Berlin, Koestler's friend Manes Sperber, and Willi Schlamm, another former communist, who became a founder of William F. Buckley's influential *National Review* magazine for intellectual conservatives. Karl B. Frank and Emmy Rado worked for the US government to assist the war against Nazism, and Reich had known them well in Europe.

Understanding the drift of Reich's fellow European colleagues in exile demands an investigation into the political shapes of pre-Nazi Germany, the pressures of emigration and immigration, wartime diplomatic intrigue and international conspiracy, and the overlapping organizational, not to mention philosophical, allegiances of Reich and his radical European former associates.

I'll attempt to synthesize the German literature (with the help of Peter Nasselstein, who provided good references and translation for me) on Weimar-era radicalism with English histories about the postwar exiles. Examining Karl Frank's global conspiracy to reconstruct and reunite Germany, we will explore new historical grounds.

There are still a few people alive who remember those times and conditions, but they are rare and ancient. Erich Schmidt, who sprang from these roots and lives today, wrote to me when I inquired about Karl Frank. He begged he was too old to revisit what he called the "devilishly complex" story of Karl Frank and the Org.

Karl Frank and Mildred Edie Brady

Karl Borromäus Frank (1893-1969) is mentioned as a footnote in Myron Sharaf's biography of Reich, and appears nowhere in any other book about Wilhelm Reich. We fail to find his entry in any encyclopedia, German or English. He surfaces once in the documents Reich collected for his legal dossier, *Conspiracy: An Emotional Chain-Reaction,* in a letter from Frank, who assessed Mildred Edie Brady as a hard-line Stalinist. Frank told Reich that he had met Mildred Brady and her husband in 1936, and that Mildred made it obvious to him that she was an affirmed Communist. So, who was Karl Frank? What kind of relationship did he have with Reich? Why did Reich place any credibility in him? And why would Reich ask this man, of all people, about Brady's political sympathies?

The answer: Karl Frank had known Wilhelm Reich since 1919 in Vienna. Reich maintained friendly contacts with Frank up to the 1950s, long after each had split from the Communist Party. Frank's papers are held at the Hoover Institution, where I found several letters from Wilhelm and Ilse Reich to Frank that documented this mutual friendship. The OSS had employed Frank during the war. If Karl Frank could not determine the true political skew of an individual, then no one could.

How did Reich suspect that Mildred Edie Brady was a Communist, in the first instance? He inquired about her with Karl Frank only to confirm it. At first, he "smelled it." Brady's article dripped with signs and codes for the Party faithful. Reich noted her mention of the Stalinist periodical, *Science and Society,* and her demands for state intervention, should the American Psychiatric Association find itself unable to curb Reich's activities. Yet Reich's biographer, Myron Sharaf, argued that Reich's suspicions of a "Red Fascist Conspiracy" were founded on too little evidence:

The Org: a New Beginning

"The idea was growing in his mind that Brady was more than a fellow traveler; she was a Stalinist, and may well have been acting on direct instructions from the Communist Party. Brady's Stalinism became a firm conviction. The evidence about Brady was scant–the tone of her article, its appearance in *The New Republic* under Henry Wallace's editorship, and some hearsay on her politics. A friend of Reich's wrote him that in 1936 Brady was 'in sympathy' with the Communist Party, but later information was not available."[2]

It's startling that nobody among Reich's circle of American colleagues was able to remember anything about Mildred Edie Brady and her husband, Robert A. Brady. They each had been denounced in Congress in the early forties, and lost their government positions; the Bradys were relatively open Stalinists. Their exposure had been front-page news. Biographer Myron Sharaf painted a portrait of Reich leaping to conclusions. It is unclear to me why he was ignorant of the Bradys' public record, because the sources for this information were no more obscure than the Congressional Record. Those American colleagues around Reich, and upon whom he relied for information about the social scene in their native land, evidently didn't follow national events.

The quotation from Myron Sharaf's biography may have typified Reich's American colleagues' attitude concerning Reich's fears of a Communist hatchet-job to destroy him. "The idea was growing in his mind"—i.e., Reich's fantasy—"that Brady was more than a fellow traveler; she was a Stalinist, and may well have been acting on direct instructions from the Communist Party," Sharaf wrote.

I'll break Sharaf's speculation down to reality in a separate chapter on Mildred Brady. For now, let's focus on Reich's friend, unnamed in the text of *Fury on Earth,* but in a footnote, revealed to be one Karl B. Frank.

This nameless friend, to whom Reich turned for reliable information about the Bradys, was indeed Karl Borromaeus Frank (*nommes de guerre*: Willi Müller, Paul Hagen). Frank was the wartime leader of a disciplined, clandestine, socialist cadre of anti-Nazi activists called "Neu Beginnen" (in English, "New Beginning" or "To Begin Anew.")

Frank found the Bradys while touring California in search of support for European refugees and his organization's German opposition to

Hitler. He wrote, in reply to Reich's inquiry, on December 15th, 1947: "Professor Brady, her husband, gave me the impression of a well-meaning American liberal, impressed somewhat by 'leftist' doctrines. The lady of the house, Mildred, did not leave any doubt in her conversation that her communist sympathies were stronger than her husband's."

This was not "hearsay," as Sharaf wrote, but direct eyewitness testimony from an individual who was extremely close to Reich, who knew him in 1920 in Vienna, and also to Reich's wife Ilse Ollendorff, who addressed Frank in a letter in intimate terms (as friends and comrades)—using the informal "Du."[3] Furthermore, Frank was in an excellent position to parse sectarian political affiliations.

I've taken great pains to search out the biographical details of the persons in the shadows of Reich's story, and in the case of Karl Frank, this has been particularly productive. Through Frank's story, we can better understand the context in which Reich wrote his Sexpol polemics, and discover a whole generation of German exiles who had been influenced by Reich's ideas on political psychology. Far from dying off after Reich left politics behind, these "professional revolutionaries" played a pivotal role in the war, and then in the reconstruction of Europe, even in the formation of the COI, then OSS, and finally the CIA.

To use the analogy of a human body, the body politic of left-wing Europe in the early thirties contained a stomach (Marxist economics) and genitalia (Freudian psychoanalysis). Between the gut and groin lay a throbbing bag of intestines, an intertwined mass of peristaltic movement. Groups like Wilhelm Reich's Sexpol and Karl Frank's Org filled the void between economic determinism and psychological fantasy-analysis with an ambitious experiment in applied social psychiatry. Reich credited Siegfried Bernfeld with the first synthesis between Marxism and Psychoanalysis, and wrote that "Bernfeld began it in 1925, but he dropped it."[4]

"A brave revolutionary politician"

Reich knew Karl Frank when he was a student in psychology, biology, and philosophy at the University of Vienna. Frank, son of a small factory owner, began his university study in 1913. After a time in the artillery cadet school in Traiskirchen, he left. Despite his growing pacifism, he served in the Austrian army in 1914 and 1915. In 1916, he declared himself a conscientious objector, and was disbarred as sick. He returned to Vienna and completed his Ph.D. degree in psychology in 1918.

Frank, an "Aryan" Austrian, belonged to the almost exclusively "Jewish" youth movement circle gathered around Siegfried Bernfeld: Rudolf Dreikurs, the famous Eisler siblings,[5] Otto Fenichel, Paul Friedländer, Hilde Geiringer, Ernst Krenek, Paul Lazarsfeld.[6] In her "autobiographical notes," Ruth Fischer, one of the Eisler siblings, recalled the time, in 1913:

> "I joined, of course, the Freideutsche Jugendbewegung and was a leading member. A special inner secret circle was formed called 'The Hydra,' a society of eleven young people including me, my brother Gerhart, my husband, Dr. Paul Friedländer (he remained in France in the Vernet Camp), Dr. Siegfried Bernfeld (in this country Paul Hagen/Karl Frank) [sic], etc. The aim of this secret society was to change the world entirely, from the head to foot. If a member should drop out or die, immediately a new head, as in the Greek fable, should grow to replace the lost one."[7]

Paul Friedländer married Ruth Fischer in 1915.

In October of 1913, Siegfried Bernfeld was one of the main organizers of the Meissner Mountain rendezvous, a conference to unify the youth movement as a whole. Bernfeld was remembered as a indefatigable organizer of innumerable projects; Freud called him "an overwhelming speaker." Bernfeld was a star in this firmament of young Austrians. Karl Frank, Otto Fenichel, Annie and Wilhelm Reich where just a few of the people, among the youth movement, who would be drawn to psychoanalysis as a career.

As a close friend of Siegfried Bernfeld, Frank was a leading figure of the Viennese youth movement. These were the days of the Wandervögel, or "roaming youth," a time that bore similarities to the counterculture of the "back to the land" movement of the sixties and seventies in the US. Urbanized young people took to the rolling hills on long hikes, sang folk songs, and set the tone for romantic youth culture in the twentieth century. Even before the war, young people were seeking to define themselves as a social force in distinction to the Germanic establishment. Educational reform, socialism, Zionism, free schools, and free self-expression were placed on the public agenda. Sexuality played no small role in the movement, and the wooded area just south of Vienna—the Wienerwald—provided the kind of seclusion, privacy and freedom the kids couldn't find at home.

After the war, Frank was one of the founders of the Free Association of Socialist Students (*Freie Vereinigung Sozialistischer Studenten*).[8] In 1918 he was one of the founders of the "junior high school movement" (*Mittelschuelerbewegung*), and served as a delegate from the University of Vienna in the revolutionary workers' council of Vienna. He joined the Austrian Communist Party (KPÖ) in 1919, and was briefly a member of the editorial staff of the party organ, *Die Rote Fahne*. In December 1919, he joined the Central Committee of the KPÖ.

Men admired Frank's political courage, and women were melted by his movie-star beauty—short, wiry, and rugged, Frank had soft, imploring doe-eyes set in his careworn face. In his private diaries, parts of which were published posthumously as *Passion of Youth,* Reich recalled that in 1919 Karl Frank had separated from a young woman, a kindergarten teacher named Lore Kahn. "She was lively, clever and somewhat 'messed up' because she had no proper boyfriend," Reich wrote.[9] "She was in love with a brave revolutionary politician who no longer belonged to the romantic Youth Movement. She had attached herself to him and slept with him." She worked in Siegfried Bernfeld's experimental, exclusively Jewish children's home *Kinderheim Baumgarten,* which operated from October 1919, until the Spring of 1920, with nearly 300 children. Lore Kahn came to Reich for psychoanalysis in the winter of 1919-1920, specifically to work out her conflicts about her relationship with Karl Frank, who was leaving Vienna to serve in Berlin's new Comintern office.

Karl Retzlaw (1896-1979) remembered when, at the youth center Alte-Jakob-Strasse in Berlin around 1920, the young Karl Frank from Vienna read aloud to Communist youths from the works of Sigmund Freud and Auguste Forel. When he championed "free love," his enthusiasm had to be curbed and "we" (that is, the Communist activists around Karl Retzlaw) had to tell him that the youth center was not an appropriate place to speak on sexual matters. "We" wanted to remain political.[10] Walter Löwenheim, the future founder of the Org, worked with Frank in this same youth center in Berlin in 1920.

Back in Vienna, as Reich and his patient, Lore Kahn, sorted out her feelings for Frank, the newly-minted analyst must have learned a great deal about this "brave revolutionary politician," and Reich's evident admiration and respect for Karl Frank continued throughout his life.

At the conclusion of her analysis, Kahn told Reich she wanted a sexual relationship with him. At first Reich resisted, but she said she would wait. After a time, Reich surrendered to Lore's request. "After all," he wrote, "she was no longer a patient. And it was nobody's business. I loved her, and she grew very happy."[11] She wanted to have Reich's child, but they did not want to get married so young. Tragically, Lore Kahn died of sepsis in 1920. Reich believed that her illness was due to their nights in an unheated room, the only place the young couple could find to be together. Kahn's mother fantasized that Lore had died as a result of complications from an abortion. Sharaf had speculated that the rumors of an abortion were true, but when the trustee of Reich's estate published Reich's memoirs, a different picture emerged.[12] The entire abortion scenario was a fantasy conjured by Lore Kahn's distraught mother. Judging from his memoir, *Passion of Youth*, Lore was Reich's first true love.

Frank in Berlin

Karl Frank met his first wife, Alice Herdan[13], then eighteen years old, in Vienna during the spring of 1919. Soon Alice and "Karli" married and lived in the Ninth District. He went to Bavaria and participated in the abortive Bavarian Soviet. On orders from the Comintern, they relocated to Berlin, where Frank lived in the underground as "Willi Müller" because of his phony passport. His Austrian friends followed him to Berlin: Otto Fenichel in 1922, and Siegfried Bernfeld in 1925.

In Berlin, Karl Frank worked in the secret offices of the Western European Section of the Comintern and traveled widely—mainly to Moscow. The Berlin branch of the Comintern was a top priority, as the crucial first step in what the Bolsheviks, and Lenin and Trotsky in particular, predicted would be the inevitable victory of international Communism, sweeping across the world in five or ten years, and starting in Germany. In Berlin, at the age of twenty-three, Frank had every reason to consider himself as a professional revolutionary standing at the very forward point of the entire Communist movement. He answered directly to a man known only as "Comrade Thomas"—an old Bolshevik, a personal friend of Zinoviev, Karl Radek, Lenin and

Trotsky. "Comrade Thomas" had been sent to Germany by Lenin, with huge financial resources, to foment the inevitable revolution in the most highly developed country in Europe. We will return to "Comrade Thomas" later in this chapter. For now, we can establish that the entire operation was completely covert and had no overt connection with the German Communist Party.

It was a deadly serious mission. The Franks shared housing with Frank's close friend, Gerhart Eisler [1898-1968]. Frank's wife Alice described him in her memoir: "Gerhart Eisler was small and ugly. In later years he became a 'cleanser' [Aufräumer], this means he traveled into half-ripe Communist countries with the assignment to put an end to people who were in the Party's way."[14] His own sister, Ruth Fischer, called him "the perfect terrorist type."[15] She later accused her brother of carrying out purges in China, and with responsibility for the murder of Bukharin. She said, "In the inner circles of the Comintern, it is well known that Eisler denounced to the GPU many anti-Nazi refugees living in Moscow. He is responsible for the death of the German Communist Hugo Eberlein [..] and of Nikolai Bukharin."[16] Arthur Koestler, who knew all three Eisler siblings, Ruth, Gerhart, and Hanns (the composer), recalled in his book *The Invisible Writing* that Gerhart Eisler had solidified his standing in the Comintern when he went to China in 1929 "to liquidate the Party's revolt against Stalin's pact with Chiang Kai Shek (the Stalin-Chiang pact of 1927 was a forerunner of the Stalin-Hitler pact of 1939)."[17] "Until this time" Ruth recalled, "[Gerhart Eisler] was an ordinary Communist in the sense that he carried out political tasks and that he fought, in his way, for what he thought was the best policy in Germany. Then he was taken into the apparatus in Moscow in 1929. He was trained in the carrying out of terrorist orders. He became another man. He was transformed. He was conditioned by his participations in GPU activities in China. When he came from China he was another type of man from the man sent from Moscow in 1929."[18] Arthur Koestler and Gerhart Eisler were later interned in the same camp in France during the war. The US State Department secured Eisler's release and brought him to America.

In his youth, Gerhart Eisler had married Hedwig Tune, later known as Hede Massing. Eisler subsequently married Hede's younger sister. From 1933 until 1937 Hede Massing had been a Soviet spy in the US. She ended her secret Party work because of the Moscow trials, and in 1948 she was major witness against the disgraced former State Department official, Alger Hiss.

Gerhart Eisler had joined the KPÖ in 1918. Like Frank, he joined the German Communist Party (KPD) at the end of 1920. In the 1940s, Eisler (a.k.a. Hans Berger) became a celebrated cause for the American left when he was charged with being the boss of the Soviet's illegal networks by the House Un-American Activities Committee. Louis Budenz testified in Congress in 1946 and said, "This man never shows his face. Communist leaders never see him., but they follow his orders or suggestions implicitly." For a time Eisler loudly proclaimed his innocence, and claimed to be the victim of red-hunters. "Free Gerhart Eisler" rallies became a focal point for Party recruiters. He addressed the crowds in 1947: "There has never been a Communist Party in the world that would allow anyone, even a foreign communist, to run its business." He refused to testify before HUAC, whereupon Richard Nixon made his maiden speech as a congressman with a motion that Eisler be charged with contempt. While his case was being argued in the Supreme Court, Eisler jumped bail, fled to East Germany in 1947, and there took charge of the Office of Information.

The Eislers, Frank, Bernfeld, Fenichel, and Reich, all sprang from this same world-revolutionary milieu in Vienna.

Between 1920 and 1928, Karl Frank held leading positions both within the KPD and the KPÖ. After two and a half years in Germany, Karl and Alice had a daughter, Michaela Frank, who was born in Munich, where Karl was serving a two-month prison sentence for his part in the short-lived Bavarian Soviet Socialist Republic. For the next twenty-five years, Frank was either on the road, or in jail. In 1923 he was arrested in Bavaria, where the Kahr government had banned left-labor organizations. After serving his sentence in a fortress, Frank was expelled from Bavaria. He returned the following year, was arrested again, and this time was sentenced to six months. He staged a hunger strike lasting 21 days, but was able to escape from prison and return to Vienna. In 1925 Alice left Frank and married Carl Zuckmayer, the famous Austrian playwright. In 1926, Frank took a six-week trip to Moscow at the invitation of the Soviets.

In 1929, Frank made a final break with the Communists after being released from prison, this time for kidnapping a Socialist speaker and substituting a Communist member of the Reichstag at the public radio station in Berlin. Finding the Stalinization of the German Communist Party complete, he "prepared [a] statement formulating criticism of the undemocratic rule within the party, the 'splitting tactics' employed in the trade unions, and the dependence of the party leadership on the dogmatic rule of Moscow through the Communist International."[19]

In 1931, Frank reapplied for admission to the Austrian Social Democratic Party, and was admitted with the intervention of Otto Bauer, its leader. (Bauer's daughter was a patient of Freud's, and given the pseudonym Dora in Freud's paper on hysteria.)

In 1930, Siegfried Bernfeld married Elisabeth Neumann, who then had an affair with her old friend, the notoriously attractive Karl Frank. In the summer 1932, Bernfeld returned from Berlin to Vienna. The Bernfelds had divorced because of Frank but the couple remained friends.[20] Siegfried Bernfeld and Karl Frank remained friends, as well as political comrades. With the view of making a career in psychoanalysis, Frank decided to enter a training analysis at the Berlin Institute . Max Seydewitz reports that Frank was formally trained by a friend he had known in Vienna.[21]

During the Bernfelds' separation, Elisabeth underwent psychoanalysis with Jeanette Lampl de Groot. Karl Frank meanwhile went for analysis to Jeanette's husband, Hans Lampl.[22] When Otto Fenichel visited Vienna a few years later, he wrote, "Federn and the Lampls treated me a bit like a 'repentant sinner': 'Now you finally see that Reich is crazy'."[23]

The Org

The Org was founded by Frank's old friend, Walter Löwenheim (1896-1977), in 1929. Since the early 1920s, Löwenheim, alias "Miles," had been a member of the KPD, and a student of the Communist Paul Levy. He left the party in 1927, and rejected sectarian parties. Löwenheim came to the conclusion, in the late 1920s, that Stalin's "planned economy," which had supplanted Lenin's New Economic Policy, spelled ruin for the socialist movement as a whole. Equally ruinous were the reformist policies of the Social Democrats. He sought to renew the workers' movement by secretly organizing Social-Democratic and Communist Party members, to wrest control of the socialist movement from Moscow. His cadre did not take a formal name, as it did not view itself as a mass party.[24] Löwenheim was not a member of any party, but he gathered a small group of young KPD members, KPO members (Communist Party—Opposition, the "right-wing") and revolutionary SPD members.

The Org: a New Beginning

Until 1933, the secret *Kreis* (or leading circle) of the Org was composed of Walter Löwenheim, his brother Ernst Löwenheim, Wolfgang and Eberhard Wiskow, and Walter Dupré. After Hitler's takeover, Franz Schleiter and Karl Frank joined them.

The Org was founded on the principle that it is counter-revolutionary to establish one splinter party after the other. The quasi-revolutionary fascist movement is the main enemy (not the "Social Fascists," nor big industry). The old methods of illegal work are useless. Instead revolutionaries must work for long-term conspiracies. Structurally, the Org was Leninist: strictly centralist and conspiratorial. Löwenheim was "the German Lenin."

In German, there are two words for "conspiracy." One is *Konspiration*, a positive, aware and rational caution that provides safety from an extreme, oppressive authority. The other sense of "conspiracy" in German, *Verschwörung*, implies an offensive, anti-democratic suppression of freedom. This is a good distinction to make, between rational privacy and irrational gangsterism.

From the start, this group worked clandestinely, following Lenin's model of the cadres. Seeking to infiltrate and control *all* socialist organizations,[25] each member had to successfully complete a course in ideology, ciphers, covert tradecraft, social agitation, and survival. The group, internally known as "the Leninist Organization," was informally known as "the Org," or simply, "the O." In 1933, after the mass arrests of socialist cadres by the Nazis, the Org was in position to represent the revolutionary underground in Germany. Instead of propaganda and sabotage, the Org[26] concentrated on strict, disciplined secrecy to continue functioning within Germany after Hitler had taken power.

The so-called "Psychoanalysis Debate" in the Org took a central place in the group's history before 1933.[27] Sergei Feitelberg (1905-1967) came to the Org from the KPD. Feitelberg (a.k.a. "Werber," meaning "canvasser" or "recruiting officer") was the key figure among those intellectuals inside the Org who propounded psychoanalysis as a key to the formation of ideology. But the Org's *Kreis* opposed Feitelberg's "factional" activities because of his theoretical deviations, the danger of psychoanalytic therapy posed for *Konspiration*, and also because of the "unfortunate moral influence of psychoanalysis on revolutionaries." Löwenheim later poked fun at Feitelberg's Freudo-Marxist position.

Siegfried Bernfeld was, according to Löwenheim, the center of the

"Conspiracy Debate" ("*Konspirations-Debatte*") that took place within the Org in 1932—a debate about the character and the form of the conspiratorial methods of the group. Comrade Bernfeld demanded infiltration, i.e. *Konspiration,* against the Weimar Republic and preparation for the inevitable illegality in a fascist state by planting Org comrades into fascist organizations as future moles when the Nazis took over. Bernfeld's position was not adopted by the Org's board. They took the view that *Konspiration* should not be an end in itself, especially when Weimar democracy, with its workers mass movement, still survived, therefore *Konspiration* would apply to the relationship of the Org to the other Socialist and Communist parties, but not to the Weimar state apparatus *per se*. Also, a decision was taken that Org members should operate within the various socialist and communist parties to form a true united workers front—it would have nothing to do with the Bernfeld approach, to send Org members to operate in fascist or bourgeois organizations. Bernfeld submitted to the decision of the Org board at the time, but later, when Frank replaced Löwenheim as the Org's leader, Bernfeld's original position would be adopted.

In 1929, Bernfeld and his friend Sergei Feitelberg, both in Berlin, initiated an effort to give psychoanalysis a natural scientific basis through "libidometry."[28] They sought to demonstrate experimentally that there is no special "psychic" energy and prove that psychic life obeys the laws of mechanical energy. They tentatively equated the death instinct with the law of entropy.[29] Later, when Bernfeld returned to Vienna in 1932, he continued this research with the help of Frank's psychoanalyst, the physiologist Hans Lampl, and the physicist Franz Urbach. In 1935 this research was dropped, having lead nowhere. It was an interesting sideline, one that predated Reich's own investigations into a biological basis for social and emotional life. Perhaps, "libidometry" anticipated Reich's work, but was dropped completely just when Reich had obtained positive results in the opposite direction. This "libidometric" faction of the Org, Bernfeld and Feitelberg, who fought against orgone energy ("psychic energy") even before Reich had actually discovered it, may have later played some part in the suppression of the discovery of the orgone in the US.

Born in Moscow, Feitelberg came to Germany after WWI, and became an engineering graduate at the Berlin College of Technology in 1928, and then studied medicine. In 1934 he went to Austria, and in 1937 to Switzerland, where in 1939 he got a post as a physician in

The Org: a New Beginning

Lucerne. In the same year he went to the US. From 1939-1967 a member of the Mount Sinai Hospital, in 1966 he became professor of medicine at Mt. Sinai College of Medicine. In World War Two he joined the US Army. From 1942-1967 he was a member of Columbia University. His research focussed on radioactive nuclei and isotopes. He was a member of several sub-commissions of the AEC and of the radiation protection agency.

Wilhelm Reich, after mentioning Fenichel's rumor-mongering about Reich's "insanity" in a diary entry, continues: "A foul-mouthed individual named Sergei Feitelberg announced at a party that I was running an illegal brothel. He is the type of man who boasts in public about how many women he has laid and how often."[30] Some five years later, on Dec. 5, 1947, in a letter to his attorney Arthur Garfield Hays, Reich mentioned this incident again: "As a matter of fact, a physicist, Sergei Feitelberg, who is close to hostile psychoanalysis, once called me a secret brothel keeper in a private gathering." Reich gave little indication that he knew Feitelberg very well. Yet, in Karl Fallend and Johannes Reichmayr's book, *Siegfried Bernfeld oder Die Grenzen der Psychoanalyse,*[31] there is a photo from the summer of 1928 at Grundlsee featuring Pepa Kramer (the sister of Elisabeth Neumann), "Mädi" Olden (a lay psychoanalyst), Siegfried Bernfeld, Richard Kramer, Wilhelm Reich, Sergei Feitelberg, Elisabeth Neumann (an actress, wife of Bernfeld), each, save for Bernfeld, wearing swimsuits. Reich is resting his elbow on the shoulder of his 22 year-old pal Feitelberg.

Bernfeld's animus for Reich was on the record in 1932, when he published his "counter-article" against Reich's article on masochism.[32] In his "Rundbrief" of Nov. 25, 1940, Otto Fenichel remembers the events as follows: "In the year 1931, when I was in the editorial office of the 'Zeitschrift' [*Internationale Zeitschrift für Psychoanalyse*], Freud, after reading the galley proofs of Reich's article 'Der masochistische Charakter,' ordered that a publication of this article should only be allowed with a footnote by him, which publication (i.e. the publication of the footnote) would have been very unwelcome to the socialist analysts. For this reason I called together the 'leftist' analysts of Berlin, to discuss with them what to do. We tried to persuade Reich to make certain changes (in the article), which he rejected strictly. Later this matter was settled: Freud renounced to publish his footnote on the condition of the publication of Bernfeld's counter-article."[33]

On January 17, 1932, Freud wrote to Jeanette Lampl de Groot, a close friend of Bernfeld, that he (Freud) had to "make a stand against the bolshevist aggressors Reich, Fenichel."[34] Two days later, Freud wrote to Max Eitingon that he, Eitingon, was not the only one with an objection to the Freud note declaring Reich a Communist operating under orders from Moscow. Also Jekels in Vienna and Bernfeld in Berlin had advised him, Freud, not to publish the note. Bernfeld had said that "we should not preempt the Soviets with a declaration of war."[35]

Wilhelm Reich did not know during this time that Bernfeld, together with quite a few other psychoanalysts and former Communists like Karl Frank, was secretly engaged in the revolutionary organization, the Org.

In October 1933, a manifesto for the Org was published by the exiled Social-Democratic Party Committee. An edited version was republished, under Löwenheim's pen-name "Miles" and 5,000 copies of the pamphlet were smuggled into Germany under a false cover with the title: "Schopenhauer—On religion." By 1934, translations in English and French were widely distributed, making the so-called "Miles group" or "New Beginning group" the most widely-known German resistance group in the pre-war years. Reinhold Niebuhr's review of the pamphlet, "Socialism's 'New Beginning'" was published in *The Nation*.[36] In the USA, the pamphlet was titled "New Beginning—a Secret Manifesto," and was known to have made an impression with the Roosevelt administration. As conditions in Nazi Germany precluded the formation of a mass resistance movement, the cadre of Org operatives inside Germany had to be masters of espionage and conspiracy. Outside Germany, Org operatives, working secretly within the SPD and the KPD in exile, had to be extremely cautious and faced the possibility of being exposed by the Moscow-directed Communists as "deviationists."

According to one participant, the "New Beginning" manifesto was published abroad "so that its historically new analysis of fascism and communism should be known at the same time as the existence of determined forces inside Germany, ready to act if and when the situation might offer new chances."[37] This witness took a dim view of the "intellectuals, former communists, members of 'the Org' (and among them Karl Frank, living abroad)" who broke away from Löwenheim and a few older members of the group in 1935 to form what came to be known as New Beginning. They had "tried to involve the Org in 'mass actions' and put under pressure to give up such utopian and

The Org: a New Beginning

dangerous conceptions, broke away from the 'Org' with the immediate effect that members of this new 'splinter group' who had stopped respecting the basic rules of disciplined behavior were arrested by the Gestapo. Later on, Löwenheim and the other responsible leaders of the original 'Org' decided to dissolve the organization."

Walter Löwenheim emigrated from Germany in 1935 and settled in London. Frank's own view of this rupture was very different. By his account, the older members of the Org wanted to have all members go into exile, effectively eliminating the possibility that this underground network might be reactivated when conditions changed, such as in the event of a foreign liberation. Over sixty years later, Org cadres still debate these issues. Although Löwenheim emigrated in 1935, and apparently ceased political activity while he lived in England, the younger group known popularly as "New Beginning" continued the underground activity along the lines of the original program, so we could as well call this more active continuing group "the Org" and "New Beginning" interchangeably. The main goal for the underground activists remained: to simply survive in a kind of deep-freeze while overwintering for the duration of the war.

After the shelling of the Viennese worker's district in 1934, the Austrian government took its cue from Hitler and began rounding up leftists and imprisoning them in concentration camps. The more resourceful and fortunate were able to escape. Karl Frank was able to operate from Prague while illegally reentering Germany and Austria repeatedly, under harrowing conditions and many close scrapes. In one famous example, he tried to enter Germany over high mountain passes, disguised as a tourist skier, but got caught in a blizzard and fell unconscious. He was awakened by Nazi paramilitaries who brought him to a chalet, where he recovered sufficiently and escaped before they realized who he was.

For the most part, the Org's members came from the intellectual cream of Europe. Among the Germans and Austrians, hand-selected over the years, each cadre was given intense training to prepare them for long years of underground work. Karl Frank, along with the rest of the members, completed the "F-Course" (F[ortgeschrittenen]- Kurs) of instruction.

When Hitler took power, the Social Democrats fell into disarray, and Frank's network of young anti-Nazi activists held the center, providing communications as well as funding. While known leftists were arrested or forced into exile, Frank created an international organization of underground clandestine and exiled foreign opposition. From Prague to London and Oslo and New York, Frank raised funds to support the fight against Nazism on its home turf.

In June 1935, this organization of "younger elements from different sections of the German labor movement" gave itself a name from the title of a publication of an earlier manifesto, "New Beginning." In his account for the US government, Frank described the group as having been culled from the Socialist and Communist parties. These Marxists were critical of both Nazism and Stalinism, and had pointed differences with the old Social Democrats and capitalism. He claimed that the Org and New Beginning were completely separate organizations, albeit rooted in a common history.

In Frank's papers, there is an historical outline of the Org, written for the US State Department and the OSS. According to Frank, the fundamental principles of New Beginning in 1935 were:

"(a) German Nazism has beaten the forces of democracy. A long period, perhaps a decade or more, may be needed to overthrow it. There is no use trying to keep mass movements alive; they will be liquidated by the totalitarian regime and its police.

"(b) Nazism will most probably lead to war. Therefore, the only visible chance for a new liberation will come with the military defeat of Nazi Germany.

"(c) In order to survive it will be necessary to select younger active elements of the former labor movement, proved in character and qualifications and able to withstand the pressure of a long period in which these groups will be isolated, persecuted and have no support other than their own inner strength.

"(d) Networks of such groups should be developed underground all over Germany; the most advanced elements should be kept informed and kept politically alive and kept in contact with the European democratic labor movement through an organization abroad."[38]

The mass emigration of activists brought together disparate groups ranging from the centrist Social-Democrats, to more left-wing groups like the Revolutionary Socialists, the KPD and KPÖ, the SAP (Socialist

The Org: a New Beginning

Workers Party, including Willy Brandt) as well as the KPO and the Trotskyists. The left-wing sectarianism that had contributed to Hitler's victory loosened once activists found themselves thrown together and isolated from the masses in Prague, Paris and Zurich. As Otto Bauer, a leader of the SPD, put it, "...the political emigration has only one chance to escape the fate of becoming jetsam of history, namely to help the illegal underground movement and to undertake those functions which can only be carried out abroad."[39] As the Org was founded on the central principle of covert activity within existing political organizations for the purpose of unifying the socialist movement and preserving its interests in the post-war era, we can assume that the Org succeeded, because it did survive the war.

Otto Bauer headed the SPD's committee-in-exile in Prague, while Karl Frank was the head of the Org's Prague *Auslandbüro* (Foreign Bureau). Contacts centered in Berlin and throughout Germany were maintained for a surprising duration, until 1938; a few cells, such as those in Bavaria and Berlin, even survived the war's end. Frank's closest friend in Prague was Joseph Buttinger (1906-1992), who had married Anna Freud's friend, the American heiress, psychoanalyst, and leftist Muriel Gardiner. Gardiner was a source of money for both the Org and the Revolutionary Socialists. (She is widely believed to be the model for the woman immortalized by Lillian Hellman in her play, *Pentimento*, later made into a film starring Jane Fonda and Vanessa Redgrave, in 1977.) Joseph Buttinger was a liaison between the SPD leadership and the Revolutionary Socialists, a group of militants who had gained distinction for their organization of the defense of Vienna's working class districts when they were shelled by Dolfuss' Heimwehr troops in February, 1934. (It was in this context that a young Briton, Kim Philby, entered covert life as a courier for the Revolutionary Socialists as well as for the NKVD, the Soviet intelligence apparatus.) While the younger, more radical left-wing elements (such as the Revolutionary Socialists and the New Beginning group) blamed the Social Democrats' timidity and Moscow's stupidity for the success of fascism in Germany and Austria, there was close cooperation and overlapping memberships between all of these groups in exile. The Revolutionary Socialists disbanded in 1938 and most of its membership concentrated on the Social Democratic Party in exile, to redirect it from within.

In Prague, an entire cohort of Freudo-Marxists found themselves in exile from the Nazis. Banned in Austria, *Der Kampf*, the theoretical organ of Austrian Marxism, was published in Czechoslovakia. Its editor was the famous Otto Bauer, the leader of SPD in exile.

For Freud's eightieth birthday Der Kampf published two articles. Karl Frank was the author of one of them: "Sigmund Freud and Revolutionary Socialism."[40] According to Michelle Weston, Frank's daughter, Frank had previously been in close contact with Freud.[41] Frank wrote that until then there had been no Marxist analysis of psychoanalysis, due both to the bourgeois background of psychoanalysis and to the narrowness of contemporary Marxism. There was a Freudian philosophy: Freud's death instinct. It shares common roots with Oswald Spengler's philosophy: that of the decaying bourgeoisie. Nevertheless, Frank argued, Freud created a modern psychology by pointing out the social causation of psychological facts and thus developing a biological-materialist basis that was useful towards an understanding similar to Marxist historical materialism. The basic accomplishment of psychoanalysis, Frank says, is this contribution to an understanding of the social superstructure.[42] All of this agrees clearly with Wilhelm Reich's ideas about the relationship between Marxism and psychoanalysis at that time.

Some Org members participated in the Spanish Civil War and in 1937 Karl Frank went to Spain to investigate the disappearance and probable murder by Stalinists of the group's member Mark Rein. At that time, Frank had already made the first of many visits to the US, where he lectured to liberal groups about the Nazi threat in Europe, and raised funds to support the clandestine resistance by Org cadres within Germany. On many of these occasions, he appeared, dramatically, in a black hood for security reasons.[43]

The advocates of New Beginning included Germans and Austrians as well as a number of important English and American citizens. Sir Stafford Cripps, British Ambassador to the Soviet Union, and Chancellor of the Exchequer, served as Treasurer to the British section of the Org as early as 1934. Joe Buttinger ("Gustav Richter"), head of the Revolutionary Socialists, the Austrian resistance group "very closely allied to the Org," helped found the International Rescue Committee, and served for 42 years. Over the years, the Committee's highest award, the Medal of Freedom, would be awarded to Winston Churchill, Willy Brandt, Wild Bill Donovan, Lech Walesa, and Joseph Buttinger himself.

The Org: a New Beginning

Joe Buttinger

In the late 1930s, Joseph Buttinger—with Klaus Ehring and Karl Frank—wrote a pamphlet for a "united German revolution" [i.e., Germany including Austria]. Buttinger left Prague and went to the US in November, 1939. In 1975, still living in the US in Pennington, New Jersey, he said: *"Karl Frank and I succeeded—with the help of Mrs. Roosevelt—to make the President issue emergency visas for the emigrants who had fled to the south [i.e. Latin America]. We had to threaten the President that we would let the refugees cruise up and down the East Coast with the ships until the entry permit would be granted. We have gotten all visas."*[44] In New York he founded a library for the history of the workers movement, that would include Buttinger's own archives.

Joseph Buttinger and Karl Frank were the only members of the Org who were denied permission by the SPD to return to Germany after the war. They remained in close contact with each other until the end of their lives. In 1946, shortly before Mildred Brady's article was published, Frank wrote to Buttinger about his idea for a book on political psychology; it was never published, and probably not finished:

> "One difficulty is, of course, also the restriction to the essential theme of the book, the contribution to the political psychology. At the beginning, the more imprecise the thoughts are, the easier one holds forth in generalities about epoch-making perspectives. I know already that for sure this time I could not and would not write the 'applied Fenichel.' That is a life's work or, expressed more appropriately, a life's work in a well run big institute, which does not exist yet. But what I concentrate on is to work out the dogmatic and semi-theological features of the so-called progressive politics. This is only one chapter in the analysis of the superstructure. But it is what interests us most and which maybe one could also—sell best. Probably also that part, one could handle best."[45]

The following year, when Frank was still doing research for his book on political psychology, he wrote to Buttinger and told him that he was studying almost constantly on the subject of his book. But he is not satisfied with the results of his work thus far. He sees much confusion

Wilhelm Reich and the Cold War

in the (then) current Theoretical Psychology: everybody opposes everybody else in everything. He finds a dozen theories about fascism, "among them even such [theories] which have the good presentiment that aside from biological factors also cultural factors play a role." Even the most fundamental things are not sure and Frank himself loses confidence in his own working hypotheses. He lacks first-hand observations and so he takes two "patients" in "superficial analysis." Since he has no license for this work he has to keep it secret. Therefore in order to publish his book, he needs to persuade one of the psychoanalytic factions to take him as a member. In all of this, his political life experience isn't of much use, Frank writes to his friend Buttinger. But that experience will be fruitful again after he has gained a firm theoretical foundation in psychology. And this means that he must achieve something to straighten out the chaos in psychology, which the most productive psychologists, the psychoanalysts, could not achieve.[46]

A Soviet spy in the US State Department, Lauchlin Currie, sponsored the visa application of "Paul Hagen, also known as Karl Josef [sic] Frank, a German national later alleged to be a Soviet agent"[47]—by his old comrade Ruth (Eisler) Fischer. It didn't look good on the resume, and Frank was persona non grata in the 1950s as far as the State Department was concerned. That Currie sponsored Frank's visa remains the strongest piece of evidence that Frank may have been a Soviet agent, but I have been unable to find any supporting evidence. As far as we know, Frank's name did not turn up in the decrypted Soviet cable traffic known as VENONA, and on the Russian side, he does not appear to figure in the KGB archives released to date. To question Frank's ultimate allegiance is understandable, given the nature of the Org's ideology.

In the 1960s, Buttinger began a series of books on the Vietnam War, four in all, the first titled *Vietnam, a Dragon Embattled*.

Erich Schmidt, former leader of Berlin Socialist Youth, returned to the USA and pursued his work as a printer, and worked on his political memoirs. He lives in Massachusetts today.

Karl Schmidt, the Org's Berlin leader in the 1930s, wrote in his memoirs:

> "I myself have no memories or documents about the then events in the Prague emigration and only can marvel at Jan Foitzik's patience and sense in drawing out the major lines of a

political labyrinth, which looks today almost archaic. My own information I gained partly through official reports submitted to me and by my active correspondence with our Prague foreign office; in frequent intervals I also spoke directly with the leader of the Neu Beginnen foreign work Karl Frank (alias Willi Müller and Paul Hagen).

"Frank, born Austrian, doctor of psychology, was already since the 1920s an active member of the KPD and since at least 1931 in contact with the 'Org', but Org's leading group kept him at arm's length. 'He thinks he belongs to us', explained 'Kurt Menz' (this is Walter Löwenheim). Frank's status, after Frank in one of the F classes (Fortgeschrittenen-Kurse = advanced courses) which we attended, probably had criticized one of the theses of Miles [Walter Löwenheim]. But it would have been hardly very wise to simply discard a man like Frank. He knew already too much and maybe it would be valuable to straighten out things with him. He was a thoroughbred politician, intelligent, brave, as the times required him, more charismatic than the whole leadership of the 'Org' together, a debater with much diplomatic talent, who even could handle retreats with his quiet Austrian charm. But Frank had also this somewhat mysterious, impenetrable character trait, which made it sometimes very difficult to know what to make of him. For example there where always some people, who wondered how amazingly unhindered Frank was able to travel through and inside Germany, which could not only be explained with Frank's intelligence and courage and also not with luck.[48]

"[...] Frank was a coolly calculating professional revolutionary, who with his melting eyes of a deer looked more like a popular star of the French movies of those years, a factor he used to his advantage.

"In the inevitable competition with the groups and with the Sopade [SPD, the German Social Democratic Party], Frank could expect not much mutual understanding and sympathy. Nevertheless there was no other emigré politician, who was so much trusted and admired like Frank. Paul Hertz, member of the Sopade board of directors, the Grenzsekretäre [border officials, officers at the outer border of Germany,] Erwin Schöttle, Waldemar von Knoeringen, Franz Bögler, Emil Kirschmann, as well as Otto

Bauer and other members of the Revolutionary Socialists of Austria, especially their young talented leader Joseph Buttinger and naturally Friedrich Adler, Robert Grimm, Ernst Reinhard and our numerous Swiss comrades—they all esteemed him. 'When will Willi [Willi Müller, i.e., Frank] come back?' Ernst Bütikofer, an editor of the Berner Tagwacht [a Swiss newspaper in Berne] always asked me, when Frank was away any long stretch of time.

"Via Frank I received shortly before the publication of the pamphlet Neu Beginnen! an order from the Org-board to write about the expulsion of the Berlin SAJ-board members by the board of the Berlin SPD and the SAJ general board. The article legitimized the 'Org' among the SAI (Socialist Workers International). (...) It was to inform them about the role of the "Org" especially in the Berlin SAJ (Socialist Workers Youth).

"As an early Org member I was a major witness of the practical success of a series of new fascinating ideas about the problems of the time, the mission of the workers class and the capability of a group of people who left the KPD or were expelled from the KPD to save us from irreversible mistakes, e.g. to follow sectarians or to become Communists. [...]"

Karl Schmidt concluded:

"In its Prague Manifesto of January 1934 the Sopade wrote: 'In the fight against the National Socialist dictatorship there is no place for compromises, for reformism and legality.' Exactly these were the major points of the young Berlin SAJ comrades in their quarrel with the [SPD] party. The Manifesto spoke about an 'elite of professional revolutionaries' which we would need under the new circumstances. This was one of the main ideas of the Neu Beginnen! brochure. [...] The Sopade Manifesto was a radical break with the policy the SPD had before the Nazi takeover."

Franz Neumann (member of the Org) wrote *Behemoth-Structure and Practice of National Socialism* in 1942.[49] His widow married his best friend, Herbert Marcuse. (Marcuse himself was an OSS and CIA employee.) Robert Bendix, President of the American Sociological

The Org: a New Beginning

Association was in the Org. Edith Jacobson, later the President of NY Psychoanalytic Society, and an analytic colleague of Reich's, was an underground Org "illegal" in Berlin. British Labor Party leader Michael Foot supported the Org very actively.

In wartime London, the Org staff included: Richard Löwenthal, Waldemar von Knoeringen, Erwin Schoettle, Paul and Evelyn Anderson (the Org's representatives to the British Labour Party). Karl Frank mentioned a Mr. Burney, with the British Ministry of Information, who helped him with the Adam von Trott zu Soltz affair.[50]

In Oslo, Norway, Bernhard Taurer lead the Org's operations. Bernhard Taurer (born 1905), formerly an office equipment mechanic, had been a student of the College for Politics in political science in Berlin (1931-1935.) In 1931 he joined the SPD, and in 1932-1933 became the chairman of the university group of the Socialist students. After the Nazi takeover, he joined Neu Beginnen and worked in the underground. In September, 1935, he emigrated to Czechoslovakia, then to Great Britain, where he studied at the London School of Economics. In July, 1936 he went to Oslo; in April, 1940, to Stockholm, Sweden; May 1941 to the USA, where he worked for the OWI, and became a commentator in the German section of the Voice of America.

Some others who fell into the Org's orbit: H. N. Braisford, Franz Carsten (author of *The Rise of Fascism*), Dr. Norbert Elices, Alba Lerner, Tibor Scitovsky. Gerhard Bry ("Paul Kemp") later wrote "we had access to high industry and to the Diplomatic Corps."[51] His unpublished memoir, "Resistance Recollections from the Nazi Years," relates to his experience in the "Leninistische Organisation/ Neubeginnen."

The Org also included Richard Löwenthal ("Paul Serling") who had been leader of KPO students at University of Heidelberg (and, later, European correspondent of the *Observer*, postwar, later Professor of Political Science at the University of Berlin, Mannheim, and in the USA). Löwenthal, along with Arthur Koestler, wrote for *die Zukunft* (The Future) a German-exile weekly paper published in Paris, in the Autumn of 1938. *Die Zukunft* was published by Willi Münzenberg, who had left the Comintern and sought a reunification of the leftist exiles under the banner of the KPO, or Communist Party—Opposition.

Org member Anna Caples, Karl Frank's second wife, later married an American supporter of the Org, Jim Loeb, the cofounder of Americans for Democratic Action and a US ambassador to Peru. With

Wilhelm Reich and the Cold War

Thurgood Marshall, she raised critical funds for the NAACP's legal defense committee, and with Muriel Gardner, Joe Buttinger's wife, fundraised for environmental action, including the New Land Foundation. According to her niece, Caples taught Freud's son English. Anna Caples died in 1996.

Another American supporter of Frank's group was Esther Caukin Brunauer, who sat on the National Policy Board of the Committee to Defend America by Aiding the Allies. The chairman of this group was Lewis W. Douglas, whom we will meet again in a later chapter. The Friends of German Freedom—the Org's front-group in New York—had Reinhold Niebuhr as a very active chairman, with Richard Storrs Childs, John Lovejoy Elliot, Mary Fox, Paul Kellogg, and Norman Thomas as members of the board in 1939.

Frank also listed the physicist Friedrich Adler, a close friend of Albert Einstein, as a New Beginning supporter who, Frank noted in a 1942 letter to Calvin Hoover of the US State Department, "became a symbol of the anti-war wing of the labor movement when he shot [and assassinated] Austrian Prime Minister Karl Stürgh in 1916." Adler had been the leader of the Second International.

Mrs. and Mrs. Leopold Ullstein, from the major publishing family in Berlin for whom Arthur Koestler worked for many years, travelled with Frank from Prague to Germany by car during one of Frank's illegal adventures for the Org.

The list of the Org's powerful proponents goes on: George Silver (Manpower Division of US Army), Theo Thiele, Kurt Mattesh, Horst Mendershausen (Economics Division, then senior member of Rand Institute), Fritz Erler (later Vice President of the postwar SPD in West Germany), OSS-men Otto Schoettle, Waldemar von Knoeringen, Boegler (all three became provincial chairmen of SPD after the war and held various elected positions), Karl Mattech (served in the post-war Bundestag), Ossip Flechtheim (later, Professor of Political Science at University of Berlin), Paul Hertz (later, Senator in Berlin), Stefan Eliasberg (later, senior member at German desk of Voice of America), Vera Eliasberg (researcher for National Bureau of Economic Research), Henry Ehrman (later, Professor of Political Science at University of Colorado, then Professor of Government and Chairman of the Government Department at Dartmouth), "Ted" Schmidt (later, Director of Educational Institute of Learning and Research), Fritz Schmidt, George Garvy (later, VP and Senior Advisor to New York's Federal

The Org: a New Beginning

Reserve Bank), Fred and Elli Sandersen (worked for US government in France), Jerry Jeremias (later, mutual funds).

Wilhelm Reich's wife had a connection to the Org. Ilse Ollendorff's brother, Robert Ollendorff, was an Org operative who stayed inside Germany as an underground worker until 1935, when he left for East Africa (where he conducted research into sleeping sickness for the British Government, and, according to Ilse, became "totally apolitical." After 11 years, he moved to England, where he became involved with the anarchists in 1947-8.) Both Ilse and her brother knew Karl Frank "quite well" and "visited him shortly before his death" in 1969. Ilse described Frank as an "absolutely charismatic character, very alive and charming" and told me that it was a tragic situation to see him in a debilitated state at the end of his life, "a shadow of what he used to be." Robert Ollendorff was very close to him, and when he visited his sister in America in the sixties, they shared fond memories and stories about Frank and decided to pay him a visit. "My brother, who was not a very sentimental person, after he left Karl, he started to cry. It was a shocking, sad change. He could barely speak, barely walk."

Ilse Ollendorff was relatively unaware of the Org's activities even while she worked in the SAP's offices in Paris. She only learned in the 1990s, after meeting him through a mutual friend from the old socialist movement, that Erich Schmidt, who also lives today in Massachusetts, was a member of the Org. She confirmed the high level of secrecy used by the Org: Erich Schmidt, who was "very much involved in Neu Beginnen" did not know about her brother, Robert, who was very active in the group at the same time.

The Ollendorffs had both known Karl Frank since their early days in Germany; Ilse remembers meeting Frank in either 1929 or 1930. She remembered "very vividly when we met him in Breslau when he was staying with Hans Eckstein and his family, and we all went together to the circus." She also recalled that she and her brother had gone to Berlin at one time, where Robert met with Frank. Ilse Ollendorff left Germany in 1933, because of her own political activity but mostly because she was married to Konrad Reisner, a man who was deeply involved in the SAP and also the Friedensgeschellschaft (Peace Society) and the League for Human Rights. In Paris, she found fulfilling work as a secretary for the American-German Joint Distribution Committee, an organization dedicated to helping German-Jewish children escape by "paying a sort of ransom" with money that was collected in America.

Wilhelm Reich and the Cold War

In Paris during the early thirties, Ilse knew Karl Frank, but only as a fellow member of the SAP, not as an Org leader. "Actually, my brother put Frank up with some relatives who were quite innocent and didn't know who he was," Ilse Ollendorff recalled. She also remembered Willy Brandt as a "very devoted comrade, very loyal comrade" of Karl Frank's. Ilse came to the US in January of 1939.

Although as secretary of the Orgone Institute she wrote him letters, Ilse did not meet with Karl Frank personally in America until shortly before his death in 1969, because Reich "did not want any contact with those old [socialist] friends. He broke all contact with the psychoanalysts and he didn't want any contact with former comrades. I did not have any contact with them until I separated with Reich. After that, I met with Gertrude Gaasland again [Wilhelm Reich's laboratory assistant in Oslo], and I met Willy Brandt in New York. He was giving a lecture—he was at that time the Reichschancellor [Bundeskanzler of West Germany] and he was invited to speak at Columbia University. Afterwards I wanted to meet him, and here I was, I hadn't seen him since I came to this country, and we still said 'du' to each other like old comrades would. He came up to me in the hallway and said 'Wie geht's dir, Ilse?'—he recognized me immediately."

By the end of 1943, an organization for all German exiles was founded in New York, with author Thomas Mann as chairman, and the direction and support of Karl Frank's Org members. The Council for a Democratic Germany became influential with the parts of the US government, which was already making plans for a post-war reconstruction of Germany. The Council was open to Communists, and closed to Nazi supporters, but all-in-all it was a remarkably broad coalition of the full range of exile groups.

"But from the beginning," one historian wrote, "it was apparent that the left-wing 'Association for a Democratic Germany' (a new name for the former American Friends for German Freedom) and its Research Director Hagen [Karl Frank] were the 'dynamic element' in the Council."[52]

To evacuate the many political refugees that had escaped to unoccupied areas of southern France, Karl Frank and the American Friends of German Freedom founded the Emergency Rescue Committee (Emerescue). The American Federation of Labor (AFL) convinced President Roosevelt to issue 800 emergency visas. Frank, using the pseudonym "Paul Hagen," toured the US on speaking engagements,

The Org: a New Beginning

raising money to pay for the French bureaucratic graft. Emerescue worked closely with other groups such as the American Friends Service Committee. Frank toured Europe and America, raising funds to support the remnant underground groups still inside Germany. In the US, Frank, along with fellow New Beginner Paul Hertz, founded American Friends for German Freedom, with Reinhold Niebuhr as chairman and Adolph Held (chairman of the Jewish Labor Committee) as treasurer.[53] As a fund-raiser for this group, Karl Frank traveled to California and met Mildred Edie Brady and her husband Robert A. Brady. At the time of the US entrance into the war, there were 105,000 German and Austrian refugees in America.[54] This could not have been possible without a delicate but determined effort to unify the fractious exile groups to support refugee immigration.

After ten years in the Comintern, and another ten years as a clandestine "professional revolutionary," Karl Frank was not easily pushed aside in political activity. A complicating factor is that Frank was working for the State Department and the OSS. Frank had been introduced to Allen Dulles by his friend, Arthur Goldberg, the future Supreme Court Justice. Indeed, an OSS counter-intelligence official, textile executive David Seiferheld, was the secretary of the American Friends of German Freedom.[55]

Frank had enormously influential support. But other exiles were working within the OSS at that time, including Emmy Rado, who opposed Frank's influence.

Emmy Krissler Rado

Swiss-born Emmy (Krissler) Rado was the wife of Hungarian psychoanalyst Alexander "Sandor" Rado.[56] Reich said, "Rado married a patient, Emmy."[57] On Sandor Rado: "His third marriage was to Emmy Chrisler [sic], the mother of his second son, Peter, lasted many years, though they lived separate lives. Emmy Rado was active in civil rights and she worked for the Office of Strategic Services during World War II. She died of cancer in 1961 [...] For a number of years during his mother's illness while she lived in Mexico Peter remained with his father [...] Peter followed in the footsteps of his father's aborted path: he became an accomplished estate lawyer."[58] Peter Rado, the son, later became the lawyer for the New York fur merchant Motty Eitingon's estate.[59]

45

Wilhelm Reich and the Cold War

Emmy Rado provided crucial assistance to the OSS during the war, and worked directly for Allen Dulles' "psychological operations" against the Nazis. In a building in New York City at 630 Fifth Avenue, on the twenty-fifth floor, guards checked Rado's identification badge, worn on a necklace, as she passed through the doors to a section of military intelligence called Biographical Records. She had been working with Dulles since September, 1941, before the formation of the OSS. She has been described as the kind of woman that Dulles enjoyed having in his offices. S. Peter Karlo, an OSS veteran, said that Emmy Rado was a marvelous combination of beauty and brains. "'And she loved to swim in the nude in the Danube,' he recalled with a smile."[60] Dulles, a famous womanizer, brought Emmy Rado to Berne to join him in Switzerland during the war's final year.

Several sources report that there is a likelihood that Emmy Rado had an affair with Allen Dulles. Of Emmy, it has been written, "There was also an attractive young woman whom Dulles had brought into Germany with him and upon whom he relied heavily for advice and suggestions when it came to dealing with Germans. She was an American of Swiss origin, Mrs. Emmy Rado, who knew the Dulles family well through her close association with the World Council of Churches. She dazzled and warmed everyone she met, including Allen, and, as Frank Wisner subsequently remarked, 'She was really a shrewd cookie about what was going on in Germany, long before everyone else got off their white horse.'"[61]

Dulles' close assistant, Rado had vivid memories of Reich to share with the founder of the OSS. She had known Reich in Berlin in 1930, and probably before then. Reich had gone to her husband, Sandor Rado, for analysis in order to determine whether his (Reich's) conflicts with Freud had a neurotic basis. Reich saw Sandor Rado for several sessions in early 1931 before Rado left Germany soon thereafter. In an interview, Freud's curator, Dr. Kurt Eissler, asked Reich,

> "And Rado—what was Rado's part?"
>
> Dr. Reich: "Oh, that—I am not sure about depositing that, but I will tell you the story. It was this: Emmy, his wife, and I had very strong genital contact with each other. Never anything like full embrace happened between us, but we danced a lot together and we had very strong contact. And Rado was jealous."
>
> Dr. Eissler: "And then he started an intrigue against you?"

The Org: a New Beginning

Dr. Reich: "Yes. He was the one who started the rumor in 1934. He began the rumor that I was schizophrenic. He was the one. And Fenichel picked it up...So, to begin with, it was Rado because of Emmy...The rumor preceded me by a year in the United States. Everyone thought I was psychotic. That was my punishment for the discovery of the orgasm function."[62]

For his part, Sandor Rado privately told Annie Reich that Wilhelm Reich had an "insidious psychotic process" and suggested that she leave her husband, according to Edith Jacobson.[63] On the other hand, psychoanalyst Helene Deutsch wrote to her husband Felix in 1935 that Sandor Rado himself "was in the beginning phase of a psychosis."[64]

Emmy Rado came to the United States with her husband in the fall of 1931 and they settled in New York, where she did "research work," some of it for Eleanor Roosevelt. While they remained married in New York, Emmy and Sandor Rado lead separate lives.

In 1941, Otto Fenichel noted in one of his *Rundbriefe* letters, that the US government, through an agency called "Coordinator of Information" (COI) "was seeking to enlist psychoanalysis in the war effort, and Fenichel noted, it was 'sad' it had little to offer."[65] Fenichel was out of the OSS loop; but many German and Austrian psychoanalytic exiles would work for the OSS. "Through government contacts," one source recorded, "she learned about a new secret government agency called the Coordinator of Information (COI), organized by General William Donovan in July, 1941." How Emmy Rado had come to develop "government contacts" in the first place is not clear. Using her State Department connections, she had been able to obtain entrance visas for Karl Frank's ex-wife, his daughter and his ex-wife's second husband (the playwright, Carl Zuckmayer) in 1939.[66] Emmy had also been able to obtain visas for Annie Reich and her two daughters; she met them at the dock, clothed them and sheltered them for a few weeks. At the same time Emmy Rado was joining the US pre-war effort, Sandor Rado, her husband, was beginning a battle with the orthodox Freudian psychoanalysts (between 1940 and 1947). Ultimately, Dr. Rado started a neo-Freudian analytic training clinic, along with Abram Kardiner, at Columbia-Presbyterian Hospital.

Emmy Rado spoke five languages: German, French, Italian, Switzerdeutsch (a Swiss-German dialect), and English. In the fall of 1941 she went to work for a branch of the COI in New York known as the Oral Intelligence Group under Colonel Edward G. Buxton, who had served with William Donovan in World War I. This section, Oral Intelligence, later became called Biographical Records, and Emmy Rado's job was to interview exiles from Europe for military intelligence. With the help of her psychoanalyst husband, Sandor Rado, she developed guidelines for interviewing the exiles. Rado's section interviewed over 4,000 exiles during a 13-month period during the war, and compiled dossiers of over 5,000 individual German officials.[67]

Many German political exiles were hired by Shortwave Research Company, "a front organization for the COI Foreign Information Office that was interlocked with both Emerescue and the American Friends of German Freedom."[68] Colonel William Donovan had read Arthur Koestler's *Scum of the Earth*, and had been moved by the plight of anti-Nazi Germans.

"At that time OSS was just a nebulous idea in the back of the mind of General Donovan. [...] Mrs. Rado, in on the very start of the organization, watched it grow in purpose and scope during the four years she served."[69]

Altogether, Rado served close to five years in US intelligence services. She worked with the COI from late 1941 to mid-1942; she was in the OSS for more than three years, from July, 1942 until October, 1945; and six months in the SSU until March, 1946.

As a reference point for Emmy Rado's importance within the OSS hierarchy, one incident serves to place her quite near the center of things. A letter from a network of German military officers seeking to make a separate peace with the western powers came to Donovan in New York in late 1943.

> "The letter said that the opposition could not guarantee that the entire western front would remain inactive in the event of an Allied invasion; it had, however, sufficient influence with important German Army and some Luftwaffe commanders to ensure that counter-measures against an Allied landing would at least be delayed. If this were to happen the Western Powers should be prepared to negotiate with a German government formed by the opposition after a coup [...]That night the document [...] was taken to Washington under strict guard by Mrs. Emmy Rado, General Donovan's assistant (who later worked for OSS in the US Military Government in Berlin, with the rank of Colonel).

["X"] later heard from Mrs. Rado that, based on his opinion, Donovan had accepted the offer but that President Roosevelt had flatly declined to negotiate with 'these East German Junkers'."[70]

Mrs. Rado was one of the first women to enter Germany with a SHAEF pass after the war's end in Europe.

Willy Brandt and Mot Dag

Two of Norway's leading literary figures, Sigurd Hoel and Arnulf Øverland, were friends of Reich's work and also his patients in Norway. Each of these two men had belonged to an obscure Norwegian socialist sect called Mot Dag,[71] in English, "move towards daylight," or "move toward better times." Founded in 1920 by Erling Falk (1887-1940), Mot Dag not only shared a similar name with Neu Beginnen, but also shared the Org's practice of infiltration of other groups with the purpose of directing them from within. Mot Dag concentrated its organizing efforts on intellectuals, professionals and leading cultural officials. As with the German Org, many of its members assumed key government positions in Norway immediately after the war, and continued to hold them as late as the 1980s.

In the 1920s, Sigurd Hoel, one of Norway's most prominent writers of fiction, edited the group's journal, *Mot Dag*. It was "probably the best written political periodical in Norway. Its contents were often malicious, but always with a strong emphasis on form and style. Its main weapons were irony, wit and erudition."[72] Hoel was one of Reich's most reliable supporters in Oslo, defending his work publicly and donating money for laboratory equipment. Hoel was in therapy with Reich for two and a half years. Trygve Braatøy, M.D., who studied with Reich in Berlin, and later with Otto Fenichel in Oslo, was a Mot Dagist. Still later, Braatøy came to America to work at the Menninger Clinic in Topeka, Kansas, from Jan. 1949 through Feb. 1951.

The young German leader Willy Brandt (1913-1992) worked in Norway as a professional underground activist for the Socialist Worker's Party (SAP). During the pre-war years, most of the SAP's leadership was exiled in Paris, where Ilse Ollendorff worked as a secretary for the party office. She did not know her future husband, Wilhelm Reich, at that time, but was familiar with him by reputation since hearing a talk given by a SAP comrade in Paris who had attended one of Reich's lectures at the MASCH in Berlin, years before.

Wilhelm Reich and the Cold War

Brandt was joined in Oslo by his nineteen year-old girlfriend, Gertrude ("Trudel") Meyer in the summer of 1933. She had obtained Norwegian residency papers through a fake marriage with Gunnar Gaasland. The young couple became friends with Reich while he lived in Oslo. Gertrude "Gaasland" became Reich's secretary in Oslo, and her boyfriend, Willy Brandt, became a volunteer subject in Reich's bioelectrical experiments.

Brandt was been introduced to Erling Falk, the founder of Mot Dag, by Jacob Walcher, an official of the Socialist Worker's Party. (Walcher later worked with Karl Frank in New York during the war, as a member of the Council for a Democratic Germany.) When Gertude Meyer (Gaasland) fled Germany after her arrest there, she worked for Erling Falk as a maid in October 1933. At the same time, she served as SAP chairman in Norway. She was the representative of the Oslo branch of the SAP at the party's 1936 convention in Czechoslovakia in 1936. She made repeated, illegal entries into Germany while carrying messages to and from the socialist underground.

It should be noted here that the Oslo socialist community was a key element in Karl Frank's Org. In 1935, with financial assistance from French socialist leader Leon Blum, the Org developed contacts within the Olso-based Norwegian Labor Party, which took in "special refugees from our group to Oslo, brought us in touch with Norwegian labor people, who went as couriers for us to Germany and became among our most helpful supporters," Frank wrote, in 1942, in a letter to Calvin Hoover of the OSS.

Max Seydewitz, a former Reichstag official who had been expelled from the SPD, was cited as a reference by Karl Frank in his communication with the OSS. In Oslo, Seydewitz lived in the residence of Willy Brandt and Gertrude Gaasland, while Brandt was in France. Seydewitz was posing as a member of the SAP, while in reality he was a secret member of the Comintern. Brandt said that only three members of the KPD Politburo and the KPD representative in Scandanavia, Hermann Matern, knew of Seydewitz's true allegiance. Years later, Brandt himself would resign as Germany's Chancellor when his top aide was exposed as a Soviet spy.[73] After the war, Seydewitz became the prime minister of Saxony, in Communist East Germany. Brandt's allegation was true. Seydewitz, a covert Comintern agent, was living with Wilhelm Reich's secretary in 1938, when Reich was on the verge of the discovery of orgone energy.

The Org: a New Beginning

Willy Brandt wrote that he was a member of Mot Dag for a time, 1933-1935. In June, 1934, he became a member of the group's board, and sales manager of the journal *Mot Dag*. Along with another member of Mot Dag, the historian Johan Vogt, Brandt wrote the first Norwegian translation of Marx' *Das Kapital*.

In his 1982 memoir, Willy Brandt described the Mot Dag group as a sort of "monastic sect," a commune with about one hundred members in Olso and a few dozen more in Trondheim and other Norwegian cities. Its membership was drawn almost exclusively from the intellectual bourgeoisie, while Brandt himself came from a destitute working-class family. Members of Mot Dag were lawyers, physicians, architects, secondary school teachers and scientific assistants who all tithed a substantial part of their income to Mot Dag. Originally Mot Dag was exclusively male, but later a few females joined. Strict discipline was maintained. Everyone had his special duty assigned. Mot Dag had an office floor with a library and a kitchen, were many Mot Dagists ate, even breakfast. Leisure time was spent together. Those members who needed extra money had to ask for an allowance. Under Falk's authoritarian, but also inspiring leadership, this small cluster of people accomplished a great deal. They issued a periodical, had a publishing house, controlled leading students' associations, maintained a night school. New ground was broken with a journal for sex education.

In 1966, Brandt remembered Mot Dag (at the time of his 1933-35 involvement) as having had large influence in the academic life of Norway, but at first not so much influence on the political landscape, until its members joined the Norwegian Labor Party (NLP) in 1936 and were promoted to the highest positions. Other sources paint a more complicated picture of Mot Dag's relationship with Norway's socialist parties. In March, 1922, the editorial staff of *Mot Dag* had declared itself as "The Communist Association Mot Dag—member of the Norwegian Labor Party."[74] In 1923, Falk went to Moscow as the NLP delegate to the Comintern. He spoke "American with a Chicago accent" and was remembered as having the temerity to treat the Bolshevik leaders as equals.[75] When this party split in 1923 after voting to leave the Comintern, a new Norwegian Communist Party (NKP) was formed by members who wished to retain ties to Moscow. Mot Dag joined the NKP. In 1925 the entire Mot Dag group was formally and collectively expelled from the social-democratic Norwegian Labor Party. Only later, when the Norwegian Labor Party won the national elections in

1933, and then became the governing party in 1935, did Mot Dag rejoin the Labor Party in 1936. Significantly, the NLP refused to admit Mot Dag's founder, Erling Falk, though Mot Dag members were accepted and soon took many positions of administrative authority in the NLP. As Karl Frank told his contact with the OSS in 1942, it was during this time (1935-1936) that the Org developed special contacts with the Norwegian Labor Party. Mot Dag was formally dissolved in 1936, yet the political influence of its former members only grew larger.

In 1935, several Mot Dag associates, including Hoel and Øverland, actively campaigned for a Nobel Peace Prize to be awarded to Carl von Ossietzky, who languished in a German concentration camp. Willy Brandt was one of the more active campaigners for the award to Ossietzky as a protest against the Nazis. When Ossietzky was awarded the prize in 1936, Hitler decreed that no German would accept the Nobel in the future. At the same time, Leon Trotsky, who had been given asylum in Norway in 1935, suddenly became a persona non grata in that country and was interned in 1936 after the Moscow trials by Minister of Justice Trygve Lie of the Norwegian Labor Party. Although Lie later denied it, communications with Moscow resulted in the Norwegians denying Trotsky asylum. Few, if any, Mot Dag members spoke against the deportation of Trotsky. Trotsky predicted that in less than five years these same bureaucrats would all become exiles, and so it would be, after the German invasion. It was also Lie who denied Wilhelm Reich's permission to stay in Norway. Lie would go on to become the first Secretary General of the United Nations.

Among left organizations in Norway, Mot Dag was the most regimented. For example it was a matter of course to give one's inheritance to Mot Dag, and members had to participate in dubious financial transactions for the benefit of the group. Its leading members held many key positions in front organizations. The radical student-organization "Clarté" was one such front. Reich lectured to the Clarté group in 1933; one attendee remembered Reich's speech in 1972, and said "I have never listened to a more ingenious and more stimulating speaker, and just as brilliant was his art of response in the discussion which followed."[76] Brandt was unique because he came from the working class and Mot Dag, with its elitism, could not hold him too long. In spring 1935 he separated from Mot Dag. Brandt compared Mot Dag's elitism to the Bolsheviks and their French forerunners, and also to some ideologues of the New Left of the late 60s.

The Org: a New Beginning

Erling Falk, the man who had founded the group in 1920, was born in northern Norway and attended three different high schools but did not graduate. He lived in the US for eleven years (1907-1918) and worked as a laborer in Minnesota, Chicago, and Montana. Gradually, Falk worked his way up to white-collar work, and had established his own business by the time he returned to Norway.

In America, Falk read widely from the radical literature of the times, from Karl Marx to Friedrich Nietzsche. He was influenced by the revolutionary-syndicalist Industrial Workers of the World (IWW) and came into close contact with William ("Big Bill") Heywood and Vincent St. John. But Falk was more inspired by the strong leadership Lenin offered to the revolution in Russia, rather than the anarchist-populist tendency of the IWW.

In Oslo, Falk attended local socialist meetings of various groups, but did not participate in the discussions. From the outskirts he listened to the speakers, while quietly identifying potential supporters. Soon he chose seven core members for his inner circle: Viggo Hansteen, John S. Hazeland, Trond Hegna, Arne Ording, Aake Anker Ording, Axel Sømme and Johan Vogt. Whether Mot Dag should influence the Norwegian Worker's Party, or whether it wanted to form the core of a future party, these were questions that Mot Dag left open. In Brandt's recollection, Mot Dag was absolutely certain that the world was to be formed according to the ideological notions of Mot Dag and especially that the worker's movement needed to be rescued from its reformist decline.

Willy Brandt described Falk as a lanky, frail, ugly man, with the neck of a vulture, the head of a featherless bird, and the eyes of an eagle. He was in total command of his disciples. Brandt learned then how sadistic inclinations are sublimated and masochistic desires are fulfilled in a political cult. According to Brandt, sexual inhibition bred gifted haters and schemers. Politics as a substitute for love camouflages itself as altruistic absolutism. Falk was the "intellectual high priest" and "ascetic guru"—very neurotic, which one could see through his uptight and twisted relationship towards the other sex. While Brandt was a lover of women, Falk feared them. The same problem troubled the party leader of the Norwegian Worker's Party Martin Tranmäl, although Tranmäl and Falk did not like each other.

It was through Mot Dag, that Brandt initially came in contact with many writers and artists, including Sigurd Hoel. The Norwegian Poet

Laureate, Arnulf Øverland, was politically active and proved fearless during the German occupation. Brandt met him again at the end of the war, when Øverland returned drained and embittered from the German concentration camp at Sachsenhausen. He gradually moved far away from his radical convictions and became a vocal anti-communist. Even so, in 1960, Brandt, then mayor of West Berlin, remained friends with Arnulf Øverland, as well as with Sigurd Hoel.

Brandt saw Falk for the last time as Falk lay dying in 1940 in a Stockholm, Sweden hospital. In view of the world catastrophe, Falk had reconciled with his adversaries in the socialist movement. The very last advice Falk gave Brandt from his deathbed in Sweden: since the whole of Europe will go fascist, an elite of Scandinavian socialists should go to the USA for survival, so that one day they might bring back to Europe the socialist idea.

Karl Frank is not mentioned in Brandt's autobiography. Yet Willy Brandt and Karl Frank remained in personal contact through the end of Frank's life in 1969.

In 1958, Willy Brandt wrote an official endorsement for Frank, probably aimed at getting Frank readmitted into Germany, where he was unwanted by the SPD leadership.

Statement from Brandt to the State Department, Feb. 27, 1958:

> "I met Dr. Karl B. Frank for the first time personally only at the end of 1936 in Prague. But he was known to me through many common friends already before the Nazis took over. I can attest that he then was a leading figure in a circle of younger Social Democratic students and workers and that he took an active position as a journalist and speaker in the confrontation with Nazism.
>
> "After the so-called Machtübernahme Dr. Frank was forced to leave Germany, since otherwise he most probably would have not only lost his freedom but also his life. As a political refugee he first lived in Prague and finally in New York and continued his activities against the Hitler regime—in close contact with friends working in the German underground.
>
> "The circumstances of emigration and the activities Dr. Frank was involved with resulted in his having to live under difficult material circumstances. There is no doubt that under normal circumstances he would have gained a respected position with a considerable income."[77]

The Org: a New Beginning

Later in life, Willy Brandt mentioned his association with Wilhelm Reich with fondness and complete comprehension of Reich's contributions to the revolutionary movement of his youth. In the film, *Viva Little Man*, he was interviewed by Digne Meller Marcovicz, and Brandt offered:

> "The idea that authoritarian power also had something to do with sexual repression was for many new and strange. This invites narrow-minded people to deceive themselves about their own stupidity and intellectual capacity by making silly comments, as I put it lately, writing on that subject [in Brandt's autobiography]. It clearly distinguishes him [Reich] that due to the impression of Hitler's rise to power he saw the connection, as I just said, between authoritarian power and the lack of freedom people experience in their personal lives, that he directed all his attention to these correlations and that he did not deal with the question of sexuality alone, as is familiar to us through the literature of the time or of the previous century, as a question pertaining only to the bourgeois classes, but that he considered it, as we all know today, to be the central predicament in the life of the masses, one that occupied, troubled the minds of young working-class boys and girls—and that, indeed, was quite a step forward from the classical issues of psychoanalysis."[78]

As an epitaph of sorts to the discussion of Willy Brandt and his political career, newly leaked information might indicate that Brandt was himself an asset of Soviet intelligence. A recent article published in the German periodical *Focus* gave new details, provided by an ex-KGB agent, Vasili Mitrokhin, known as "Curb," that would indicate that Brandt had been recruited by the Soviets in Sweden during the early forties. Brandt had signed a receipt for money from the NKVD. "Curb" said the Soviets later used this document in an attempt to blackmail Brandt when he was an established politician in postwar Germany.[79] Brandt refused. Because the actual receipt had been destroyed in the Soviet archives, it was only a bluff. As it turns out, Brandt had indeed given the Soviets, along with the British, information about the deployment of German forces in Norway. He quite openly visited the Soviet Embassy to do so.[80]

Wilhelm Reich and the Cold War

Willi Schlamm

For another viewpoint on Karl Frank, we will turn to one of Wilhelm Reich's old friends from Vienna who followed the development of the Org and of its leaders who went on to hold leading positions in politics and society after the war. Willi Schlamm, a founder of the American conservative journal *The National Review*, would write his recollections about the Org near the end of his life, after he had returned to Germany. Schlamm was never a follower of the Org, but was in a position to describe its leadership and ideology.

William Siegmund Schlamm (1904 -1978) was born in Przemysl, Galicia (Eastern Austrian Empire). His father was a well-to-do businessman.[81] His pacifist and socialist idealism led him, as a high school student in Vienna, to join, in 1919, the *Kommunistische Jugend-Internationale* (Communist Youth International). At the tender age of 16, in 1920, Schlamm visited Moscow. In 1922, after receiving his high school diploma, he joined the KPÖ-organ *Rote Fahne* (Red Flag) in Vienna, where he was editor until 1929.

In 1927 he married Stephanie Kohaut. In 1929 he broke with Communism after Moscow took proceedings against him for forming a "rightist" opposition within the Austrian Communist party. Schlamm was expelled at the same time Reich officially joined the Party. Reich was, at the time, in complete agreement with the Stalinist line against "social fascism," whereas Schlamm saw the need for cooperation on the left against fascism.

Schlamm remained a leftist and joined the magazine *Jugend* (Youth), as well as the satirical magazine *Simplicissimus* in Munich, Germany. In 1931 he got in contact with the Berlin liberal *Weltbühne*, published by the famous Carl von Ossietzky who later became a Nobel laureate. In 1932 Schlamm took over the Viennese edition of the *Weltbühne*. When in 1933 the Gestapo arrested von Ossietzky, Schlamm became, until 1934, director of the *Neue Weltbühne* in Prague, Czechoslovakia. From 1934-1937 he published his own magazine, the *Europäische Hefte,* in Prague. In 1937 he published the book *Diktatur der Lüge* (Dictatorship of Lies), a thorough denunciation of Stalinism. This book was positively reviewed in Reich's Sexpol journal in Oslo.[82]

In 1938 Schlamm emigrated to the US, where in 1941 he published the book *This Second War of Independence* about Hitler's victories over the Western democracies. Ilse Ollendorff, Reich's wife-to-be, met

The Org: a New Beginning

Schlamm when she first came to America in 1939, through a mutual friend. She remembers him fondly from those days, but told me that Willi Schlamm and his wife warned her not to get involved with Reich after she started seeing him. The Schlamms were friends with Annie Reich, "and that was where the animosity against Reich came in," Ollendorff said. "It was probably much more personal than political." In 1944 he became an American citizen. From 1944-1950 he was assistant to the director of the magazine group *Life*, *Time*, and *Fortune*, of Henry Luce. Schlamm became close friends with Whittaker Chambers at *Time*. In 1949 he came back to Europe as a correspondent of *Fortune*. In 1951 he was among the founders of the conservative magazine *National Review*. He was also contributor to *American Mercury*, *The Nation*, *Freeman*, and *The New York Times*.[83]

In 1957 he visited Germany and wrote the book *Die Grenzen des Wunders—Ein Bericht über Deutschland* 'Limits of the miracle; A Report on Germany.' With a German edition of 100,000, he became famous in Germany as "Cold Warrior No. 1."

In 1966 he published the book *Vom Elend der Literatur—Pornographie und Gesinnung* 'About the misery of literature.: Pornography and attitude.' Attacking the contemporary "sex literature": the modern author fleeing from the failed social revolution runs into the "sexual revolution."

In the twilight of his life, Willi Schlamm remembered his friend, Wilhelm Reich, in a bittersweet eulogy that revealed the depth of his feeling for his "friend from my youth:" "Wilhelm Reich, the father of the 'Sex-Revolution.'"[84]

Confronted by the resurgent popularity of Reich's writings among the New Left in 1970, Schlamm argues that the "Sexual Revolution" is a continuation of Revolution by other means, but that a break-down of inhibitions makes the individual incapable of disciplined actions necessary for any revolution.

Schlamm, an old man by 1970, completely misunderstood Reich's theories, and describes them in much the same way Mildred Brady had in 1947. Schlamm wrote that Reich's ideas could be traced to his unresolved hostility to his "master" Freud, and Freud's bourgeois world. "I know it, because Wilhelm Reich was a friend from my youth." They had met in 1919, but it wasn't until the late 1920s that Schlamm became warm friends with Reich—paradoxically, Schlamm said, because it was at that time Schlamm broke with Communism, while Reich was just joining the Party. Schlamm said he had no doubt that Reich would soon leave the Party as well, since he was too intelligent and decent to tolerate that ghetto of stubborn fanaticism. Schlamm recognized Reich's contribution to psychoanalysis, and said that Reich was the only young analyst that Freud took seriously, but Schlamm could make no sense of Reich's later work, beginning with Sexpol and ending with what Schlamm called "Orgone caskets."

Schlamm had only sadness for Reich: never before had a man died so lonely, who had so believed in the vegetative togetherness. "For his sake (among other reasons) I hope that there is a life after death—and that it can be a life in peace for my friend from my youth Willi Reich."[85]

Mimicry

In his later years, Schlamm drew upon his experiences in the socialist movement to sketch some of the leading figures in post-war German politics. Schlamm had followed the career of Willy Brandt, and he considers him "the most helpful servant of a certain historical movement of socialist revolutionaries." He described its political program as "mimic play-acting." In an article written for his own right-wing journal, *Zeitbühne,* in 1972, Willi Schlamm claimed that the German Chancellor Willy Brandt simply did what he had learned in the 1930s in the group *Neu Beginnen.*[85] Previous to Hitler's takeover, the following elements met in Neu Beginnen: the Sozialistische Arbeiter-Partei (SAP, Socialist Worker's Party, of which Brandt was a member), disappointed Communists, and some Marxist scholars. Dr. Karl Frank, one of the leaders of the secret organization, was a Viennese friend from Schlamm's youth. Its purpose was to infiltrate and takeover *all* parties to the left of Hitler. Neu Beginnen tried to achieve this by "mimicry," pretending to agree with whomever one is with at the moment. Instead of expending the limited forces of revolutionary socialism with the dangerous strategies of the Communists, Neu Beginnen wanted first a takeover of all existing leftist parties and second with the same secret strategy a takeover of all of society as a whole. "Revolution by adjustment," socialism in alliance with big industry. At the time, Schlamm told Frank that this would be a bizarre program.

Schlamm recalled Frank as a very attractive and fascinating man, and that Willy Brandt became one of his most devoted followers. Schlamm described the success of Neu Beginnen as phenomenal: its

The Org: a New Beginning

agents sat in the outer office of Chancellor Brüning, in the English House of Commons, in the Norwegian Worker's Party, in the office of the American President, in the main office of Himmler's Gestapo, in the secretary's office of the SPD in exile, and in the secretary's office of the League of Nations. The leadership of Neu Beginnen (with Karl Frank and his faithful coworker Willy Brandt) remained a dense and closed secret organization with a solid socialist identity despite all of that mimicry.

After Hitler's fall the disciplined cadres of Neu Beginnen returned to Germany as cadres of the SPD, including Fritz Erler (deputy leader of the SPD) and Willy Brandt. Norbert Ernst was also a member; he returned to Germany in 1945 as a UN official involved in "re-education." Other "new beginners" became managers of the German unions, public health insurance organizations, party newspapers, cooperative banks; all of these institutions were dominated by the SPD. They also became university professors, Bundestag representatives, administrative undersecretaries. Schlamm estimates that in 1972 a quarter of the German establishment was controlled by Neu Beginnen and its inheritors.

The most successful of Neu Beginnen members was, of course, Willy Brandt, the most devoted and receptive pupil of Karl Frank. But, Schlamm wrote, Frank himself was in the end disappointed by his life's work. Brandt remained true to Neu Beginnen and its cadres. His consultants in the administration and the lead of the SPD are almost completely old cadres of Neu Beginnen. Schlamm warned that, as Chancellor, Brandt continued to believe in the omnipotence of mimicry, of infiltration, of penetration, and of the secret takeover. In the company of bourgeoisie Brandt is a bourgeois, in the company of patriots Brandt is a patriot, in the company of statesmen Brandt is a statesman, and in the company of revolutionaries Brandt is a revolutionary. Schlamm described this as the essential characteristic of Neu Beginnen. Brandt, Schlamm accused in 1972, was determined to realize socialism with the help of big industry, to establish friendly relations with the Soviet Union using the cooperation of the USA, to free the people from the "opium" of religion with cooperation of the Protestant church, to achieve German unification by the splitting of Germany into West-Germany, East-Germany, and West-Berlin, and to get rid of the freedom of press with the cooperation of the newspaper publishers. Schlamm wrote that Brandt considers all of life and politics

Wilhelm Reich and the Cold War

as nothing more than a play with cast roles—a play conceived by intellectuals and performed by actors—and Brandt was the leading man.

Schlamm's assessment of the Org's methods remind one of Edgar Allan Poe's famous detective story, "The Purloined Letter," where police fail to find a stolen document because the document was on the table for everybody to see. This is the secret trick of all conspiracies: they are always public. And this is the conspiracy of Neu Beginnen: a public play.

Other European exiles who worked with the OSS, including Ruth Fischer, took a dim view of Frank and denounced him to the OSS and the State Department as a covert Stalinist. (She had known Frank when she worked in the Berlin's secret Comintern office with "Comrade Thomas.") Yet Frank's influence survived at the OSS, within bounds. The Council for a Democratic Germany called for a unified post-war Germany, the dissolution of large landholdings, nationalization of heavy industry, and the destruction of the "military caste."

Since the Org was a secret society, we can't be certain what the ultimate goals of the operation were. We can only take them at the words of their manifesto. The wide range of members in positions of authority make it obvious that the Org had influenced the shape of post-war Europe. The Org's central tenet required a disciplined cadre of stateless spies working toward a common goal, the unification of the socialist movement. The Org's interactions with the State Department, the OSS, the NKVD—and limitless other diplomatic, military, and intelligence groups—were laced with manipulation, duplicity and intrigue.

In his foreword to the 1936 edition of *Die Sexualität im Kulturkampf* (*Sexuality in the Cultural Struggle*, later retitled *The Sexual Revolution*), Reich replied to friends who worried about his involvement in "dangerous politics," as follows: *"Who, in a house that is on fire, would leisurely write aesthetic treatises on the color sense of crickets?"*[86]

It's unlikely that Wilhelm Reich was himself a formal member of the Org. They kept no written membership records. Still, there are a number of direct points of agreement between the Org's program and Reich's own assessment of the situation in Europe, as he wrote of it in his book, *Mass Psychology of Fascism*. While in Oslo, Reich seriously considered a political career, raising the question as to whether he had been

The Org: a New Beginning

inspired, informed or provoked by the Org program. Reich mentioned "Miles'" famous pamphlet, *Neu Beginnen!* in the essay he published in his *Zeitschrift für Politische Psychologie und Sexualökonomie*: titled "What is Class Consciousness?"[87] This essay was written during two weeks in late 1933, while Reich was newly in exile, and published in June, 1934, in Denmark, shortly after the "Night of Long Knives," the failed "second revolution in Germany" in which "left" elements of the SA were liquidated. In some ways it is a direct response to the message in *Neu Beginnen!* Reich agreed with parts of the program, but pointed toward a more psychologically astute avenue of organization, and Reich rejected "Miles'" assertions that the masses were incapable of spontaneous revolt. Reich tried to show that until revolutionaries understood that "class consciousness" sprang from the unconscious psychic life of the people, all their efforts would be stillborn. Reich had been meeting with Org members, as well as members of Trotsky's Fourth International, in Paris and Zurich, Switzerland. He later described the attitude of these exiled politicians in disparaging terms.

During the period between 1933 and 1937, Reich published several articles in the *Zeitschrift für Politische Psychologie und Sexualökonomie* that made reference to the founding principles of the Org and its manifesto, *Neu Beginnen!* In 1934, "Towards a Critical Analysis of Communist Policy in Germany" (attributed to "Karl Teschitz" but written by Karl Motesiczky[88]), the author issues a Leninist analysis of the Party's errors leading up to and immediately following Hitler's assumption of power. The author criticized the Comintern policy of doing battle with the "social fascists"—the Social-Democrats—rather than against the Nazis. "Teschitz" also ridicules the Comintern fantasy that the German masses were on the verge of a "revolutionary upsurge," and he cites and agrees with "Miles" (i.e., Löwenheim, the author of *Neu Beginnen!*) for having described this fantasy as "subjective idealism."[89] In an afterward, Teschitz notes that many of the criticisms he made when writing the essay had since been adopted by the Comintern. He noted that a top party functionary had recently spoken about the "defeat" of the German working class; "only six months ago, one could be expelled as a counterrevolutionary for calling it this (e.g. as in the case of Wilhelm Reich.)"

Another article in the *Zeitschrift für Politische Psychologie und Sexualökonomie* by Karl Teschitz responded to several published critiques of Sexpol's founding principles from anarchist groups in

Holland and Spain. While acknowledging the anarchists' natural inclination to support Sexpol's open criticisms of the Comintern, Teschitz held to a strictly Leninist viewpoint and insists that a centralized cadre of leaders is necessary for any revolution. He notes that French anarchist trade unions marched in favor of the government's involvement in WWI, and pointedly remarks that the Spanish anarchists were unable to defeat Franco's fascists. "They might therefore be a little less high and mighty about Lenin."

Having distanced himself from the anarchists, the author takes a step closer to the principles of *Neu Beginnen*. "Anarchism, too, has fallen prey to authority—namely the authority of its own doctrine."

"Might it not, therefore," wrote Teschitz, "be best to work towards the revolutionizing of the great mass parties from within them? (cf. the appendix to *What is Class Consciousnes).*"[90]

That reference is interesting because there Reich advocated working with, and within, Nazi mass organizations as well as all socialist parties. Karl Frank pointed out that, in order to camouflage themselves, underground workers inside Germany adopted the life and lifestyle of the Nazis after 1933.[91]

As we can see, there was a fundamental agreement between many of the political statements of Sexpol and the Org. First, a Leninist cadre of dedicated leaders must understand the current political realities. They must be able to unify the support of the working masses by understanding their needs, without concern for their superficial party allegiances, be they SPD, KPD, or Nazi. They must understand fully the defeat of socialism in Germany, be patient and work for the long-term, "with the purpose of reinforcing the struggle of the proletariat against the bourgeoisie, of overthrowing the latter and establishing a political system based on soviets, which will be able to resolve the urgent matter of the sexual life of the young within the framework of a total proletarian revolution. Long live the Proletarian Revolution."[92]

Reich later wrote of his essays during this period:

"Thus I sought a new social revolutionary organization which was willing to learn productive lessons from the catastrophe. Many of my political-psychological essays of that time were based on this. Further experiences were needed to free me from these illusions completely and to make me realize that the problem of human organization per se was at issue."[93]

The Org: a New Beginning

Edith Jacobson

Edith Jacobson (she sometimes spelled it Jacobsohn) (1897-1978) was an Org supporter who was arrested by the Gestapo in 1936. The daughter of a doctor, born in Upper Silesia, she took a medical degree in Munich and completed psychoanalytic training at the Berlin Institute in 1929. She had been a part of the Reich-Fenichel circle of Marxist analysts in Berlin, where they formed the so-called Kinderkreis or "Children's Circle" of younger doctors. She joined Reich to work in Berlin's "Sexpol" clinics, counseling adolescents and working women on sexuality.[94] "She was a woman of exceptional intelligence and deep humanity," Reich recalled.[95] After Hitler's takeover, she met with Reich and others in Malmö and Oslo, and decided to return to Berlin. Russell Jacoby[96] states that she returned out of concern for her patients; Ellen Siersted wrote that she was concerned about her family.

Edith Jacobson was the Org's house psychoanalyst [*Haus Psychoanalytikerin*] and she used the analytical couch to protect endangered comrades. Jacobson became Karl Frank's psychoanalyst. She was arrested in October of 1936, not because she was a psychoanalyst, but because she was a member of the Org.

Jacobson was incarcerated in a local German prison where conditions were not so bad as in the concentration camps. Reich wrote later in *People in Trouble*, calling Dr. Jacobson "a diligent participant" in the Berlin group of "dialectical-materialist psychoanalysts." Then he added suddenly and parenthetically, "(Her later misfortune in having to spend two years in a German prison lies heavily on my conscience.)"[97] This comment may make reference to the decision taken by Reich and Elsa Lindenberg to obtain an abortion in Berlin in 1935, a dangerous undertaking assisted by Edith Jacobson and her contacts within the "German resistance movement against Hitler."[98] Did Jacobson's arrest in 1935 weigh heavily on Reich's conscience because of the abortion, or was Reich himself a member of, or sympathetic to New Beginning? Did Reich advise Jacobson that she should return to Berlin? We do know that their relationship deteriorated after Reich's expulsion from the IPA in 1934. Reich wrote, in 1935, about his disappointment in Jacobson, who he felt had not adequately defended him and the cause of Freudo-Marxism they shared against the psychoanalytic establishment: "I had a letter from Edith I didn't like. I won't answer it, either. I 'promised'—to let myself be quietly slaughtered. Just because it would have been more convenient to the Ediths and their ilk, to spare them any pangs of conscience..."[99]

Reich's own bad conscience was not unique. Otto Fenichel's *Rundebriefe* continuously discussed plans to obtain Jacobson's release.[100] Annie Reich's new husband, Thomas Rubinstein, orchestrated Jacobson's escape and smuggled her into Prague. Rubinstein, who was by 1936 in Prague and closely associated with the Org, would have been motivated by a concern to protect other Org activists inside Germany.

Edith Jacobson emigrated to the US, where she later served as the President of the New York Psychoanalytic Society. Myron Sharaf described Jacobson as having turned against Reich by 1935, the same year Karl Frank ascended to the leadership of the Org, and Walter "Miles" Löwenheim retreated to England in resignation. Openly and publicly, Wilhelm Reich had left the psychoanalysts and the politicians by 1935. Edith Jacobson, Otto Fenichel, Annie Reich, Sandor Rado and Karl Frank all moved to the US and hid their Marxist backgrounds in their new home. Sandor Rado shed his Freudian background and started a psychoanalytic school of his own in New York. Wilhelm Reich was an unwanted witness to their past lives.

Comrade Thomas

There remains one life-story to be told, of a man who entered the Org's orbit around 1934, as the former head of the Comintern's office in Berlin. Until recently, virtually no one knew his given name. Perhaps his wife in Prague and America knew it. His legal pseudonym in the US was *Arnold Rubinstein* (1882?-1955).

"He was an enigma," said Ilse Ollendorff. "He was supposed to be a very influential person in Russia at the beginning of the Soviet Union. No one knew his real name, no one knew what had happened to him, and everyone was very cagey about it." We do know that "Thomas Rubinstein," as he called himself, was the second husband of Annie (Pink) Reich. They met in Berlin in the early thirties. Rubinstein emigrated from Germany and lived with Annie Reich and her children in Prague from 1933 until 1938.

"I know that he knew Lenin and Trotsky," remembers his stepdaughter, Lore Reich, who lived with him in Prague as a child, and the US as a young adult. "According to my mother he [participated in] the 1905 revolution in Russia and then fled to Switzerland," and later Rubinstein

The Org: a New Beginning

was active in the 1917 Russian revolution.[101]

Until the release of the CPSU archives, the only published information about Thomas came from a few memoirs, like Karl Retzlaw's, and two conversations Thomas had in 1935 with Boris Nikolayevsky in Prague.[102] In 1993, Alexander Watlin published a history of the Comintern, including a chapter on "Comrade Thomas,"[103] providing new details about his life.

Rubinstein's party name, or *klichka,* was "Thomas." He also used the aliases "James Thomas," "James Reich," "Gordon," and "Dieckmann." His nickname was *der Dicke* ("the fatty," or "the pudge"). Lore Reich recalls someone telling her that his real name was James Reich. He was short, plump and smoked a pipe. According to contemporaries, he was one of the "key figures of the most secret activities of the Bolshevik leadership before the October Revolution" and was considered a "master of conspiratorial work."[104] In French and American books, he appears under the name "Yakov S. Reich." And, indeed, the real name of "Thomas Rubenstein" was Yakov Samojlovic Reich, born May 23, 1886 in Lemburg (Lvov), a Russian-occupied part of Poland. Lvov is now part of Ukraine, and is approximately 200 kilometers from Chernowitz, the boyhood home of Wilhelm Reich. If these two Reichs were relatives, it has never been documented.

Yakov Reich moved to Warsaw when he was nineteen and helped to establish the underground illegal press. As a student, he was a member of the Polish Socialist organization *Promien* (Ray), and later to the group *Zjednoczenie* (Association). During the first Russian revolution in 1905, he was a member of a "combat organization" and was wounded twice. He was one of the bomb throwers at the assassination of the Russian governor of Warsaw. Thomas's then close friend Pilsudski was involved, the later head of state of the independent Poland.[105]

Lore Reich, Wilhelm Reich's second daughter, who grew up with Thomas Rubinstein in the US, took issue with some of these points: "The idea that he ever threw a bomb is absolutely ludicrous. He was so unathletic he could not have thrown a ball."[106] Yet, with the release of documents from the Comintern, the evidence points toward her stepfather as a youthful bomb-thrower, if not an athlete.

After the failed 1905 revolution, Yakov Reich emigrated to Switzerland, by way of Germany, in 1907. In Zurich, he directed a chemical laboratory, experimenting in high explosives. Between 1906 and 1912, he studied jurisprudence, social sciences and, remarkably, pedology (educational reform) at the university. According to Alexander Watlin, he worked for socialist free-school publications like *L'Ecole Nouvelle* (The New School) and founded *Wolna Szkola* (The Free School) in 1908. His first wife, Beta Osipovna Reich, gave birth to a daughter in 1914 in Switzerland, but we know nothing of their fates. He served briefly in the Austrian Army (1914-1915) but was demobilized due to heart disease. He returned to Zurich, where he taught in schools. He published an anti-war pamphlet under the name "James Reich."

After the October Revolution, Comrade Thomas served as a member of the first Soviet delegation in Berne, Switzerland. In addition to being the editor-in-chief of the *Russian News*, in German and French, he prepared related propaganda material through Promachos-Verlag in Belp, publishing German translations of Lenin, Trotsky, and Karl Radek. (Some sources allude to other duties of Comrade Thomas: laundering money and jewels looted from Russia.) His press was shut down by the Swiss, and the members of the Soviet delegation were deported in late 1918. Thomas was imprisoned by Swiss authorities until January. He went to Moscow in February 1919. Three of his fellow exiles in Switzerland, Grigori Zinoviev, Nikolai Bukharin, and Karl Radek, had become his high-ranking patrons in the new Soviet government. He was a close friend of Radek's, who, like Thomas, had been born in Lvov.

As an employee of the People's Commissariat of Foreign Affairs, he was chosen to organize an international conference—held during the first week of March 1919—at which the Comintern was "spontaneously" founded. American journalist John Reed was a co-signer there to the Comintern Manifesto, which declared war on all capitalist countries. Comrade Thomas assisted in the publication of the periodical *Kommunistische Internationale* and wrote an article about the situation in Germany in its first issue, under the pseudonym "James Gordon."

Thomas' main duty was to keep undesirables away from the Comintern-founding conference, including notable functionaries of the German Communist Party (KPD) such as Paul Levi and August Thalheimer. Thomas belonged to the Central Committee of the Comintern, where he was known as "James Thomas." Lenin personally gave Thomas the assignment to organize the West European Section of the Comintern in Berlin. He arrived in Berlin in the autumn of 1919.

Some of the earliest financial records of the Comintern, first published in 1995, show that Comrade Thomas had received valuables and hard currency, worth no less than three million dollars, directly from Moscow during the 1919-1920 calendar year.[107] During this period, "the Comintern conducted covert operations and acted as the Soviet Union's foreign intelligence agency," notes historian Harvey Klehr.[108]

Gerd Koenen described those heady days:

> "The attitude of the Bolshevik party leaders, in the years of their victories in the civil war, was futuristic indeed. The world was one big playing field where one could make one's entries. In 1919, they tried to throw a 'red bridge' across the Baltic to Germany, because there, too, was also a civil war. They wired to Budapest and to Munich that bourgeois hostages should be taken, and a Red Army formed, just as they had done it themselves in Russia. In Berlin, where the Spartakus uprising had been smashed, 'Comrade Thomas' founded, in Autumn 1919, a 'West European Bureau' of the still-virtual Communist International, founded in Moscow in the Spring of 1919. In Vienna, the old capital of the Hapsburg Empire, a 'South-East Bureau' was installed, to further the fermentation in the various Balkan states"[109]

In his Comintern function, Thomas monitored meetings of the KPD and sent his reports directly to Lenin, Zinoviev, and Trotsky. Initially, German Communists such as Paul Levi, August Thalheimer, Herman Remmele, Willi Müzenberg, Clara Zetkin and Eduard Fuchs worked with Comrade Thomas, but he quickly assumed a dominant conspiratorial role. His closest co-worker, Ruth "Österreich" Jensen (1894-1943), alias "Ruth Gebhardt," served as Thomas' wife and secretary. They had a daughter, also named Ruth.[110] Communist parties throughout Europe had to report to Thomas. Of the office and its work, Lore Reich remembered, "I was always told it was the Russian Information Service."

Immediately after his arrival in Germany, Thomas organized a secret meeting in Frankfurt, for representatives from the Communist Parties of Europe, including Clara Zetkin of the German KPD, Mieczyslaw Bronski from Poland, Valerian Marcu from Rumania, Sylvia Pankhurst from Britain, and Karl Frank, representing the Austrian Communist Party. Thomas saw something in Karl Frank that he liked, and invited him to move to Berlin.

Karl Retzlaw worked with Thomas in the Western European Section of the Comintern.[111] Retzlaw reports that Thomas never appeared at a single KPD meeting personally, never once entered a KPD office, and was unknown to all but a handful of KPD officials. He conducted his business out of a bookstore. Also working in Thomas' staff in the Berlin Comintern office, according to Retzlaw, was a young Austrian, Karl Frank.

Thomas' assignment in Berlin was not to organize a political center in Berlin, but to provide an economic link between Moscow and Germany as a springboard to the rest of Europe. The Western European Section would disburse funds to European Communist parties, organizations and publishing houses. Vast sums of money would have to be laundered and the office set up business-fronts for its activities, which included providing false passports to illegal Comintern agents. Thomas also provided Moscow's new rulers with Western food delicacies (after all, there was a famine in Russia), books and periodicals, stationery, and sundry necessities. Like a corporate executive officer, he chartered airplanes, signed leases, traded stock in shell-companies, established tax-dodges. He bought the Hamburg publishing house Carl Hoym, which became the official publisher of Comintern literature. While Thomas ran the Hamburg operation from a second-hand bookshop in Berlin, the bookshop's accountant, Elena Stasova ("Hertha Sturm," "Lydia Wilhelm") was a member of the Russian Communist Party and Secretary of the Central Committee of the Comintern.

Thomas' publishing venture failed from an economic point of view: from the ten million reichsmarks invested, the receipts brought in a half million. Meanwhile, his conspiratorial activities made him enemies with the leading German communists, who astutely viewed Thomas as a Russian commissar sent to spy on them. While the German party had no funds for publishing, Thomas' products flooded the market. The KPD leadership had little knowledge of Thomas' activities, and what they knew they didn't like. Still, they depended on him for funds. They appealed to Moscow for a new commissar.

Thomas' Moscow patrons stood by their man, sending him millions in cash and valuables looted from the bourgeoisie. In 1921, Thomas brought 25 million marks in cash and 37 million in valuables to finance the "March Action"—a stillborn German revolution that wasted many revolutionary lives. Fifty million were earmarked for the "Frankfurt Fund," start-up capital for financing Europe's Communist parties. Thomas lost half that when he failed to transfer the fund into stable

The Org: a New Beginning

currency. In 1923 Thomas gave 7 million marks to the German Party for the October Revolution, with more lives lost.

Weimar-era inflation got the better of Comrade Thomas, and soon the financial affairs of his enterprise were in disarray. In August of 1921, a new man, Pyatnitsky, took over the budget of the Comintern back at the home office in Moscow and was shocked to find that there was absolutely no accounting for the millions being handled by Thomas in Berlin. He folded his books and marched to the Central Committee, where he charged Thomas with lining his own pockets. Thomas' patron, Comintern chief Grigori Zinoviev, stood up for him, but questions lingered until 1923, when a special commission in Moscow voted that Thomas should be kept away from Communism's cash-box. In Berlin, another special commission acquitted Thomas of most of the charges, except one issue: the fate of nearly nine million that Thomas claimed he had stuffed into some furniture and forgotten. Still, Bukharin, Zinoviev and Radek all vouched for Thomas' character. In August, 1923, a decision was made to retain Thomas as head of the publishing ventures, but that his secretary, Elena Stasova, should take responsibility for financial affairs. Thomas was promoted, and elevated to the publishing commission of the Executive Committee of the Comintern. Meanwhile, he remained in Berlin.

Elena Stasova, too, came under the scrutiny of Pyatnitsky the Comintern accountant. In 1924 Moscow made the New Economic Policy ultimatum that the Berlin operation support itself financially without subsidies. Thomas' patrons in Moscow were already losing influence, and the "Trotskyist" Karl Radek was removed from his important Comintern posts in early 1924. Thomas handed in his resignation in the autumn of 1924, but the bankruptcy of one of his publishing houses threatened to become a scandal. Thomas was recalled to Moscow, and after postponing his return several times, he finally arrived on April 23, 1925. The old charges of malfeasance were renewed, and on this occasion, even Zinoviev could not help.

The verdict of the International Control Commission was delivered on May 30, 1925: the finances of Thomas' publishing houses were "wrong and chaotic." The Commission decided that Thomas had never technically joined the party, and therefore he was no longer a "Comrade," but a "Spez," a specialist. On this technicality, the issue became closed because Thomas no longer existed as far as the Party was concerned. It appears that Radek, and the others, retained enough power to save Thomas' life and arranged for his safe departure from the Soviet Union.

Thomas had run the West European Section of the Comintern from the fall of 1919 until April, 1925. For the next eight years, he lived in Berlin, and the Comintern archives contain no records on him during this period. Whether he survived on money stashed in the furniture, we do not know. How he remained unmolested and living openly in Berlin in the late nineteen-twenties, we cannot say. Did he have documents, perhaps, as life insurance? Thomas officially took the name "Rubinstein." He made one last trip to the Soviet Union in 1926, but apparently lived a "legal" life and began work on a book on the Russian revolution. In 1928 he edited, along with Communist intellectuals Alexander Slepkov and Valentin Astrov, the book, *Illustrierte Geschichte der Russischen Revolution* (Illustrated History Of The Russian Revolution). Thomas joined the Communist Party-Opposition (KPO) in 1928, and in 1929 he completed another book, *Illustrierte Geschichte des Russischen Bürgerkriegs* (Illustrated History Of The Russian Civil War).[112] Both books were published by Willy Münzenberg's Neuer Deutscher Verlag.

Lore Reich tells a story of how Rubinstein was detained by the Nazis, while in possession of a working draft of his book about the Russian Revolution, which, if discovered by them, would have certainly resulted in his death. They let him go, and upon leaving them and his never-published magnum opus behind, one of the Nazis called him back. "You forgot your gloves," the Nazi told Thomas.

In 1932, together with the minority faction of the KPO, Thomas went over to the Socialist Workers Party (SAP), which elected him to a leadership position. Secretly, he also became a member of "the Org."

In 1933 Thomas Rubinstein emigrated to Czechoslovakia where he lived with the former wife of Wilhelm Reich, Annie (Pink) Reich. In 1934, his application to the Social Democratic Party was refused. In 1935 he resigned from the SAP. Next he openly associated with Neu Beginnen. Both Neu Beginnen and Revolutionäre Sozialisten had foreign bureaus in Prague. Rubinstein was now in Prague, where he was in control of the illegal work of the SAP inside Germany. Thus, Thomas was the SAP counterpart of Neu Beginnen's Karl B. Frank, who was responsible for the Org's *Auslandbüro* in Prague. There were 20 or 30 SAP members in Prague.[113] Along with the *klichkas*, the factions, and the overlapping associations between socialist groups, historians have yet to unravel the unstated and hidden conspiratorial agendas of covert operators within this period, when German and Austrian leftists were scattered across Europe by Hitler's putsch.

Wilhelm Reich met Rubinstein in Grundlsee, Austria, where Annie and Thomas were vacationing in August of 1935. Reich and Rubinstein did not get along well. "They had little pet-names for each other," said Reich's daughter, Lore Reich, "Thomas called my father 'the skunk' while my father referred to him as 'the snake.'"

In January 1936, Rubinstein, along with Annie Reich and Otto Fenichel, participated in the organization of a "Prague Marxist-Psychoanalytic Study Group." Karl Frank, who was also involved in this project around that time, was in California, raising money for the Org, and meeting with Americans such as Mildred Edie and Robert A. Brady.

From Prague, Rubinstein somehow orchestrated the 1937 escape of fellow New Beginner Edith Jacobson from a German prison hospital, where she had been moved since developing diabetes during her imprisonment, Jacobson having been arrested in 1936 for her activities with underground units of New Beginning.

Rubinstein, like most old Bolsheviks, did not believe in legal marriage, but he had fathered a child with Ruth "Österreich" Jensen during his early Berlin days. The daughter was also named Ruth Österreich. Thomas abandoned this family. The daughter, age 15, was imprisoned by the Nazis in Belgium, where she was impregnated by a German soldier. If Rubinstein's daughter, Ruth Österreich, is still alive, she would be seventy-six years old today. Lore Reich spent some effort tracking down these relations, to no avail.

In 1938, Thomas moved to New York, with Annie Reich and her children, Lore and Eva Reich. Rubinstein began to recompose his history of the Russian revolution.

According to Eva Reich, her mother supported the family on her income as an analyst in New York. Thomas cooked for the family, and Eva remembers him making yoghurt in the kitchen. She recalled that Thomas had once lent her a book on the French Revolution. She was reading it in the bathtub and accidentally dropped the book in the bathwater. When she told Thomas about the mishap, he said to her, "I was testing you. You *failed the test*." Thomas never learned English and never acclimated himself to life in America. He sat in his study, trying to reconstruct a book that had been lost to the Catastrophe.

Wilhelm Reich arrived in New York the next year, in August of 1939. He found he could communicate with his children only with great difficulty. He, Annie and Thomas Rubinstein and family, all lived in and around the New York City area.

Wilhelm Reich suspected his ex-wife, along with Rubinstein, of involvement in his arrest as an "enemy alien" on December 12, 1941, by the FBI, who picked him up at 2 a.m. and delivered him to Ellis Island, where he was imprisoned for nearly a month.[114]

It is unlikely that Rubinstein, living under an assumed name, hiding in the United States, would have directly contacted the FBI, or permitted Annie to do so. However, Thomas having been the former head of the Comintern's Western European apparatus, the possibility that Wilhelm Reich might have indirectly denounced Rubinstein to the FBI cannot be discounted.

There is no evidence that Reich denounced Rubinstein, or anyone else, for that matter, who he knew to be Communists. Given the mistrust between Annie and her ex-husband, and given the absolute secrecy with which Thomas conducted his business in Berlin during his Comintern days, Reich may well have been ignorant of Rubinstein's importance. Rubinstein had already left his Comintern work long before 1930, when Reich moved to Berlin, and Rubinstein's work had been well concealed.

In 1941 Rubinstein worried far less about the FBI than Stalin's henchmen. Eva Reich vividly recalled her stepfather hiding in a locked room each time a patient knocked on the apartment door for an analytic appointment with Annie. Unanticipated visits to the residence sent Rubinstein into a panic. "They've come for me," he'd say. Rubinstein knew the reach of Stalin's retribution.

Eva Reich remembers another anecdote about Thomas that speaks to his skills in counterintelligence. He became furious with her when she once disturbed the papers on his desk. He had been alerted to the slight rearrangement because she had disturbed "some dust" he had carefully placed there before his absence.

"Everybody Goes to Rick's"

The immortal film, "Casablanca," starring Humphrey Bogart, Ingrid Bergman, Claude Rains and Paul Heinreid, still offers us a glimpse into the world of the Org, in the Hollywood style of 1942. Ronald Reagan

The Org: a New Beginning

was considered for the lead. The film was based on a play by Joan Alison and Murray Burnett, highschool teachers who visited Vienna and Europe in 1938. The film opens with a newsreel map of the "refugee trail" taken by thousands of exiles from Europe, from Paris to Marseilles, across the Mediterranean to places like French Morocco in December 1941. (*Casablanca* begins its time frame comfortably *after* the German invasion of Russia: in July, 1941.) The film is more historically accurate than has been credited by film critics. The luscious and dangerous feel of the film helps us capture that time today.

Victor Laszlo, played by the Austrian actor Paul Heinreid, is a veteran Czech underground fighter against the Nazis, who has eluded them time and again. Laszlo arrives in Tangier with his young wife, Ilsa Lund from Oslo, Norway. Ingrid Bergman—I must insert my personal feeling here—plays the role of Ilsa Lund perfectly and is the most beautiful woman I have ever seen on the screen.

Humphrey Bogart plays the role of Rick Blaine, a former arms smuggler ('35, Ethiopia; '36, Spain) who camouflages his leftist heart with a character attitude of indifferent egoism. "I'm the only cause I'm interested in," Rick tells the Vichy French police inspector, played by Claude Rains. "I stick my neck out for nobody," Rick tells the French Inspector, who retorts, "An excellent foreign policy."

Casablanca was directed by Michael Curtiz (1888-1962)—he was born in Hungary and learned filmmaking in Austria. The screenplay was written by three Jews, Howard Koch, Julius J. and Philip G. Epstein. Many of the "Nazis" in the film were portrayed by Jewish actors. It won Oscars for Best Picture, Best Director, and Best Screenplay for 1942.

Two endings for the film were shot, and the actors were kept guessing throughout production where the plot might lead. One version, with which we are all familiar, had Rick giving up the girl for the greater good of all mankind. There was an alternate ending, one not chosen, after sample audiences reacted negatively in one of the first uses of "focus groups" in Hollywood. This version had Rick leaving Tangier with Ilsa Lund, and leaving the heroic Victor Laszlo in the dust.

Karl B. Frank was a dead ringer and perfect model for Victor Laszlo. The eyes, the hair, the simple dignity of Frank, along with his career in the underground, his modern attitude toward his wife's infidelity, his personal risks, his resigned dedication, cannot help but remind us of Victor Laszlo in "Casablanca." And Rick Blaine, in his tormented sense

Wilhelm Reich and the Cold War

of abandonment by a woman, was like Wilhelm Reich.

We see the deeper truth of *Casablanca* as its characters develop through the movie. Just as the actors did not know how the movie would end when they performed the individual scenes, nor did the world know how the war in Europe would end in 1942. The players are confused, torn between dialectic pulls from four directions. In the end, they win, but they didn't feel like winners.

Notes

1. See *Reich Speaks of Freud,* and *People in Trouble.*

2. *Fury on Earth*, p. 366. Reich's second source, aside from Karl Frank, was Dwight MacDonald, via Myron Sharaf. In *Fury on Earth,* Sharaf writes that "in 1946, the most laudatory and insightful review (of Reich's writings) was published by Paul Goodman in Dwight MacDonald's periodical *Politics*, which reached a small but influential audience" (p. 268). Sharaf writes that he himself led Reich to the inevitable conclusion: "Regrettably I, too, contributed to the loose political characterization of Brady by repeating to Reich a statement heard from Dwight MacDonald. MacDonald had casually mentioned something about the fellow-traveling or Stalinist sympathies of Mildred and her husband, and Reich exaggerated the significance of this vague remark" (p. 366). Dwight MacDonald was a coeditor of the *Partisan Review* (1937-1943), member of the Trotskyist Socialist Workers Party (1939-1941), editor of the leftist magazine *Politics* (1944-1949), a frequent contributor to *The New Yorker*, and author of *Memoirs of a Revolutionary* (1957). He most certainly would have known Brady's political line.

3. Letter from Ilse Ollendorff Reich to Karl B. Frank, June 10, 1948. Karl B. Frank Papers, Hoover Institution Archives, Palo Alto, California.

4. *Reich Speaks of Freud,* p. 115.

5. Gerhart Eisler's brother Hanns was a noted composer who scored screenplays and who had worked with Bertolt Brecht. He wrote "The Comintern March." He was admitted into the US in 1939 through the personal intervention of Eleanor Roosevelt, who wrote a handwritten appeal to Michael Straight's housemate and superior at the State Department, Sumner Welles. Hanns Eisler taught at the New School For Social Research in 1939, the same year Reich arrived in New York to teach there as well. Straight's mother was a founding donor of the New School. In *American Odyssey,* Reich comments that Hanns Eisler taught that music has nothing to do with sexuality.

6. Johannes Reichmayr, *Spurensuche in der Geschichte der Psychoanalyse*, Frankfurt, Nexus Verlag, 1990.

The Org: a New Beginning

7. Ruth Fischer/Arkadij Maslow, *Abtrünnig wider Willen. Aus Briefen und Manuskripten des Exils*, edited by Peter Lübbe, München, R. Oldenbourg Verlag, 1990, p. 443.

8. Arnold Deutsch, the NKVD recruiter of Kim Philby, was a member of this group as a young man in 1920. See *The Crown Jewels,* by Nigel West and Oleg Tsarev, Yale 1999.

9. *Passion of Youth* p. 124.

10. Karl Retzlaw, *Spartakus.* Aufstieg und Niedergang. Erinnerungen eines Parteiarbeiters, Frankfurt, Verlag Neue Kritik, 1974.

11. *Passion of Youth* p. 125.

12. *Passion of Youth*, p. 137-8.

13. Alice Herdan's full name: Alice Henriette Alberta Herdan-Harris von Valbonne und Belmont.

14. Alice Herdan-Zuckmayer, *Genies sind im Lehrplan nicht vorgesehen,* Frankfurt, Fischer Taschenbuch Verlag, 1981.

15. Goodman, Walter. *The Committee.* FS&G, NY, 1968, p. 201.

16. Eric Bentley (ed.), *Thirty Years of Treason.* Excerpts from Hearings before the House Committee on Un-American Activities, 1938-1968, New York 1971.

17. Koestler, *The Invisible Writing.* Beacon, Boston, 1954. p. 39.

18. Eric Bentley (ed.), *Thirty Years of Treason.* Excerpts from Hearings before the House Committee on Un-American Activities, 1938-1968, New York 1971.

19. "Autobiographical Data" from Karl B. Frank, submitted to the OSS in July, 1942. Hoover Institution Archives, Karl B. Frank Papers.

20. Karl Fallend, Johannes Reichmayr (ed.), *Siegfried Bernfeld oder Die Grenzen der Psychoanalyse.* Materialien zu Leben und Werk, Frankfurt, Stroemfeld/Nexus, 1992.

21. Max Seydewitz, *Es hat sich gelohnt. Lebenserinnerungen eines alten Arbeiterfunktionärs,* (East) Berlin, Dietz Verlag, 1976.

22. Elisabeth Neumann-Viertel, *Du mußt spielen,* Wien, Edition S, p. 96. Lampl was a close friend of Freud's son Martin. In 1921 he went to Berlin to study psychoanalysis, and after Hitler's take-over he went back to Vienna. Together with Walter Marseille and Franz Urbach he was involved in Bernfeld's and Feitelberg's "libidometric studies" in Berlin and Vienna between 1929 and 1935.

23. Otto Fenichel, *119 Rundbriefe,* Frankfurt, Stroemfeld, 1998, p. 280.

24. Soell, Hartmut, *Fritz Erler. Eine Politische Biography.* Verlag J.H.W. Dietz Nachf., 1976.

25. Hoffman, *History of the German Resistance,* p. 21.

26. It is interesting to note that the Nazi intelligence officer Reinhart Gehlen created a spy network in Eastern Europe, which survived the war under the American authorities, and it was similarly called the "Gehlen Org."

27. Walter Löwenheim, *Geschichte der Org (Neu Beginnen) 1929-1935*. Eine zeitgenössische Analyse (written in Autumn 1935). Berlin, Gedenkstätte Deutscher Widerstand und Edition Hentrich, 1995, pp. 116-118.

28. Karl Fallend, Johannes Reichmayr (ed.), *Siegfried Bernfeld oder Die Grenzen der Psychoanalyse*. Materialien zu Leben und Werk, Frankfurt, Stroemfeld/Nexus, 1992.

29. See Kapp, Reginald O., Comments on Bernfeld and Feitelberg's "The Principle of Entropy and the Death Instinct." In *International Journal of Psychoanalysis* 12, 1931.

30. Nov. 6, 1942, *American Odyssey*.

31. Frankfurt, Stroemfeld/Nexus, 1992.

32. See also translator's note at the beginning of "The Masochistic Character" in *Character Analysis,* Orgone Institute Press, pp. 208-210.

33. *119 Rundbriefe*, p. 1383f).

34. Karl Fallend/Bernd Nitzschke, *Der "Fall" Wilhelm Reich*, Frankfurt: Suhrkamp, 1997, p.181f.

35. ibid., p. 34.

36. *The Nation*, 8/1/34, pp. 135-136 (issue 3604).

37. email from John Wilder, 4/27/98.

38. Hoover Institution Archives, Karl Frank Papers; Memoranda from Karl Frank to the US State Department, July 13, 1942. "New Beginning Group."

39. Bauer, Otto, *Die Illegale Partei*. (Paris, 1939) p. 142.

40. *Der Kampf* III (1936), p. 157.

41. Interview with Michelle Weston by John Wilder, 3/3/98.

42. Ernst Glaser, *Im Umfeld des Austromarxismus; Ein Beitrag zur Geistesgeschichte des österreichischen Sozialismus,* Wien, Europaverlag, 1981, p. 271.

43. Emmy Rado, working for the OSS, recorded an event where "Paul Hagen" (Frank) appeared at a luncheon meeting of the American Friends of German Freedom at the Biltmore Hotel in 1942, sharing the dais with Michael Straight, the publisher of *The New Republic.* National Archives, Military Records (OSS); memo to Lithgow Osborne from Emmy Rado.

44. Helene Maimann and Heinz Lunzer (editorial staff), *Österreicher im Exil 1934 bis 1945.* Protokoll des Internationalen Symposiums des Österreichischen Exils von 1934 bis 1945, Abgehalten vom 3. bis 6. Juni 1975 in Wien, Vienna, Österreichischer Bundesverlag für Unterricht, Wissenschaft und Kunst, 1977.

45. Letter from Frank to Buttinger, Nov. 20, 1946, Hoover Institution, Karl Frank Papers. Translated by Peter Nasselstein.

46. Letter from Frank to Buttinger, March 24, 1947. Hoover Institution, Karl Frank Papers.

The Org: a New Beginning

47. Klehr, Harvey and Radosh, Ronald, *The Amerasia Case; prelude to McCarthyism.* UNC Press, Chapel Hill, 1996, p. 160.

48. Similar sentiments were expressed by Max Seydewitz: Seydewitz, Max, *Es hat sich gelohnt. Lebenserinnerungen eines alten Arbeiterfunktionärs,* (East) Berlin, Dietz Verlag, 1976.

49. Identified as a "source of information" for the Soviets in the VENONA decrypts, given the code-name "Ruff." See p. 194-5 in *VENONA; Decoding Soviet Espionage in America,* by Haynes & Klehr. 1999, Yale.

50. At the beginning of the war, Frank was sent from London to the US, since the English political administrative bodies had a certain problem: Adam von Trott zu Solz. Trott zu Solz, a former Rhodes Scholar and personal friend of many important Englishmen from his student days, was a German diplomat. The English did not know whether they could trust him, thus they decided to send Frank to interview him. Karl Frank came to the then-neutral US with the mission to meet Trott zu Solz and to determine whether he was reliable. Frank concluded that he could be trusted, but the English paid little heed to his recommendation. (Richard Löwenthal, *Die Widerstandsgruppe "Neu Beginnen."* Beiträge zum Thema Widerstand 20, Dokumentationszentrum Berlin, Gedenk- und Bildungsstätte Stauffenbergstr, 1982.)

51. Leo Baeck Institute.

52. Nichols, Anthony, and Matthias, Erich, *German Democracy and the Triumph of Hitler; essays in German history.*

53. Ibid., p. 247.

54. Ibid., p. 248.

55. Smith, Harris, *OSS.* Berkeley, 1972, p. 218.

56. Apparently Emmy Rado's husband was unrelated to another Alexander "Sandor" Rado, but it has not been determined conclusively. The "other Sandor Rado" was a cartographer and an agent of the NKVD in Switzerland. This Rado employed Arthur Koestler in his offices. See Arthur Koestler's *Invisible Writing.*

57. *Reich Speaks of Freud* p. 105.

58. Roazen and Swerdloff, *Heresy: Sandor Rado and the Psychoanalytic Movement,* page 180.

59. The Eitingon family estate was built on an international cartel based on Russian furs. When Motty's cousin, Waldemar (Vladimir) Eitingon had died, Arthur Garfield Hays served Waldemar Eitingon's estate lawyer. (Later, Hays served as Wilhelm Reich's lawyer.) This fortune became Max Eitingon's grubstake in Berlin, where he funded Sigmund Freud's International Psychoanalytic Association.

60. McIntosh, Elizabeth, *Sisterhood of Spies; the women of the OSS.* Naval Institute Press, Annapolis, 1998, p. 83.

61. From Mosley's *Dulles*, p. 226.
62. From *Reich Speaks of Freud* pp. 111, 112.
63. Sharaf, *Fury on Earth*, p. 193.
64. Roazen, Paul, *Helene Deutsch: A Psychoanalyst's Life*. Garden City, NY: 1985, p. 282-3.
65. Russell Jacoby, p 163.
66. John Wilder's communication with Frank's children.
67. McIntosh, Elizabeth. *Sisterhood of Spies; the women of the OSS*. Naval Institute Press, Annapolis, 1998, p. 82.
68. Smith, Harris, *OSS* Berkeley, 1972, p. 405, note 33.
69. Winks, Robin, *Cloaks and Gown: 1931-1961, Scholars in the Secret War*. Morrow & Co., New York, 1987, p. 175. From this same source comes the information that Karl Frank was once considered for use in an assassination attempt against Hitler.
70. Hoffman, Peter, *The History of the German Resistance 1933-1945*. MIT Press, 1977, page 227.
71. The author thanks Peter Nasselstein of Hamburg, Germany for his information on the Mot Dag group. His sources in the German literature included:

- Brandt, Willy, *Mein Weg nach Berlin*, München, Kindler Verlag, 1960.

- Brandt, Willy, *Draußen*. Schriften während der Emigration, München, Kindler Verlag, 1966.

- Brandt, Willy, *Links und Frei*. Mein Weg 1930-1950, Hamburg, Hoffmann und Campe, 1982.

- Brandt, Willy, *Erinnerungen*. Frankfurt, Propyläen, 1989, p. 100.

- Lorenz, Einhart, *Willy Brandt in Norwegen*. Die Jahre des Exils 1933 bis 1940, Kiel, Neuer Malik Verlag, 1989.

- Schröck, Rudolf, *Willy Brandt*. Heyne Biographien, München 1991, pp. 70-73.

72. Jens A. Christophersen, "'Mot Dag' and the Norwegian Left," *Journal of Contemporary History*, Vol.1, No. 2, p139. London, 1966.
73. For a discussion of the mole within Willy Brandt's administration, Günter Guillaume, from the standpoint of the former chief of state security of Eastern Germany (also known as the Stasi), see Markus Wolf, *The Man Without a Face*, Public Affairs, New York, 1997 chapter 9. The scandal was widely viewed as a complete blunder on the East German side.
74. See http://www.multinet.no/~falk/Falk7.2.1.htm. Information drawn from *Erling Falk and Mot Dag,* written by Mot Dag member Tryggve Bull, Cappeln House, 1955.
75. Jens A. Christophersen, "'Mot Dag' and the Norwegian Left," *Journal of Contemporary History*, Vol.1, No. 2, p141. London, 1966.

76. Helmut Müssner, *Exil in Schweden*. Politische und kulturelle Emigration nach 1933, München, Carl Hanser Verlag, 1974, p. 470f.

77. Statement of Brandt about Frank, Feb. 27, 1958. Hoover Institution; Karl Frank Archive. (Translated from German by Peter Nasselstein).

78. Meller, Marcowicz Digne, *Viva Kleiner Mann; über Wilhelm Reich*. Nexus, Frankfurt, 1987, p. 28-9.

79. Josef Hufelschulte: Deckname "Polarforscher." Exclusiv: Wie der sowjetische Geheimdienst KGB Willy Brandt mit peinlichen Agentendossiers und einer Quittung erpressen wollte. In: FOCUS, Nr.15, 12th April 1999, pp.88-95.

80. Andrew and Mitrokhin, *The Sword and the Shield*, Basic Books, NY, 1999, p. 440-445.

81. The name Schlamm, in German, literally means "mud." Jews had been forced by Austrian authorities to obtain family names, and the rich could buy beneficent names like "Adler," "Rubinstein," "Einstein," "Reich," "Freud," "Goldstein," "Rosenberg," which sounded either like nobility or had a good meaning; for instance, Reich means, literally, rich. Apparently Schlamm's ancestors could not afford to bribe the Austrian officials.

82. *Zeitschrift für Politische Psychologie und Sexualökonomie*, Vol. IV(3)(14), pp. 230-231 (Sept. 1937) by H-P (Harry Proell), Review of "Willi Schlamm, Diktatur der Lüge."

83. Source of biographical material on Willi Schlamm, *Munziger-Archiv/Internat.Biograph.Archiv* 7.10.1978 - Lieferung 40/78 - K - 8929.

84. Schlamm, Willi. *Welt am Sonntag*, April 5, 1970, "Der Vater der 'Sex-Revolution.'" Source: William S. Schlamm, *Am Rande des Bürgerkriegs*, Berlin, Zeitbuch Verlag, 1970, pp. 341-348.

85. William S. Schlamm, *Zorn und Gelächter. Zeitgeschichte aus spitzer Feder*. Ausgewählt von Kristin von Philipp, München, Georg Müller Verlag, 1977, pp. 324-327 (from the *Zeitbühne* 1972).

86. *The Sexual Revolution*, 1962, p. xxxiii.

87. Reich, Wilhelm; Baxandall, Lee, ed. *Sex-Pol Essays, 1929-1934*, Random House, NY 1976, p. 285. Ernest Bornemann was born in 1915. He was a young follower of Reich's Berlin Sexpol organization. In 1933 he emigrated to London, and when Reich came to London in 1933, Reich met Bornemann's university professor Bronislaw Malinowski. Bornemann claims that Reich's article "What is Class Consciousness?" was actually written by Bornemann for Reich's *Zeitschrift für politische Psychologie und Sexualökonomie*. Reich, according to Bornemann, had rewritten the article and had, without Bornemann knowing, published it under the pseudonym "Ernst Parell." But Reich then assured Bornemann that "Ernst" stood for Bornemann's first name and "Parell" meant that they are a collective (from French *pareil* = uniform). For example Ola Raknes would

trade under the name of "Ola Parell" and he, Reich, as "Wilhelm Parell." In reality only Reich used the pseudonym "Ernst Parell." (Marc Rackelmann: "Wilhelm Reich und der Einheitsverband für proletarische Sexualreform und Mutterschutz: Was war die Sexpol?" *Emotion*. Beiträge zum Werk von Wilhelm Reich, Nr. 11, 1994).

88. Christiane Rothländer has written a diploma dissertation on "Austrian Psychoanalysis During National Socialism" (1938-1945). In this connection she came across a certain Graf (a German count or earl) Karl von Motesiczky. She was moved by the story of his life and tried to reconstruct it. Her research was recounted in an article by Ella Lingens in the *Sigmund Freud Bulletin* (1985): Motesiczky was born May 27, 1904 in Vienna, to a rich family. His great-aunt Anna Tadesco was one of the first patients of Freud. As a patient and student of Reich, Motesiczky went to Berlin and followed Reich also to Sweden and Norway. He was very upset by Reich's expulsion from the German Psychoanalytic Society. According to Ella Lingens, Graf Karl von Motesiczky went back to Austria after Reich left Norway for America. But Anni Urbach, the daughter of Paul Federn, writes in a letter that he came back to Austria shortly after the Nazi occupation of Austria, to his hereditary estate near Vienna. According to Ella Lingens, Motesiczky was a friend of the conservative psychoanalyst August Aichhorn. He continued his psychoanalysis in Vienna with Aichhorn. He also attended Aichhorn's introductory seminar to psychoanalysis held in Aichhorn's apartment. As a "half-Jew," Motesiczky could not become a psychotherapist officially. In the winter of 1940 Motesiczky joined the Wehrmacht on a *voluntary* basis ("half-Jews" were not required to become soldiers), thus he hoped to escape racial discrimination. During his military basic training he notified his superiors that a sergeant came back to the barracks too late. For this "offense," because he was a "Jewish traitor of his military companions," he was expelled from the Wehrmacht. In 1939 he had met the couple Kurt and Ella Lingens at Aichhorn's training seminar. They lived together at Motesiczky's estate. In the summer 1942, they tried to hide Polish Jews at the estate to prepare for their escape from Germany to Switzerland. They were betrayed by one of their representatives. On Oct. 13, 1942 they were arrested by the Gestapo, were Motesiczky confessed everything, after the Gestapo said they had all the "evidence." On Feb. 16., 1943 Motesiczky was deported to Auschwitz together with Ella Lingens. Four months later, on June 25, he died of typhoid.

89. Socialist Reproduction, *Selected Sex-Pol Essays, 1934-1937*. London, April, 1973, p. 83.

90. Ibid. p.112.

91. "An Early Anti-Nazi" by James A. Weschler. undated (early 40s?) periodical unknown, Hoover Institution Karl B. Frank papers.

92. Socialist Reproduction, *Selected Sex-Pol Essays, 1934-1937*. London, April, 1973. p. 58. Translation of the "Manifesto of the Zeitschrift für Politische Psychologie und Sexualökonomie," Verlag für Sexualpolitik, Copenhagen, *Zeitschrift für Politische Psychologie und Sexualökonomie*, Vol 3/4, 1934.

93. *People in Trouble*, p. 212.

94. Sharaf, *Fury on Earth*, p. 136.

95. *People in Trouble*, p. 227.

96. Jacoby, Russell, *The Repression of Psychoanalysis; Otto Fenichel and the Political Freudians.* Chicago, 1983.

97. *People in Trouble*, p. 227.

98. Sharaf, *Fury on Earth*, p. 245-246.

99. Letter to Lotte Liebeck, January 15, 1935, reprinted in *Reich Speaks of Freud*, p. 207.

100. Jacoby, Russell, *The Repression of Psychoanalysis; Otto Fenichel and the Political Freudians.* Chicago, 1983, p. 101.

101. Letter to the author from Lore Reich, M.D., January 2, 1998.

102. Boris Nikolaevsky, Les Premieres Années de l'internationale Communiste - D'après le récit du "camarade Thomas." In *Contributions à l'histoire du Comintern*. Ed. Jacques Freymond, Genève 1965, pp. 1-28.

103. Watlin, Alexander, *Die Komintern 1919-1929*. Mainz, Decaton Verlag, 1993.

104. Vorholt, Udo. *Die Sowjetunion im Urteil des sozialdemokratischen Exils 1933 bis 1945. Eine Studie des Expilparteivorstandes der SPD, des Internationalen Sozialistischen Kampfbundes, der Sozialistischen Arbeiterpartei und der Gruppe Neu Beginnen*, Frankfurt, Peter Lang, 1991.

105. Karl Retzlaw, *Spartakus. Aufstieg und Niedergang. Erinnerungen eines Parteiarbeiters*, Frankfurt, Verlag Neue Kritik, 1974.

106. Letter to the author from Lore Reich, M.D., January 2, 1998.

107. Klehr, Harvey; Haynes, John; Firsov, Fridrikh. *The Secret World of American Communism.* Yale, New Haven, 1995, pages 22-24.

108. ibid., page 20.

109. Gerd Koenen, *Utopie Der Säuberung. Was war der Kommunismus*, Alexander West Verlag, 1998, p. 120. A related question: who was the "Comrade Thomas" of the Vienna Office? Was it General Walter Krivitsky? In 1938 Willi Schlamm met Krivitsky, the former West European chief of the GPU, in New York. Krivitsky had defected in 1937 and published a lengthy, confessional book. Schlamm was surprised by Krivitsky's incredible detailed knowledge and an even more incredible power of recollection. He could remember the most intimate details of Schlamm's life in 1919 (details even Schlamm himself had meanwhile forgotten), even though Schlamm had never met Krivitsky before. Krivitsky replied that in 1919 he was the chief of the Viennese Secret

Office. Schlamm then asked if Krivitsky could tell him why the Russians never ordered him into the "abyss," i.e. underground work. Krivitsky laughed and replied that they never really trusted Schlamm. "When you came in 1919 to the Party and said that it can have you, save for two evenings in the week for concerts, we did not trust you, because Mozart was apparently more important to you than the Party." (William S. Schlamm, *Die jungen Herren der alten Erde. Vom neuen Stil der Macht*, Stuttgart, Seewald Verlag, 1962).

110. Until 1928, Ruth Österreich Jensen was a member of the KPD; she joined the KPDO in 1929. At the end of 1931 she went to the SAP. In 1933, she had to emigrate to Prague, were she became the secretary of Willi Schlamm. In 1933, on emigration to Prague, she worked there at the Spanish Embassy as an agent of the Secret Service of the Spanish Republic. In 1938 she went to Paris and on to Brussels, where she and her little daughter were arrested by the Gestapo on April 21, 1941. According to Eva Reich, Österreich's daughter was released after being impregnated by a German soldier. Her mother was convicted by the *Volksgerichtshof* (People's Court) in Berlin, sentenced to death for high treason, and executed in Plötzensee on June 25, 1943. Her daughter was still alive in 1986 when she told Theodor Bergmann her story. (See Theodor Bergmann, *"Gegen den Strom" Die Geschichte der Kommunistischen-Partei-Opposition*, Hamburg, VSA-Verlag, 1987, p. 410, and, Herbert Crüger, *Verschwiegene Zeiten. Vom geheimen Apparat der KPD ins Gefängnis der Staatssicherheit*, Berlin, LinksDruck, 1990, p. 241).

111. Karl Retzlaw, *Spartakus*. Aufstieg und Niedergang. Erinnerungen eines Parteiarbeiters, Frankfurt, Verlag Neue Kritik, 1974.

112. Reprinted in 1970, Frankfurt, Verlag Neue Kritik..

113. Bremer, Jörg, *Die Sozialistische Arbeiterpartei Deutschlands (SAP). Untergrund und Exil 1933-1945*. Frankfurt, Campus-Verlag, 1978; pp104-107.

114. Sharaf, *Fury on Earth*, p. 271.

Chapter Three: General Eitingon and Doctor Eitingon: The Sword and the Shield

Surrounded by "spies," and "class enemies," the new Soviet state formed a "Sword and Shield" of the Party, known as the Cheka. During the Russian civil war of the years the Cheka proved itself valuable in its violent suppression of counter-revolutionaries within and without Russia. Over the years, the Cheka, and its inheritors, the NKVD and the KGB, carried out innumerable political assassinations across the globe, of which the plot against Trotsky's life is most infamous. Of all the officers who served in the ruthless security apparatus, Stalin had a high regard for the agent who directly orchestrated the Trotsky assassination, Naum Isakovich Eitingon.

Surrounded by "enemies" and "distorters of his teaching," Sigmund Freud formed a "Secret Ring" of analysts upon whom he could rely to defend psychoanalysis from distortion, embarrassment, and corruption of its ideals from within. No person was more important in this role than Dr. Max Eitingon (1881-1943), the Russian-born analyst who lived in Berlin and had provided most of the funding for the publications of the International Psychoanalytical Publishing House. Eitingon functioned like a political professional, a bureaucrat. Jung joked to Freud that Eitingon would likely be elected to the Duma (the Russian parliament) one day.

In Berlin, Eitingon set up an analytic, literary and cultural salon that attracted the brightest and most charming of Russian emigrés, as well as leading figures of German society. Enormously wealthy by virtue of his family's fur-trading company, Eitingon had no need to see patients and wrote few if any psychoanalytic papers; he was nonetheless highly valued by Freud, who would come to view Max Eitingon as a son.

Wilhelm Reich and the Cold War

Reich's Expulsion from the
International Psychoanalytic Society 1933-1934

"*Eitingon,*" Reich wrote in *People in Trouble,* "president of the Association, had already asked me in October 1932 not to admit any candidates to my technical seminar, which was attended by approximately twenty practicing Berlin analysts. I rejected this unjustified request and he vetoed my election to membership in the Berlin Training Institute. Nevertheless, I gave lectures at the Institute which were very well attended."[1]

When Reich first looked into moving to Copenhagen, he asked Eitingon whether his training candidates would be recognized by the IPA. Eitingon replied rather wormily that due to "differences in opinion, my [Reich's] candidates were to be subjected to stricter examination."[2] At the same time Eitingon officially approved the teaching credential for a Dr. Harnik, whom Freud had assessed as a "manic-paranoid."[3] Reich's Danish colleagues wrote a letter of protest directly to Freud, arguing that the old man "was scarcely in a position to judge from such a distance how much Harnik has damaged the psychoanalytic movement here. [...] Let me just say this: people in Copenhagen were greatly surprised, and still are, that a man in his condition was a member of the teaching committee of the German Psychoanalytic Association, that he was given the difficult assignment to teach psychoanalysis in Denmark, and that he was an authorized analyst at all."[4] As for Wilhelm Reich, the Danish analysts valued his work highly and wished "to keep him at all costs." Freud replied and acknowledged Reich's stature, but rejected the Danes' plea.

When Wilhelm Reich was expelled in a secret meeting from the German Psychoanalytic Society in 1933, Max Eitingon was presiding. Eitingon had not made this decision alone—certainly Anna Freud and Freud had discussed it. Expulsion from the national society automatically meant expulsion from the international body. The IPA never formally announced nor explained this decision—ever. The furtive and unexplained decision to banish Reich requires that we take a closer look at Max Eitingon, taking our cues from his objective behavior.

In a letter to Reich dated December 29, 1935, after Reich had published the details of the expulsion in his own journal, Eitingon claimed that he had played no role in the expulsion of Reich from the IPA.[5] "As late as 1933," Eitingon wrote, "when I was still in Germany,

General Eitingon and Doctor Eitingon

I was against your expulsion from the German Society and kept pointing out to the Executive Committee that under my aegis a thing like that would not be allowed to happened."[6] Eitingon's recollection varied from Felix Boehm's.

Felix Boehm, a psychoanalyst in Berlin, wrote an account, dated August 31, 1934, describing the events in the German Society around the time Wilhelm Reich was expelled from the group. "Early in 1933, in various National Socialist periodicals, there appeared very disparaging articles about psychoanalysis. The associations in Germany, scientific and medical societies, changed their boards according to the race-political point of view. Immediately after the seizure of power of the new government, Dr. Eitingon informed Dr. Reich that he could no longer enter the offices of our institute, so that, if he should become arrested, that could not happen in our offices."

Boehm visited Vienna and met with Freud and a senior IPA official, Paul Federn, on April 17, 1933. He discussed with Freud what to do in Germany. Before they left, Freud issued two demands for the direction of the German Psychoanalytic Society. First: Schultz-Hencke must never be elected to the board of the Society. And secondly, Freud said, "free me from Reich."

Boehm had tried to keep the minds of academics open for psychoanalysis, but to no avail. One of his acquaintances, a Nazi official, called psychoanalysis "Jewish-Marxist smut." Boehm's account connected this Nazi's slur on psychoanalysis directly to the influence of Wilhelm Reich, as Boehm continued, "As everybody knows, Reich often appeared in public as a Communist, where he had made his opinions out to be results of psychoanalysis. Berlin had been warned about Reich in countless leaflets. This was the prejudice I had to fight against." It was as if Boehm, and perhaps Freud and Federn, imagined that the Nazis would embrace psychoanalysis if Reich had been disconnected from it.

The decision on Reich's status was made in Summer of 1933:

"In a board meeting Simmel proposed a motion to remove Reich from the membership list. (Fenichel, who was away on a trip, was absent from this board meeting.) Simmel voted for the motion along with Müller-Braunschweig and myself. Eitingon also voted for Reich's expulsion, in principle, but he asked urgently to wait with this 'cleansing action' until the next general assembly in early October when he has resigned from all his posts. The decision to notify Reich about this was not executed because we didn't consider it opportune to make any contact with Reich, who was abroad. I want to add immediately that in a later board meeting in early 1934 we asked Mrs. [Edith] Jacobson to notify Reich about this decision during the convention in Oslo, but she neglected to do this."[7]

Sandor Rado, in an interview in the US, later recalled Max Eitingon:

"His personal income did not come from his medical practice, which he did not have; but it flowed from a fur enterprise his family ran in five countries. The Eitingons were one of the biggest fur traders. They had an establishment in Russia, one in Poland, one in England, two in Germany, and one here. The old man had died, and Max Eitingon's brother-in-law ran the whole enterprise, and then came the years of the Depression, during which all this began to collapse. For a while, even under the communist regime, they had the biggest contract with the Russians for furs."[8]

Rado provided the essential facts of the matter, except that the Eitingon fur business survived the Depression, as did Max Eitingon's wealth. It's said that Freud had a pun: *"Unsere besten Fälle sind die Felle von Eitingon."* ("Our best cases are the furs of Eitingon.")

Max Eitingon: Soviet Spy?

All of this information about the Eitingon family has been cast into the shadow of another story, first published in 1988 by San Francisco reporter Stephen Schwartz, who wrote an article about "Stalin's Killerati." Schwartz presented a rogue's gallery of first-rank academics, poets, and artists who had been involved in the violent criminal conspiracies and political assassinations, including kidnapping and murder, on behalf of the Soviet Union.[9]

Schwartz was the child of a leftist California couple; his father was an ardent member of the CPUSA, and his mother joined the Communist Youth League. As a "red diaper baby" he developed an encyclopedic knowledge of labor history, while at the same time developing a keenly critical view of the Party. Schwartz' article in 1988 ignited a rare, public

General Eitingon and Doctor Eitingon

controversy about the role Max Eitingon may have played in the kidnapping and murder of a White Russian general, Yevgeni Karlovich Miller on Sept. 22, 1937 in Paris.

Schwartz also listed a number of Stalinist intellectuals, such as Mark Zborowski, an anthropologist who worked with Margaret Mead. Another example was the poet Pablo Neruda, who actively supported the assassination of Leon Trotsky in Mexico.

Drawing on the work of historian John Dziak,[10] Schwartz devoted the bulk of his article to the evidence that Max Eitingon, Freud's closest colleague, had a hand in the General Miller kidnapping in Paris. By Max Eitingon's own admission, however, there is some truth to the evidence that connected him to the Miller assassination.

The charges against Max Eitingon rose from the trial of Nadyezhda Plevitskaya, a beautiful Russian singer who was the wife of a Soviet spy, General Nikolai Skoblin. It was Skoblin, we now know, who masterminded the Miller operation. Skoblin escaped from Paris after Miller disappeared and made it to Russia on a Soviet freighter, but his wife, Nadyezhda Plevitskaya, was arrested by French authorities and charged with participating in the crime. Ultimately, Plevitskaya was convicted and sentenced to a long prison term in a hard labor camp. She died in prison in 1940.

Plevitskaya and Skoblin had been frequent guests at Max Eitingon's mansion in Berlin, where she sang Russian folk songs that greatly affected the Russian-born Eitingon. Skoblin, Plevitskaya's husband, served as her impresario. Max Eitingon gave them a considerable amount of money by his own admission. The court trial testimony revealed contacts between the Skoblins and Eitingon in the days immediately prior to the kidnapping. And to add a twist of mystery, Eitingon had given Nadyezhda Plevitskaya a small "green bible" that was later determined to have contained coded messages, evidently part of the Soviet communications surrounding the operation. Plevitskaya left history a diary recorded in prison; it was found in Columbia University's archives by a historian. She had written that her husband, General Skoblin, had met "with the Bolsheviks" in the 1920s in Max Eitingon's mansion in Berlin.

Schwartz' article unleashed a wolf-pack of essays and polemical letters to the editor that could not be contained within a single periodical. Eitingon's defenders included the American branch of Eitingon's family, partisans of the IPA, and other interested parties. In *The New*

Wilhelm Reich and the Cold War

York Review of Books, historian Theodore Draper, who had met the scion of the American Eitingons socially, countered Schwartz's and Dziak's charges by listing a conflation of personalities involved, and misidentified persons with similar names, caused by the practice of intermarriage among the Eitingons. Draper correctly established that Max Eitingon was not likely Naum Eitingon's brother. The NKVD general known as "Leonid" to friends in his staff of the Soviet secret service, this was the "Leonid" Eitingon who oversaw the murder of Trotsky. Max Eitingon's relatives, along with officials from the International Psychoanalytic Association, vigorously objected to Schwartz' basic accusation that Dr. Max Eitingon had been a servant of Stalin's secret army, but they were not clear as to the question as to any blood-relation between Max and Naum, nor did they fully account for the ultimate source of the Eitingon family fortunes.

Contrary to what Theodore Draper has written, Max Eitingon remained wealthy after the crash of 1929. Boris Bazhanov, Stalin's personal secretary, has written that a Russian "capitalist," who worked in Paris for the Soviet secret service, was involved in the organization of the Miller kidnapping.[11] In 1930 Eitingon-Schild Fur Company, the Eitingon's family concern, was suspected of involvement in the illegal import of Soviet funds for subversive activities in the US, according to Stephen Schwartz. The debate would continue for more than ten years.

In 1994, a grisly book by Pavel Sudoplatov became one of the more memorable of the confessions produced by former intelligence officials after the collapse of the Soviet Union. The theme of this book could be described as "How To Keep Your Nose Clean When You're Up To Your Eyeballs In Blood."

Sudoplatov, as commander of the euphemistically named "Special Tasks" section, oversaw sabotage and political assassinations overseas during the early part of World War Two (1941-1942). In his memoir, *Special Tasks,* Sudoplatov revealed that he had been a colleague and close personal friend of General Leonid Eitingon, both before and after they each fell from grace following the death of Stalin.[12] Both Sudoplatov and General Eitingon served long and miserable prison sentences in the Soviet Union after many years of handling Stalin's dirtiest work.

Sudoplatov specifically addressed the Eitingon Affair in his own memoir. While denying that Naum Eitingon of the NKVD had anything to do with the Miller kidnapping in 1937, and questioning whether Naum/Leonid was even related to Dr. Max Eitingon, Sudoplatov did

General Eitingon and Doctor Eitingon

remember Naum Eitingon mentioning his wealthy relatives in Europe and that he had tried to get these relatives to provide him with a legitimate cover as a businessman, but that they had refused him.

Sudoplatov could not recall Naum Eitingon ever mentioning a relative named "Max." It is unlikely that Eitingon would have spoken openly with junior officers about the specifics of his wealthy relatives. Pavel Sudoplatov recalled that Naum Isakovich Eitingon once joked, "There is one small guaranteed way not to end up in jail under our system. Don't be a Jew or a general in the state security service."

Eitingon and his Western relatives (or namesakes) may have had their reasons for disavowing each other. The nature of the working relationship between the Eitingon family in the West, and the Soviet Union in the East, is a more important question than whether Naum and Max were blood relations.

In 1997, the historical controversy spread to Germany, where Michael Schröter published a summary of the evidence in *Psyche*, the official journal of psychoanalysis.[13] His essay was the first to include the recently-published correspondence between Max Eitingon and Sigmund Freud, and its bearing on the Miller kidnapping.

Max Eitingon wrote to Freud that in November, 1938, he had trouble with some Paris newspapers who mentioned his name "in an absurd and sensational way" in connection with the trial of Plevitskaya. Max Eitingon had to sent to Paris "a very official statement, so that no myths around my name are created." In another letter to Freud Max Eitingon continued his account: "Our friends in Paris, the Princess [Marie Bonaparte] and [René] Laforgue, had to speak out in favor of me in the trial of that singer." The trial and the newspapers cooked up "a funny figure" out of him, his New York brother-in-law and other ingredients. The accused singer, "although it seems there was no exceptional evidence," was sentenced to a long prison term. "A terrible story," Dr. Eitingon wrote.

Max Eitingon had more to tell on Jan. 8, 1939. He became involved in the pre-trial hearings because the accused singer said, when asked about her income, that she had some friends "among them the Dr. Eitingon, formerly Berlin, now Palestine, which had helped her quite a lot." Eitingon continues: "This is partly true, especially for the early past, because Mira [Max's wife] and I knew the singer for 16 years and truly appreciated and loved her very much for her unusual talent for singing Russian folk songs. They [the Skoblins] visited us often in

Berlin, and we also saw them always in Paris, when we were there." Eitingon described the Miller kidnapping as it was related in the newspapers. "The attorneys of the civil plaintiffs tried to place the Soviet Government in the dock and for this they needed Soviet agents, and through some fantastic and infantile contaminations they made out of me some kind of an agent for a short while." But that did not last long, "and in the trial they themselves had to admit this was a 'strategic mistake.'" In contrast to his Parisian friends, Eitingon was not that annoyed about the "grotesque lies of the newspapers," but he felt very sorry for the accused singer. They had made a example of her, as a warning "so that the Russians do not take such liberties on French soil."

KGB General "Leonid" (Naum) Eitingon (1901-1981) was born in Sklov, a village in the province of Mogilov, Byelorussia. Max Eitingon (1881-1943) was born in Mogilev, (a town in the province of the similar name.) Twenty years older than Naum, Max Eitingon was not likely Naum Eitingon's "brother" as reported by Schwartz and Dziak, since Max's father emigrated with Max at least a decade before Naum's birth. Still, the towns of Mogilev and Sklov are a scant 40 kilometers apart, both in the same province of "Mogilov" and it's inconceivable that the Doctor and the General were totally unrelated. The papers of the Waldemar Eitingon estate, from whom Max received his wealth, indicates that just before his death, Waldemar bequeathed a large sum of money to a charitable institution in Sklov, the birthplace of Naum Eitingon.[14]

It is still an open question, whether Max Eitingon had any role at all with Soviet intelligence. There is plenty of circumstantial evidence, as can be gleaned from the vibrant historical controversy.[15]

I can only point out the basic facts of the issue that still remain: the Eitingon money derived from a Soviet monopoly on furs, the salon that Max Eitingon hosted for Russian exiles which was a community that was thoroughly penetrated by Soviet intelligence, Naum Eitingon's brief reference to wealthy relatives in Europe, the family scion, Waldemar (Vladimir) Eitingon, sending money back to Naum's hometown.

Alexander Etkind, in his book, *Eros of the Impossible; the History of Psychoanalysis in Russia*, offers another possible conduit between Max Eitingon and the Soviet Union. He points out that the first official diplomatic representative from Moscow to Berlin was Victor Kopp, who later responded to Trotsky's call to place psychoanalysis at the

General Eitingon and Doctor Eitingon

service of communism. Kopp had worked with Trotsky in Vienna in 1909 producing *Pravda*. In 1924 Kopp was elected to serve as vice-president of the Russian Psychoanalytic Society, whose members at the time comprised one eighth of the International Society as a whole. Etkind suggests that this official relationship with the West may have provided a cover of legitimacy for Soviet intelligence operations.[16]

General Eitingon of the NKVD

Leaving the question of the actual relationship between Doctor Eitingon and General Eitingon open, it is worth our time to take a quick overview of General Eitingon's career. The KGB-General Pavel Sudoplatov knew Leonid Eitingon personally: the assassin of Trotsky was born Dec. 1, 1899 as Naum Isakovich Eitingon, but in the Soviet secret service he was known as "Leonid Alexandrovich" to cover his Jewish origin, as was common practice. Leonid Eitingon joined the Socialist Revolutionary Party in 1917, the Red Army in 1918 at the age of seventeen, and was then brought into the Cheka, the first Soviet "security" apparatus. In 1920 he joined the Bolshevik party, and his career took off. He earned the personal notice of Cheka head Felix Dzerzhinsky when young chekist Eitingon crushed a rebellion in the city of Gomel. Leonid was sent to Bashkiria "to enforce discipline in the area," according to Sudoplatov. In 1921 he was assigned to the military academy in Moscow for training, and then sent to the Foreign Department of the OGPU, military intelligence. "His distant European relatives," Sudoplatov wrote of Leonid Eitingon, "declined his request to provide him recommendations, papers, and money for travel and cover for operational work in Western Europe."[17]

Instead, Leonid went to Shanghai and Harbin, China. He worked with fellow spy Richard Sorge, the German national posing as a Nazi journalist in Asia.

In 1930 Leonid Eitingon became deputy director of the "Administration for Special Tasks," devoted to sabotage and political assassinations abroad. In 1932 he moved over to the Foreign Department once again, where he oversaw the illegal agent network abroad.

Pavel Sudoplatov first met Leonid Eitingon in 1933, and his description of Eitingon is worth repeating here:

"Eitingon's intelligence shone from his handsome face and gray-green eyes. He had a penetrating look and a heavy head of black hair. A scar on his chin from a car accident (most people thought it was a battle scar) gave him an air of bravado. Eitingon was a charmer who recited Pushkin by heart to illustrate both the folly and heroism of everyday life. He never drank much—one glass of cognac would last him for the evening. His hobby was hunting, but he did not shoot animals; he only liked to spot and track them. What struck me immediately was that he had none of the pretensions of a high-ranking bureaucrat. His favorite weapons were humor and ridicule."[18]

In 1936, Leonid was sent to Spain to oversee guerilla operations. In 1939, he fled from Spain to Paris, where he oversaw the control of Guy Burgess, and the other Cambridge-recruited moles in London.

Pavel Sudoplatov remembered that Leonid Eitingon had been sent to the United States as an illegal agent (without diplomatic cover) and served in San Francisco "to recruit Japanese and Chinese emigrants who might be useful in military and sabotage operations against Japan."[19] Eitingon came into contact ("in the early 1930s," according to Sudoplatov) with Californian allies of the Soviet Union, a circle that included Robert and Mildred Brady. "Another chief task of Eitingon's was to assess the intelligence potential of Americans involved in Communist activities. He was shrewd enough to suggest that recruiting operations should concentrate on people who were sympathetic to Communist ideas but were not true members of the party."

Upon his return to Russia, Leonid Eitingon received a deadly assignment from Pavel Sudoplatov, who in turn had received the order from Stalin himself: kill Trotsky. Eitingon succeeded in 1940, and was rewarded in a feudalistic manner by Stalin, who told him that while he, Stalin, was alive, no one would harm Leonid, and his children would be treated like red aristocracy. Leonid would get the best of schools for his abundant progeny (by his numerous common-law marriages), lifetime pensions for his relatives, an untouchable *apparatchik* in the *nomenklatura*. Leonid Eitingon was deputy-chief of the GRU, in charge of Soviet espionage in the West, until 1952.

Generals Sudoplatov and Eitingon activated a vast network of agents during the plotting of the Trotsky assassination. Mexican and Spanish Communists who had served in the Spanish Civil War launched a failed attempt when they machine-gunned the Trotsky residence in

General Eitingon and Doctor Eitingon

Mexico City. Then, Leonid Eitingon employed the son of his lover, Caridad Mercader, and her son, Ramón, ultimately put the axe in Trotsky's brain. He didn't get a clean shot with the pickaxe he brought with him, but Trotsky died anyway. The murderer said he was "paralyzed when I heard Trotsky scream."

Soviet agents in Moscow, New York, and Mexico City worked on securing the release of Ramón Mercader from his Mexican prison cell. The FBI discovered the first mail drop in New York to belong to Lydia Altschuler, a German immigrant and an employee of Consumers Union. (Previous to working for Consumers Union, Altschuler worked for the distributor of Soviet films in the US.)[20] In 1950, Alschuler was questioned in Congress about her role in the Trotsky affair, and she declined to answer. Recently, the National Security Agency released decoded Soviet cable traffic that proved that Altschuler had worked extensively as a courier in Latin America, during the period 1941-1943, while an employee of Consumers Union. There may have been more than one courier at Consumers Union.

Soviet efforts to free Mercader failed and he served out his term until 1960. He received a medal of the Hero of the Soviet Union and a KGB pension.

Immediately after the murder of Trotsky, Leonid Eitingon and the assassin's mother, Caridad Mercader, both waiting in a getaway car nearby, were forced to flee Mexico when Ramón was apprehended. They reached Cuba and after six months took a ship to New York, where "Eitingon used his connections in the Jewish community to obtain new papers and identity," according to Sudoplatov.

"They traveled across America to Los Angeles and San Francisco and sailed for China in 1941. Eitingon took advantage of the trip to resume contact with two agents he had planted in California in the beginning of the 1930s. They were to become couriers in the network obtaining American atomic secrets from 1942 to 1945. In May, 1941, on the Eve of the Great Patriotic War, Eitingon and Caridad returned to Moscow on the Trans-Siberian Railroad."[21]

Leonid Eitingon arrived in New York City in October, 1939, and set up an import-export business as a cover for his operations. "Eitingon was joined by Itzhak Akhmerov," Sudoplatov recalled, making a direct connection between Max Eitingon's relative and Michael Straight's control officer, "Michael Green" (Akhmerov). Eitingon also organized the Amerasia spy ring, which included Mark Gayn. Michael Straight provided press credentials for Gayn, an unindicted spy. In New York, Leonid Eitingon was within hailing range of Motty Eitingon, the director of Eitingon-Schild Company, formerly the Moscow Fur Company. "Leonid was a talented man," mused Sudoplatov, "and if he had not gone into the intelligence business he would no doubt have become either a prosperous administrator or a professor in the sciences."[22]

After the death of Stalin, and security chief Lavrenty Beria's fall from power, Leonid Eitingon was arrested and sent to prison. He was not released until 1964; Sudoplatov himself was not released until 1968. Leonid Eitingon died in 1981, Sudoplatov in 1998.

The main charge brought against both Eitingon and Sudoplatov turned on their oversight of the lethal toxins project developed by Professor Grigori Moisevich Maironovsky in "Lab X." Maironovsky "had worked on the impact of lethal gasses and the use of poisons in combating malignant tumors."[23] In 1937 Maironovsky's research group at the Institute of Biochemistry was transferred to the jurisdiction of the NKVD. In addition to research, Maironovsky worked directly for the Kommandatura Service, responsible for executions and secret "liquidations." By 1947, the Soviets were conducting double-blind experiments with lethal injections that were indistinguishable from real heart-attacks. Subjects' corpses were sent to random Moscow hospitals for autopsies, and none detected the poisoning.

Follow the money: new evidence of Dr. Eitingon's dependency on the Soviets

The Waldemar Eitingon estate was settled in the early twenties, at a time when one US dollar was the equivalent of something over *four trillion* inflated Weimar marks. The lawyer for the estate was Arthur Garfield Hays, who had many clients among German nationals in the US and Germany. Aside from his lucrative practice as a corporate lawyer, Hays also championed the civil rights of radical groups such as American precursors to the Communist Party, and at times functioned as the legal mouthpiece for the CPUSA itself. He was an official observer of the trial of the anarchists Sacco and Vanzetti, who became a cause for the CPUSA. We will discuss Hays more closely in the

General Eitingon and Doctor Eitingon

chapter on Mildred Brady, because Arthur Garfield Hays served as Wilhelm Reich's attorney during the 1940s, and was a sponsor of Brady's Consumers Union before that. In terms of the Eitingon Affair, it's important to note that among Arthur Garfield's papers from the estate of Waldemar Eitingon, I found conclusive evidence that the Eitingon fur business enjoyed the steady support of the Soviet Union. This evidence, in the Hays Papers, as far as I know, has not been taken into any account of the relationship between Max Eitingon and the USSR.

Eitingon's wealth derived from the estate of Vladimir (Waldemar) Eitingon, who emigrated from Byelorussia in the 1880s and founded a fur-trading corporation, the Moscow Fur Company ("MOSFURCO"), which had offices in Moscow, New York, Paris, London, and Stockholm. When Waldemar Eitingon died in 1919, his one-third share of the business was figured at $4.6 million by the U.S. Treasury Department., the total valuation of the business being over $16 million. The MOSFURCO partnership consisted of Waldemar, Motty and Mathus Eitingon. Waldemar and Motty were brothers; Mathus is unidentified thus far in the story. Dealing in Russian furs, MOSFURCO was reorganized after the Russian revolution as the Eitingon-Schild Co., but it survived and thrived in the era when Lenin sought to centralize state monopolies of key foreign trade commodities. By 1934 the company was valued at $85 million. Waldemar Eitingon's estate papers indicate that the family claimed a total loss on hare pelts in Russia when the Bolsheviks confiscated their goods, and took a $2 million tax write-off from the estate's valuation with the U.S. Treasury Department. My surmise is that these pelts were "confiscated" and resold at a much-below market value back to the Eitingons once things got settled at the home office in Moscow. Max Eitingon's funds were dependent on the good will of the Soviet government. Had either Lenin or Trotsky wished to sever ties to Eitingon's fur interest, in the 1920s, Max Eitingon could not have provided much financial assistance to the IPA, which provided for much of the funding of psychoanalytic books and training institutes. Furthermore, the Soviet foreign trade concessions, such as the fur trade, were "massively used," according to historian Robert Conquest, as a part of Russia's overseas intelligence goals. Another example was Amtorg, as well as the far-flung businesses of Armand Hammer and his father, Julius.

Notes

1. *People in Trouble*, p. 193.
2. *People in Trouble*, p. 196.
3. *Reich Speaks of Freud*, p. 174.
4. Letter from Erik Carsten to Freud, November 10, 1933. *Reich Speaks of Freud*, p.173-4.
5. *People in Trouble*, p. 247.
6. *Reich Speaks of Freud*, p. 259
7. Karen Brecht, Volker Friedrich, Ludger M. Hermanns, Isidor J. Kaminer, Dierk H. Juelich (editors): *Hier geht das Leben auf eine sehr merkwürdige Weise weiter. Zur Geschichte der Psychoanalyse in Deutschland,* Hamburg: Verlag Michael Kellner, 1985. On pp. 99-109 the editors print a facsimile of a report (dated Berlin, August 21, 1934) by Felix Boehm on the events in the German Psychoanalytic Society.
8. Source: P. Roazan and B. Swerdloff, eds. *Heresy: Sandor Rado and the Psychoanalytic Movement*
9. "Intellectuals and assassins—Annals of Stalin's Killerati" in *The New York Times Book Review*, January 24, 1988, pp. 3, 30f.
10. Dziak, John J., *Chekisty; a history of the KGB.* Lexington, 1988. p 69.
11. B. Bazhanov, *Ich war Stalins Sekretär* (I was Stalin's Secretary), Frankfurt: Ullstein, 1977
12. Sudoplatov, Pavel and Anatoli, with Jerrold L. and Leona Schecter, *Special Tasks; the memoirs of an unwanted witness-a soviet spymaster.* Little, Brown, New York 1994.
13. Michael Schröter : "Max Eitingon ein Geheimagent Stalins? Erneuter Protest gegen eine zählebige Legende" *Psyche*, Zeitschrift für Psychoanalyse, Stuttgart, Germany, Vol. 51, No. 5, May 1997, pp. 457-470.
14. Information on the Waldemar Eitingon Estate can be found in the papers of Authur Garfield Hays, thanks to the Seeley G. Mudd Manuscript Library, Princeton, New Jersey. Box 22, File 7.
15. For further information on the Eitingon controversy, see: Theodore Draper (Princeton, N.J.): "The Mystery of Max Eitingon" *The New York Review of Books,* April 14, 1988. Revised version in T.

Draper: *A Present of Things Past. Selected Essays*, New York: Hill & Wang, 1990, pp. 265-293. See also letters to the editor: *The New York Times Book Review:* March 6, 1988 (Gay, Schwartz), July 3, 1988 and November 9, 1988 (Conquest, Draper); *The New York Review of Books:* June 16, 1988 (Schwartz, Rapoport, Draper, Laqueur), August 18, 1988 (Broué, Draper). Further works on the Eitingon issue: R. Conquest: "Max Eitingon. Another view" *The New York Times Book Review* July 3, 1988; W. Laqueur: "The strange lives of Nikolai Skoblin. Some origins of World War II" *Encounter,* March 1989, pp. 11-20; M. Moreau-Ricaud: "Max Eitingon (1881-1943) et la politique" *Revue International de'Histoire de la Psychoanalyse,* 5, 1992, pp. 66-69; J.L. Rice: *Freud's Russia. National identity in the evolution of psychoanalysis*, New Brunswick/London: Transaction, 1993.

16. Etkind, Alexander. *Eros of the Impossible; the history of psychoanalysis in Russia.* Westview, Boulder. 1997, p. 250-1.

17. *Special Tasks*, p. 32.

18. *Special Tasks*, p. 33.

19. Sudoplatov, *Special Tasks*, p. 85

20. *VENONA; Decoding Soviet espionage in America*, p. 279.

21. *Special Tasks*, p. 79.

22. *Special Tasks*, p. 36.

23. *Special Tasks,* p.269.

Wilhelm Reich and the Cold War

Chapter Four: Wilhelm Reich and the Cambridge Five

Though his writings were controversial, Wilhelm Reich could have gone on for years, working quietly, publishing privately and without fanfare. According to documents filed in court, Reich sold no more than 500 copies of any title of his many self-published books, even in his best years. He was eventually jailed on a legal technicality subsequent to public smears of "quackery," but the record shows he never personally profited from his basic experimental research. It is documented that the government's legal actions against Reich were initiated by Mildred Edie Brady's article, "The Strange Case of Wilhelm Reich," published in *The New Republic*.

Several months prior to Brady's hit-piece, *The New Republic* had published Frederic Wertham's[1] vitriolic review of *Mass Psychology of Fascism*, strangely titled "Calling all Couriers." Wertham projected onto Reich an "utter contempt for the masses." Weirder still, Wertham—"friend" of the masses—was writing for a periodical that was subsidized by the estate of a Morgan Bank partner.

Based on the evidence that Wertham was a leader in the "Friends of the Soviet Union," and that Brady was a leader of the Consumers Union, a verified "united front" organization and identified as such in the *communist* press, Reich sensed the worst: that these camouflaged "red fascists" were taking deadly aim at his work. There was much publicly available evidence of Wertham's and Brady's Stalinist sympathies, and Reich took note of it.

Standing behind Mildred Edie Brady and Frederic Wertham was Michael Straight, publisher of *The New Republic*. A decade earlier, in 1936, Michael Straight had been recruited by the Soviets into what is known as the "Cambridge Five" spy ring. These infamous Cambridge University students, including Harold "Kim" Philby, Donald Maclean,

Guy Burgess, Anthony Blunt, and John Cairncross, left a tangle of lies that is slowly being unraveled today.

It's beyond the scope of this book to present the complete history of the Cambridge Five, which is still a topic of discussion when intelligence officials brief the US Congress. One such intelligence expert explained:

> "This was the kind of operation that, in terms of the Soviet model, produced pure gold, and it took an awful lot of work, spotting apparatus, people in the international communist movement watching how the Cambridge element moved, and picking and getting the right man, pulling him out, having him denounce communism, and then enter the government, and protect him. Philby protected Burgess and Maclean. This shameless drunk, Philby got him out of there before the guillotine dropped on him.."[2]

That's a rather garbled explanation of what happened, and we know more about it today.

Sex-Pol, "Otto," and Lucky Kim Philby

A journey through the historical literature of Cold War espionage is truly a "wilderness of mirrors," mined with disinformation, ideological posturing, and an unnerving paucity of reliable fact. There are those authors, for instance, who have suggested that Kim Philby was actually working for British intelligence even after he defected; others that say he recruited CIA chief of counter-intelligence, James Jesus Angleton, as a Soviet double-agent; and the most recent books by former Soviet KGB officers suggest that Philby had significant contacts with pro-democracy elements within the KGB. For instance, Philby's KGB controller in Moscow, after he retired, began writing anti-KGB literature. He asked, "How can you assert great principles and secretly do dirty things? We're all like dumb, blond horses pulling a miserable cart into a far-off abyss, decorated, like a Christmas tree, with cheerful lights and stars."[3]

Shortly, we will explore Michael Straight's engagements with the Soviets, but it must first be pointed out that he was not the only man associated with the Cambridge Five spy ring who had an intersection with Reich's life and world.

The most well known of the Cambridge spies was Harold "Kim" Philby, the son of an Indian Raj official and Middle East explorer, St. John Philby. "Kim" received his nickname from Rudyard Kipling's novel, *Kim*, a character who was a spy. St. John Philby was a personal advisor to King Ibn Saud, and was reviled by his fellow British subjects when he secured for Standard Oil the oil concession for what is now Saudi Arabia, giving the rich contract to the American Rockefellers rather than to the British. He was seen as a traitor, and a renegade who "went native" and adopted the Muslim religion, which required a painful circumcision. He was given two slave girls by Ibn Saud, and Kim had two half-brothers by these girls. For his services St. John (pronounced "sin-gin") received a substantial commission from Standard Oil, money the elder Philby badly needed to pay his son Kim's tuition at Cambridge. Thus, in effect, Rockefeller money sent Kim Philby to Cambridge.[4]

Kim Philby's communist sympathies were well-known by those who knew him at Cambridge in the 30s. He wanted to join the British Communist Party, but older members of the Party advised Philby to avoid open Party membership, in view of the fact that he had a promising career in the civil service ahead of him. In early 1933, young Philby rode across Europe to Berlin on a motorcycle with a friend from his old school, Tim Milne. Kim's friend was the son of headmaster A.A. Milne, the author and creator of Christopher Robin and Winnie the Pooh; it's been said that Kim Philby, as a boy, was the model for Christopher Robin.[5] Philby returned to Cambridge to complete his studies. He graduated in June, 1933. Cambridge "Red Don" Maurice Dobb, Philby's tutor in economics, gave Kim the names of his contacts in Paris, officials of "the International," i.e., the Comintern, who would be able to direct Kim toward clandestine work in Europe, as the rising tide of fascism continued to grow in 1933.

Kim's contacts in Paris sent him forward to Vienna, where he lived with a politically-active Jewish family. He fell in love with the family's divorced young daughter, Alice Kohlman "Litzi" Friedman, and they worked together in the "Revolutionary Socialist" units that waged street battles with the Heimwehr forces of the Dolfuss putsch. Litzi was a dedicated communist, as well as being an activist in Wilhelm Reich's "Sex-Pol" movement. It is perhaps a loosely defined association, since Sexpol as such was more formally established in Scandanavia after

Wilhelm Reich and the Cold War

Reich's 1933 emigration from Germany, but Litzi was part of a movement that formed around Reich at that time.

Baptized by fire by the Revolutionary Socialists, Philby was recruited into the service of the NKVD at this time, either in Vienna, or in London, where he returned along with his wife, Litzi. She was described in Julius Hay's *Memoirs* (1974) as "very attractive, petite, with a nicely rounded, pneumatic, provocative figure and dark brown hair." Hay alleged that Litzi initiated a sexual relationship with Philby in order to recruit him for the Third International and "a world-wide spy network."

Just a few years before, in 1929, Wilhelm Reich had formed another Viennese action group, with the similar name: "Revolutionary Social-Democrats." He called for a united front of Communists and Social Democrats, ready to arm the workers and fight the Fascists in the streets. This gambit was an example of the belligerent game being played by the Comintern, to infiltrate the "social fascist" Austrian Social Democratic Party and gain recruits for the Communists. The Revolutionary Socialist-Democrats had an ill-fated launch: after Reich spoke at the first meeting with a mixed group of socialists, "Otto Benedikt tried to speak but a third Social Democrat attacked Reich for demagoguery and a general tumult ensued in which chairs were thrown about the room, the lights were turned out, and the meeting was closed."[6] That group ceased to exist as its members converted publicly to the Communist Party; later most of the recruits to the Communist Party from the Social-Democrats were expelled in Stalinist purges as "social fascist moles." Even so, there was a direct line between Reich's 1929 "Revolutionary Social-Democrats" and the 1934 "Revolutionary Socialists" joined by Kim Philby.[7]

Litzi married Kim Philby in Vienna in the Spring of 1934, just after Reich left for Copenhagen. Philby was able to use his British passport to act as a courier for the underground. The couple was friendly with Teddy Kollek, who would later become the mayor of Jerusalem.[8] Kim Philby saw live-fire, combat action alongside the Revolutionary Socialists against the right-wing's attacks on the workers' residential districts. On February 12, 1934, the Austrian government launched mass arrests of Social Democrats, and began artillery shelling of workers' council estates outside Vienna. Thousands of men, women and children were killed.

Wilhelm Reich and the Cambridge Five

Reich and Arnold Deutsch

As a University of Vienna medical student in the early 1920s, Wilhelm Reich needed to earn money to pay for his own education. In addition to his own studies, he tutored for cash. One of his students was a young man named Arnold Deutsch.

In 1920, Arnold Deutsch joined the Free Association of Socialist Students,[9] founded by Karl Frank. At the same time, Deutsch joined the revolutionist-intelligentsia group named after Henri Barbusse's journal, *Clarté*. Deutsch joined the Austrian Communist Youth Organization in 1922 and must have known Reich's young friend, Willi Schlamm. Between 1923 and 1928, Arnold Deutsch studied physics, chemistry and philosophy and received a Doctor of Philosophy. Deutsch joined the Austrian Communist Party in 1924, at which time Karl Frank and Johann Koplenig were provisional leaders of the Party.[10] He also worked in the International Workers Relief Organization to feed the starving Russian masses, and was a member of the Central Committee of the group's Austrian branch. A member of the Communist Student Fraction (Kostufra) at the University of Vienna, Alfred Klahr, and his friend, Arnold Deutsch, organized a group called *Freiheitskämpfer* (Freedom-fighters) around 1925. Around 1927, Deutsch worked as sex-counselor in Wilhelm Reich's clinics in Vienna, where medically sound advice on sexuality and contraceptives were distributed in working-class neighborhoods. Newly released archives of the KGB indicate that Deutsch had been expelled from the University of Vienna in 1927, although it appears he completed his degree eventually.[11] Arnold Deutsch, as we will see, intensified his commitment to the Party, as did Wilhelm Reich, around this time.

In 1928, Deustch was sent by the KPÖ to Moscow to participate in the Soviets' version of the Olympic Games, the *Spartakiad*. He then returned to Vienna and worked for three months as a chemical engineer in a textile factory. In December, 1928, he was assigned to work for the Comintern underground in Vienna, by Johann Koplenig and the secretary of the Austrian Communist Youth, identified only as KONRAD. Deutsch began working for the Comintern underground as a courier and a liaison man, traveling to Romania, Greece, Palestine and Syria. He also worked for the Soviet military intelligence, or OGPU, and "the inventive young chemist made up a few recipes for invisible ink and obtained passports for the Vienna *rezidentura,"*—and so, he knew the identities of fellow agents.[12] He married Josefine Rubel in 1929.

I wondered whether the connection between Deutsch and Wilhelm Reich was a fiction, a product of Kim Philby's imaginative disinformation work. In September, 1999, Christopher Andrew and Vasili Mitrokhin's *The Sword and Shield; the Mitrokhin Archive and the Secret History of the KGB* appeared. The book has made a tremendous international stir, since Vasili Mitrokhin, the former archivist of the KGB's most secret records, spent the decade before his retirement smuggling out and copying files dating back to the October Revolution. When he defected in 1992, the Mitrokhin files were used by the British intelligence service to corroborate the material that had been carefully released by the SVR, the successor to the KGB in the post-Soviet era. (The former Russian Prime Minister, Yevgeni Primakov, was previously the head of the SVR.) Simply put, the Mitrokhin files are the "most complete and extensive intelligence ever received from any source," according to the FBI.[13] This is quite a statement, and it bears out in truth. It must be noted here that Mitrokhin defected from the former Soviet Union, and risked his life to do so.

Christopher Andrew, a professor of history at Cambridge University, most likely had the same question I had about the Deutsch-Reich association, and he mentions Reich in his discussion of Deutsch, whose portrait hangs today in the Hall of the Great Illegals in Moscow. Andrew's account of Deutsch's career as a spy is the most complete to date, and he views Deutsch as the most talented Soviet intelligence officer of them all. Not only was Deutsch "familiar" with Reich's work in general, Andrew reveals, he had been a close associate of Wilhelm Reich in Vienna.[14]

Andrew found Arnold Deutsch's university records, which confirmed most of what has been written about him previously. He points out that Deutsch completed his Ph.D. in Chemistry after five years, and that his academic performance surpassed that of all the Cambridge Five. Andrew located Austrian police files that reflect the Vienna anti-pornography squad's records on Arnold Deutsch as a "Sexpol" worker. In 1928, while Deutsch was working openly in Marie Frischauf's and Wilhelm Reich's marriage-counseling clinics for Viennese workers, he was simultaneously working for the NKVD; yet the police were only aware of his sex-reform activities. More significantly, Arnold Deutsch *had been Wilhelm Reich's publisher in 1929.*

Reich's third book was published by Arnold Deutsch. *Sexual Exci-*

Wilhelm Reich and the Cambridge Five

tation and Sexual Satisfaction (*Sexualerregung und Sexualbefriedigung*) sold over 10,000 copies through four editions.[15] Reich and Deutsch had a best-seller on their hands by 1930, when it was translated into Hungarian.[16]

In 1930 Arnold Deutsch also published, under the Münster Verlag imprint, Wilhelm Reich's *Geschlechtsreife, Enthaltsamkeit, Ehemoral; Eine Kritik der Bürgerlichen Sexualreform*[17] which became the first section of his later book, *The Sexual Revolution*. Deutsch also published Marie Frischauf and Annie Reich's pamphlet on abortion.[18] In effect, Deutsch was director of the prototype for the Verlag für Sexualpolitik, which became the Sexpol Verlag, and finally the Orgone Institute Press.

Reich later rewrote *Sexual Excitation and Sexual Satisfaction* in view of the KPD's line, and renamed it *Sexual Struggle of Youth*, but even then, the Party delayed its publication, waiting for approval from Moscow. Reich finally published it himself, and it became wildly successful. He regretted the revisions of the earlier text, for those revisions contained what he later called "political aspects outdated and surpassed and therefore invalid"[19] and never republished it. In 1953, Reich still considered *Sexual Excitation and Sexual Satisfaction* a "valid" text, theoretically and politically.

In October 1931, Deutsch's Vienna network, "The International Relations Department of the Comintern," was discovered by the police and for months Deutsch went into hiding. Then, Deutsch was summoned to Moscow in January, 1932. There he was trained as an OGPU (military and political intelligence) illegal agent. After a long period of close examination by his superiors, and his recovery from a long bout with typhoid fever in Moscow, Deutsch became a full-fledged officer by 1933. In October of 1933, Deutsch was assigned to work in Britain as an illegal agent, but first, he went to Vienna to recruit two agents.

One of Arnold Deutsch's first recruits for espionage was Edith Tudor Hart (1908-1973). To my knowledge, her biography has yet to be written. Kim Philby was spotted by Hart in Vienna and formally recruited by Deutsch in London. Edith Tudor Hart was the daughter of William Suschitsky, a radical bookseller who agitated for sex education and contraception in Vienna. She was a good friend of Litzi Friedman. As a teenager, Edith Suschitsky trained at the Montessori kindergarten school (the same one attended by Wilhelm Reich's daughter, Eva, later in 1927-1929.) Edith went to England to work as a kindergarten teacher in 1925. She returned to Vienna in 1927, and by 1929 she was working as a Soviet illegal.[20] In 1933, Edith married Dr. Alex Tudor Hart at the British consulate, for many of the same reasons Kim Philby married Litzi Friedman. That year, Dr. and Mrs. Tudor Hart left Austria and moved to Brixton, in London.

Litzi, Kim Philby and Arnold Deutsch went to London in 1934. Deutsch took a room at Number 7, Lawn Road, in Hampstead, a flat owned by the mystery writer Agatha Christie, though they likely never met.[21] He was joined by his wife, Josefine in 1935. In London, Deutsch recruited Kim Philby and became his first Soviet control officer. (Other NKVD officers, including Alexandr Orlov and Theodor Mally, Deutsch's bosses in the London *rezidentura*, were mistakenly later credited with this crowning success.)

Contrary to earlier published reports, it was Deutsch who conceived the strategy to recruit young radical students destined for high office. His strategy was approved in Moscow. Christopher Andrew writes that "It is difficult to imagine any other controller in the entire history of the KGB as ideally suited as Deutsch to the Cambridge Five. Though four of the five graduated with first-class honors, Deutsch's academic career was even more brilliant than theirs [...] He combined a charismatic personality and deep psychological insight with visionary faith in the future of a human race freed from the exploitation and alienation of the capitalist system. His message for liberation had all the greater appeal to the Cambridge Five because it had a sexual as well as a political dimension."[22]

Philby's first assignment was to spy on his own father, St. John. Kim Philby was reported to have viewed Deutsch as a "father and tutor... the father he never had." While father Philby charted unexplored territories in the Middle East, son Kim had been raised by his mother and grandmother during the first world war. Deutsch, a student of Wilhelm Reich's work in character analysis, had developed a lifelong interest in psychology and wrote up detailed psychological profiles on all his agents, including Philby, which were transmitted to Moscow. Stalin himself took an avid interest in these profiles, and was keen to find a characterological "hook" into prospective assets for Soviet intelligence. NKVD agent Arnold Deutsch misused Wilhelm Reich's technique of character analysis to assess his agents in these reports.

In one of Deutsch's early psychological reports on Kim Philby to his superiors in Moscow, he wrote that Kim "comes from a peculiar family.

Wilhelm Reich and the Cambridge Five

His father is considered at present to be the most distinguished expert on the Arab world. He has command of several Arabic dialects and has himself become a Muslim. He is an ambitious tyrant and wanted to make a great man of his son. He repressed all his son's desires. He [Kim] has a bit of a stammer and this increases his diffidence. He is a typical armchair scientist, well read, educated, serious and profound. He is a clumsy person emotionally and does not easily get close to people. Often he is afraid to talk because of his speech defect, unwilling to make a fool of himself.

"[...] He is undoubtedly a sentimental person, but, owing to his upbringing and the whole life of the English bourgeoisie, this side of his character is rather corrupted. He is a shy person and does not know how to handle money in the sense that he does not know how to arrange his own budget. However, he handles our money very carefully. He enjoys great love and respect for his seriousness and honesty. He was ready, without questioning, to do anything for us and has shown all his seriousness and diligence working for us. He is a kind and mild person. His temperament inclines him to pessimism and that is why he needs constant encouragement."[23]

Although they were in love, Philby and Friedman separated in order to maintain Philby's cover. They did not get divorced until 1945. Deutsch, as the NKVD's most valuable illegal agent in London, controlled both Litzi and Kim in 1934. So, too, were the rest of the Cambridge Five in Deutsch's hands. Under orders from Deutsch, Philby developed contacts within pro-Nazi Anglo-German Fellowship in 1935, blurring traces of his Communism. "Otto"—the name by which Philby knew Arnold Deutsch—died on a ship headed for America during the war. Philby would recall, many years later as a defector in Moscow, that "Otto and I met regularly. And he taught me the rules of conspiracy. He hammered them into my head: how to call the necessary person on the phone, how to check, how to recognize a tail in a crowd, and other basics."[24] After Arnold Deutsch's death in 1940, Philby worked with a succession of Russian controllers, none of whom filled the fatherly role for Kim as did Deutsch. One by one, most of these subsequent Russian NKVD controllers were recalled to Moscow and shot.

Kim Philby became head of Section Nine (British counterintelligence) in 1944, and through the years he continued the practice of composing capsule psychobiographies[25] of nearly everyone he met, writing them up in tidy minuscule handwriting, and sending them to Moscow in diplomatic pouches.

In 1949, Philby came to Washington, D.C., as the chief liaison between the British SIS and the American FBI and CIA. There, he learned that the Americans would soon discover that Donald Maclean was a Soviet mole. Maclean was placed in the British Embassy in Washington, with total access to nuclear arms developments. Philby demanded that Maclean be exfiltrated by the Soviets, and that Burgess remain, since Burgess' defection would cast suspicion on Philby. Moscow Centre decided that Guy Burgess would accompany Maclean in defection to the Soviet Union in 1951. The British MI5 could not prove their suspicion that Philby had tipped his two friends, but they did force his resignation. After years of close surveillance, and a quasi-official clearance, Philby was given some spy-work once again by the British. While working in Beirut in 1962, Philby was interrogated on the basis of new evidence from a Soviet defector, and this time, Philby confessed. Somehow, he was allowed to escape to Moscow, where he became a Soviet citizen, and advised the KGB on the Western intelligence services. On the night of May 12-13, 1988, Kim Philby died in Moscow. He was given a state burial with full honors, interred at Kuntsevo in a cemetery reserved for Soviet generals.

The *New York Times* ran an article, after Philby's death, that recalled the "lingering influence" of Wilhelm Reich upon Philby even in Russia.

"His name was Arnold Deutsch and, 50 years later, in his autobiography, Philby seems still under the spell of Deutsch's magnetism, an almost sexual seductiveness. Not surprising perhaps, because Deutsch was a charismatic former sexologist, originally a follower of Wilhelm Reich, the Freudian Marxist schismatic who made healthy orgasms the key to personal as well as societal revolution. (The lingering influence of this doctrine on Philby may be glimpsed in a not entirely facetious inscription in a book [i.e., Alex Comfort's The Joy of Sex] that turned up in the Sotheby's consignment. The book was a gift to Melinda Maclean, the wife of his fellow spy, Donald; Philby betrayed his own wife to woo her away from his friend. The inscription to Melinda reads: 'An orgasm a day keeps the doctor away.')"[26]

The President's Mentor: Professor Carroll Quigley

I learned the first details of Michael Straight's role as publisher of

Wilhelm Reich and the Cambridge Five

The New Republic in a remarkable book called *Tragedy & Hope; A History of the World in Our Time* by Carroll Quigley (1910-1977). Quigley was a professor at Georgetown University's School for Foreign Service in Washington, D.C. Today, his work has benefited from renewed interest because the President of the United States, Bill Clinton, attended Georgetown University and was a student of Quigley's. The President even lauded Quigley as his "mentor" when he accepted his party's nomination in a speech at the 1992 Democratic National Convention. Quigley was a legend on the Georgetown Campus.

Angular, trim, Quigley was a Boston-born Catholic. He wore his professorial tweed blazer, elbows case-bound in leather, as if he were born in them. His deep-set eyes shone with a searching intensity. At Georgetown's School of Foreign Service, he brought an almost spiritual quest to his work to comprehend the motive forces influencing world history. As a writer, his command of fact and style draws the reader in. He was capable of shockingly original insight as well as opinionated gossip.

His students recall the Professor tossing classic books of Western Civilization out the second-story classroom window, with an exclamation of derisive contempt. He took his teaching duties as seriously as he pursued his main body of research. But by the mid-sixties, he began to bemoan the lack of critical thinking and ideology of his young charges, many of whom are working in government today. When the students called a Moratorium on classes against the Vietnam War, Quigley refused to cancel his class. He was alone, lecturing to an empty hall, when the campus activists burst in to the room and roughed him up. After that day, Quigley contracted his activity more toward research. He was never the same man again.

Quigley's main body of research has been into the global influence of a semi-secret group founded by Cecil Rhodes in Britain, which Quigley called "the Anglo-American Establishment," or "the Rhodes Group." This was a precursor to the modern-day "think-tank." It was composed of men whom, in Rhodes' day, stood at the epicenter of the British Empire, holding positions of authority in trade, journalism, and academe. Educated at Oxford and Cambridge, they had financial and matrimonial connections to the world's most wealthy and powerful. In America, their interests were handled by the Morgan Bank, and on the Continent by the Rothschilds. Rhodes (1853 - 1902) was one of the wealthiest men on the planet, with an *annual income of over $1 billion,* adjusted for inflation. Rhodesia was named after him, and when he died, his estate established the Rhodes scholarships.

What is less well-known is that Rhodes scholarships were not set up to reward gifted students with scholarships to Oxford, but rather as one component of a plan to create a world federation of governments. He was, in those times, a liberal, since he believed that the indigenous peoples within the British Empire should be given the same democratic rights and legal structures as were enjoyed by the English Islanders. The function of the Rhodes scholarships was to identify future leaders, instill them with common values at Oxford, and send them back to their native "colonies" where they would spread these acquired traits.

Today, this plan manifests itself as the Council on Foreign Relations, the Trilateral Commission and other "free-trade," "world federalist" organizations.

In his books, such as *Tragedy and Hope*, and *The Anglo-American Establishment*, Quigley points to the history of finance capitalists giving money to left-wingers and communists as but one feature of a century-old master plan to end national sovereignty, consolidate all currency, and institute the Global State. The tactic: bring together leaders in the fields of finance, journalism, education and politics. Quigley documents all this, as an insider who had the opportunity to inspect the internal papers of Rhodes' group for several years when, during WWII, the records of the group were transferred from England to New York, at the Council on Foreign Relations, for safekeeping during the Blitz.

Quigley's research encourages a suffocating sense of Commonwealth democracy being little more than a tool of control for global market planning.

Tragedy and Hope, Quigley's magnum opus, is truly fascinating, and one section that caught my eye was a long discussion of *The New Republic*.[27] The leading liberal journal in America for many decades, *The New Republic* published Mildred Brady's 1947 smear article that triggered the FDA's prosecution of Reich. Wilhelm Reich's American translator and editor, Theodore P. Wolfe, M.D., outlined the steady spread of Brady's lies through various sources in his *Emotional Plague versus Orgone Biophysics*.[28] So, I was familiar with the magazine, but Quigley provided a detailed background about *The New Republic*'s publisher, Michael Straight.

Wilhelm Reich and the Cambridge Five

There are *conspiracy theories*, and then there are *conspiracies*. It would be anti-historical, even childish, to deny that price-fixing, restraint of trade, and secret illegal deals played no role in the formation of the great financial dynasties created in the nineteenth century. It would truly be delusional to suggest that these family fortunes did not transfer to great political, social and economic power in the twentieth century. No one has ever described this influence in more clear and lucid terms than the President's Professor, Carroll Quigley.

The New Republic, Quigley explains, was founded by a full partner of the Morgan Bank. "More than fifty years ago," Quigley wrote in 1966, "the Morgan firm decided to infiltrate the Left-wing political movements in the United States." As a result of the graduated income tax in 1913, and the inheritance tax shortly thereafter, Wall Street money flooded into American tax-exempt foundations (Carnegie, Rockefeller, Whitney, Mellon etc.,) "which became a major link in the Establishment network between Wall Street, the Ivy League, and the Federal government. Dean Rusk, Secretary of State after 1961, formerly president of the Rockefeller Foundation and Rhodes Scholar at Oxford (1931-1933), is as much a member of this nexus as Alger Hiss, the Dulles brothers..."[29] These foundations in turn funded lofty-purpose magazines and public interest corporations in order to manipulate and monitor the thoughts and feelings of the left. *The New Republic* is cited as an early example.

The New Republic was founded in 1914 by Willard Straight, Michael's father. Willard had made a small fortune as assistant to Sir Robert Hart of the Chinese Imperial Customs Service before being made a full partner at Morgan Bank. His wife, Dorothy, was an heiress to a portion of the Standard Oil Trust, as well as a beneficiary of the Consolidated Edison trust. She inherited $75 million around 1910. She was a childhood friend of Eleanor Roosevelt's, and her son, Michael, was a personal friend of Mrs. Roosevelt's as well. Dorothy Straight, like Eleanor, championed the downtrodden and contributed to countless philanthropic and progressive causes. She toured the sweatshops of New York with anarchist Emma Goldman. She sponsored black Communist Paul Robeson's singing tour through Europe. She viewed her large inheritance as a "public trust," all the while retaining the full benefits of her wealth and bottom-line control over her philanthropic gifts.

At the conclusion of WWI, Willard Straight died in Paris on

Wilhelm Reich and the Cold War

December 1, 1918, while representing the United States at the Versailles peace conference. Seven years later, his widow Dorothy remarried to Leonard "Jerry" Elmhirst, an English gentleman of limited means. Dorothy, along with her children, Michael and his siblings, returned with Elmhirst to his native England, where they purchased Dartington Hall, an estate in South Devon. There they founded a utopian community and operated an alternative school for children. Influenced by the latest theories in psychoanalysis and socialist theory, especially Fabianism, the Elmhirsts encouraged the children towards self-government in running their affairs. Michael Straight, who attended the school himself, later would bemoan the emphasis on "dream-analysis" rather than classical studies. The Elmhirsts sponsored lectures by avant-garde writers, artists, musicians and educators. Michael remembered hearing A. S. Neill, headmaster of the more experimental—and more enduring—Summerhill School. Neill would become Wilhelm Reich's close friend by 1939.[30]

When Michael Straight returned to America from England in 1937, he first stopped off at the White House to inquire of President Roosevelt as to employment in the federal government. Straight had been a student of John Maynard Keynes at Cambridge, and had exceptionally close contact with him as a fellow member of an elite, private university society called the Apostles. Keynes' economic theories, of course, played an important role in the policies of the New Deal. "Keynesian" economics provided the rationale for governmental stimulus of the economy through massive deficit spending. Straight later wrote, in his autobiography, that his association with Keynes "opened every door" when he sought a job in Washington D.C. in 1937.[31] Eleanor Roosevelt helped him find a job as a "dollar-a-year man" at the State Department, where he worked on an analysis of German war-making capacity. He was best known as an ardent spokesman for Roosevelt's New Deal. Straight wrote many of FDR's pre-war speeches, and became Eleanor Roosevelt's advisor on the left-wing causes she embraced.

Lady Elmhirst of Dartington Hall (the former Mrs. Willard Straight) was sole owner of *The New Republic* until 1937 when she turned it over to her younger son, Michael Straight, setting up a dummy corporation in Canada with a grant of $1.5 million.[32]

Since 1914, it had been managed by Walter Lippmann, a member of the Rhodes Group. *The New Republic's* "original purpose... was to provide an outlet for the progressive left and guide it quietly in an

112

Anglophile direction," Quigley says. Quigley points out the close—nepotistic—financial relationships between such families as the Straights, Paynes, Whitneys, Morgans and Rockefellers. For instance, Michael Straight started *United Nations World*, which listed Nelson Rockefeller, "Jock" Whitney (Michael's cousin) and *The New Republic*—among others—as owners of record. Michael Straight and Nelson Rockefeller attended the experimental Lincoln School together in New York City, though Straight was four years younger than Nelson.

After WWII, Straight took an active interest in his family's magazine, *The New Republic*. In October of 1946, when he was barely thirty years old, Michael Straight hired Henry Wallace, the former Vice-President of the United States, as managing editor. Wallace had just been dumped from his cabinet position as Secretary of Commerce by President Harry S Truman, for breaking ranks over the Marshall Plan. Perhaps Wallace wanted something more Stalinist for Europe; his alternative to the Marshall Plan would have involved the United Nations in disbursing monies from the US Treasury to Eastern Europe under Stalin, and putting the administration of the German industrial zone in the Ruhr Valley by a Berlin-style "Big Four" commission. Wallace said that the Marshall Plan would "take the nation down the road to World War III."[33] Fired by Truman, Wallace's political career within the Democratic Party was over, apparently. The young publisher, Michael Straight, says he saw an opportunity to build his readership and that is why he offered the former Vice-President the job as editor. Soon thereafter, Straight funded an expanded *New Republic* staff, which promptly started beating the Wallace-for-President drum. The new staff functioned primarily as a campaign organization for the new Progressive Party, which received organizational support from the CPUSA and quiet direction from the Soviet Union. Michael Straight later recalled having to point out to an oblivious Wallace that each time he lectured a packed hall of enthusiastic Communist Party faithful, he was only preaching to the converted. Wallace was happy to speak to packed halls, and commended the Communists for delivering crowds.

In the fall of 1947, Wallace, Straight and Wallace's speechwriter, Lew Frank, Jr. (a leader of the Communist caucus of the American Veterans Committee, the left's answer to the more conservative Veterans of Foreign Wars and the American Legion) made a trip to the Mediterranean. They were granted an audience with the Pope; they were blindfolded and spirited to a meeting with Zionists in Palestine.

Wilhelm Reich and the Cold War

Wallace reportedly came back "a changed man" and announced his candidacy in December. Lew Frank, Jr., looking for assistance in formulating a response to the Marshall Plan and the new Cold War, joined a "Communist research group" at the home of wealthy "Wall Street Red" Frederick Vanderbilt Field.

Wallace's presidential candidacy severely split the left wing of the Democratic Party, and because they misjudged the sentiment of the broad mass of American progressives, the Communists only isolated themselves and were driven out of former "united front" organizations. What representation they had in the labor unions was crushed. The CPUSA was marginalized as a political force on the left by the Wallace campaign by 1948—long before the "McCarthy Era" (1952-1955).

Quigley, in *Tragedy & Hope*, argues that the third-party candidacy of Henry Wallace and the Progressive Party required the tacit cooperation of "Wall Street" and "the Communists." Both wanted to defeat Truman, and so they used Wallace to form a third party candidacy to draw enough support from Truman's left flank to elect the Republican, Thomas Dewey of New York. From the Stalinist viewpoint, a Republican administration would "sharpen the contradiction" between the workers and the capitalists. Michael Straight's ownership allowed many communists into the staff of *The New Republic* whose sole function was to promote the Wallace candidacy. Straight acknowledges a tightly organized presence of Communist Party followers on the staff of his magazine.

Truman squeaked by and won. After the election, Straight fired all the new "Wallace staff" off *The New Republic*. Quigley, with a grand flourish, devotes dozens of pages to Michael Straight and what can only be described as an international conspiracy to manipulate the US national election of 1948.

Wallace resigned as editor of *The New Republic* when he declared his candidacy, and Straight then resumed editorship. Over the objections of the "Wallace staff," *The New Republic* endorsed Truman for President. Straight pulled the rug from beneath the Wallace candidacy and switched his support to Truman.

Quigley sums it all up thusly:

> "The relationship between Straight and the Communists in pushing Wallace into his 1948 adventure may be misjudged very easily. The anti-Communist Right had a very simple explanation

Wilhelm Reich and the Cambridge Five

of it: Wallace and Straight were Communists and hoped to elect Wallace President. Nothing could be further from the truth. All three—Straight, Wallace and the Communists—joined in the attempt merely as a means of defeating Truman. Straight was the chief force in getting the campaign started in 1947 and was largely instrumental in bringing some of the Communists into it, but when he had them all on board the Wallace train, he jumped off himself, leaving both Wallace and the Communists gliding swiftly, without guidance or hope, on the downhill track to oblivion. It was a brilliantly done piece of work."

Quigley reported in 1966 that Straight was an anti-communist acting on behalf of the vested interests of the "Anglo-American Establishment." But Straight himself would not tell his side of the story until 1983, when he publicly revealed that he had been recruited into Soviet intelligence as a student at Cambridge University in 1935.

Michael Straight's Story

"Experience shows that there have been very many conspiracies, but few have turned out well, for whoever conspires cannot act alone, and cannot find companions except among those who are discontented; and as soon as you have disclosed your intention to a malcontent, you give him the means of satisfying himself, for by revealing it he can hope to secure everything he wants..."
-Machiavelli, *The Prince.*

In his 1983 book, *After Long Silence*, Michael Straight finally accounted for his behavior in the election of 1948. He had been an independent voice for liberals as an editor, and publicly opposed the pro-Nazi line taken by Communists during the Hitler-Stalin Pact (1939-1941). In 1940, Straight had essentially "broken Party discipline" during the Pact by publishing a long article in *The New Republic* which argued in favor of intervention by the Allies. Yet his hiring of Henry Wallace and his initial support for Wallace's Presidential campaign in 1947 was deeply conflicted, and raises many questions. In his book, Straight made considerable efforts to explain himself.

In the 1920s, both Lenin and the head of his secret police, Felix

Wilhelm Reich and the Cold War

Derzhinsky, were fascinated by British class structure and the "old boy network" which guaranteed young graduates of certain schools a place in the Civil Service. Lenin and Derzhinsky had analyzed the situation as highly favorable to a long-term plan to neutralize the powerful British Secret Intelligence Service (SIS). In the 1920s, the SIS was of paramount international importance, seasoned by decades of colonial intrigue, and was a direct threat to the new Soviet state.

In 1934, as Straight arrived at Trinity College, Cambridge University, the academic community was electrified into radicalism by events in Nazi Germany. Although immensely wealthy, Straight was a sensitive young man who realized himself in student socialist and communist movements. He studied economics with socialist professors such as Maurice Dobb, Denis Robertson, and John Maynard Keynes. Keynes in particular played a great influence on Straight's thinking.

In the following summer of 1935, Straight and a group of Cambridge students, including fellow Communists Anthony Blunt and Charles Rycroft,[34] took a long trip to the Soviet Union.[35]

"For me," Straight wrote in hindsight, "the superstructure of Communist ideology was little more than 'psychological pretense.' It was the sense of brotherhood that opened up a new life for me. I had come to deeply mistrust my own feelings after my failure with Margaret." True to his "psychoanalytic" education at Dartington Hall, Straight correlated his feelings about the communist brotherhood with his first sexual experience at the age of fifteen. With Margaret, a twenty-nine-year-old woman, he had found to his horror that he was impotent.

After the trip to Russia, in the fall of 1935, Anthony Blunt (an art historian) and Guy Burgess (who served as a "financial advisor" to the Rothschilds) took an interest in Michael Straight. Blunt and Burgess were already acting as talent-spotters and agents for the Soviets, utilizing their contacts in the homosexual underground that flourished in the British establishment. Among others, they spotted an insecure young man, Michael Straight, who was the scion of one of America's most powerful families. "I had written to my mother about the 'violent love' that I felt for my friends at Cambridge. Almost all of them were men. There was no physical element in the love I felt for them, or that they felt with me. Yet most of my friends, I came to realize, were or had once been homosexuals. Guy was an exhibitionist. Anthony, in contrast, was wholly discreet." Although heterosexual himself, Straight said that he was "sympathetic but incredulous" when a "fellow radical"

Wilhelm Reich and the Cambridge Five

invited him over to meet his lover, an eleven-year old choir boy.

A year later, Straight was invited and joined another group in the fall of 1936, a "secret society" known as the Apostles, one of the most exclusive of several such fraternities at Cambridge. John Maynard Keynes was its most influential member and had been inducted in his own university days along with "the Bloomsbury set": Lytton Strachey, Leonard Woolf, and Clive Bell. Many of Straight's dons at Cambridge were Apostles. Victor Rothschild, Anthony Blunt and Guy Burgess were also members.

"We met in Keynes's rooms. I held up my right hand and repeated a fearful oath, praying that my soul would writhe in unendurable pain for the rest of eternity if I so much as breathed a word about the society to anyone who was not a member."[36]

Guy Burgess frequently attended the Apostles' meetings. "He craved the companionship and physical love of other men, which seemed to be a binding tie of many men in the society." Burgess had left Cambridge in 1934 after a "bizarre" break with his Communist associates. (It was a recurring theme. The Soviets demanded, first of all, that young moles preparing for the underground must first sever all public and social ties with former left-wing comrades. It goes without saying that these Soviet agents would not be registered as Party members, or carry cards.) At the time, Burgess was working for both the Rothschilds and as secretary for the Nazi-sympathizing, MP Captain Jack Macnamara, while secretly serving as a "talent spotter" for the NKVD. What was Guy really up to, Straight asked himself at the time.

The Soviet moles in England and elsewhere were required by Moscow to feign allegiance to the British right wing and international fascists. Burgess worked for the pro-Nazi Macnamara, and Kim Philby rubbed shoulders with Franco's Falangists. As we shall see, Straight was given similar marching orders. This break with old friends provided the *legenda*—a Russian word meaning "cover"—for their espionage. They were introduced to the covert work of intelligence-gathering as dedicated soldiers in the war against the Soviet Union's enemies.

At Cambridge, Straight became very close friends with fellow student, John Cornford, a great-grandson of Charles Darwin. Cornford was a brilliant young poet and a dedicated communist. At his friend's urging, Straight had joined the Cambridge Socialist Society which had 200 members at the time, one in four of whom were members of "red cells." In January of 1937, Straight learned of the death of his closest friend; John Cornford had been killed in action in Catalonia during the Spanish Civil War.

Devastated, Straight described himself as emotionally defenseless when he was asked by Anthony Blunt, two weeks later, to return to America and follow in the footsteps of his father into a partnership with J. P. Morgan, and to serve as a Soviet mole in international banking.[37] Blunt suggested that "our friends" had made this assignment for Straight—Blunt's vagueness implied these friends were in the Comintern, but Straight realized much later that there was no doubt he was being recruited directly by Soviet intelligence. Straight says he had no interest in returning to America, nor becoming a mole, and that he pointedly asked Blunt, "Why on earth would anyone suppose that I'd take on an assignment like that?"

For the same reason that John Cornford went to Spain, Blunt answered. He suggested that Straight stage a break with his Communist friends, since it was known that his loyalty was to John, and not the Party.

After sleeping on the proposal, Straight met with Blunt the following day and once again refused the assignment. Blunt simply said that he would inform his superiors and that they would consider it, and ultimately they would let Straight know their decision within a week. Later, Blunt told Straight that his "appeal had been rejected." Once again, Straight protested, and Blunt offered to have his protest considered at the "highest levels of the Kremlin" if Straight would begin immediately with his staged break with known student Communists. Straight agreed and complied. As if explaining this decision, given his strong ambivalence, Straight wrote: "Do psychiatrists point out the innate submissiveness of the youngest child in each family? I expect so."[38] I was struck, time and again, with the way that Straight's apologia referred to psychoanalytic concepts and theory.

Anthony Blunt soon told him his protest had been rejected by Stalin himself. Blunt obviously had some sway over Straight, but what exactly? Straight is vague on the reason. Resigned to accepting the assignment, Straight went along with Blunt to meet with a Russian who briefly gave him instructions. Straight remarked, "He was more like the agent of a small-time smuggling operation than the representative of a new international order." This agent was most likely Arnold Deutsch, who had already recruited Kim Philby for the NKVD.

So it was, after Blunt relayed the Kremlin's decision, that Michael

Wilhelm Reich and the Cambridge Five

Straight embarked for America in July of 1937. Right away, the founder of the American Civil Liberties Union, Roger Baldwin, invited Straight on a tour of the mid-western states, where they met with John L. Lewis of the Congress of Industrial Organizations (CIO) and Philip Murray of the American Federation of Labor (AFL). On this trip, Straight wrote that he did not dare speak of his dark secret, even with Baldwin, with whom he would form a close bond. He later recalled the dread he felt should he ever be approached by his expected Soviet controller.

Soon Straight began looking for work in Washington, but that year jobs were scarce in the government. He went to the White House, personally, to ask if there were any vacancies, but President Roosevelt could not help. Straight tracked down his mother's childhood friend, Mrs. Roosevelt, to ask about a position with the Department of State.

Straight was able to locate Mrs. Roosevelt where she was touring coal fields with Doris Duke, a young heiress interested in social work. As Straight drove with her while returning from a speaking engagement, Mrs. Roosevelt startled Straight with the remark, "I would be a Communist if I thought that Russia was comparable to America." Straight offered no comment, and Mrs. Roosevelt promised him that she would inquire with Sumner Welles, Undersecretary of State. With her influence, Straight landed the job at State, which, he remarked, was "still a place of employment for gentlemen."

Straight volunteered to write a long report on Nazi Germany's war-making capacity from an economic standpoint. He was eager to do this, he wrote, since most of his friends were isolationist. For a time, he buried himself in the project, but in the spring of 1938, he received a telephone call from a stranger with a thick Central European accent. "Mr. Straight?" the voice said, "...I bring you greetings—from your friends in Cambridge University."

He introduced himself as "Michael Green." American codebreakers have now made public hundreds of decrypted Soviet espionage communications, known as the VENONA documents. The National Security Agency (NSA) has identified "Michael Green" as Iskhak Abdulovich Akhmerov. Analysts from the NSA and historians of the Cold War believe that "Michael Green" was one of Alger Hiss' Soviet control agents.[39]

Straight recounts his meeting with the man he knew only as "Michael Green" at a nearby restaurant. This would be the Soviet handler who Anthony Blunt told him to expect. Over dinner, "Green" asked him to report on State Department documents and gave him a telephone number in Brooklyn where he could be reached. Straight says he was frightened and took leave of his unpaid job for some months. In September, 1938, he returned to the State Department and finished his report on Germany. Alger Hiss called him into his office to commend him on his work. That same month, Straight met again with "Michael Green" and handed him the report to copy. Straight told "Green" that he was taking a new position with the Department of the Interior. "I had not freed myself from my own past, but in moving from a sensitive department of the government to a job that was purely political, I had gained my first, small foothold on the possibility of leading an honest and an honorable life."

This was an important juncture in Michael Straight's life, so it is worth dwelling upon. Straight had handed classified State Department documents to an illegal agent of Russian intelligence (NKVD). "Green" had gained another lever of blackmail from that, which Straight clearly realized. Although Straight now claims that the information in the report was insignificant, Straight's sword of Damocles was dangling overhead by 1938. At any time thereafter, he could have been threatened by his Soviet controllers with the exposure of his cooperation. This pressure must have been felt acutely as the Cold War heated up.

In 1939, a Soviet defector, General Walter Krivitsky, was found dead in his Washington, D.C. hotel room. Straight thought in frozen horror when he heard the news, "They've come for me."[40]

By February, 1940, Hitler and Stalin had partitioned Poland in accordance with the secret protocols of the Molotov-Ribbentrop Pact, and Hitler, with Stalin's active support, was on the verge of mastering France as well. At that moment, Germany and Russia were formal allies, since the terms of the Pact went far beyond "non-aggression" and included sharing military intelligence and trans-shipments of war materiel. In the US, the Communists promoted an anti-interventionist policy, including slowing or stopping shipments to France from US ports. Stalin assumed that with a nudge and a prod, the Fascist regime in Germany could be easily converted into another Soviet satellite state.

At this time, Straight was working directly with Eleanor Roosevelt as an advisor during the American Youth Congress (AYC), a "popular front" coalition of non-Communist and Communist organizations.[41] The AYC turned into an ideological battleground, with Communist Party members attempting to commandeer the group in the larger anti-interventionist strategy. "A Communist caucus, working under cover and in many guises, was able to manipulate the congress in ways that Mrs. Roosevelt could never comprehend," Straight recalled some forty years later. As for those covert Communists that he recognized among these groups, "It was a mark of my ambivalence that we got along well together."

In the summer of 1940, Guy Burgess visited with Straight in Washington. "Reveling in the scatological," Burgess regaled him about the time he had with three members of the French cabinet in a male brothel in Paris. "Singing and laughing, they danced around a table, lashing a naked boy, who was strapped to it, with leather whips." Straight and Burgess spoke of the Apostles Society, and Straight learned that Anthony Blunt was still recruiting selected members for Moscow's espionage apparatus within the British establishment. Burgess asked Straight if he could put him in contact with "our friends" but Straight refused. Later, after he'd driven Burgess back to Washington, Straight confessed everything to his wife and informed her about his contacts with Michael Green. She demanded, he said, that he end these contacts within a year. Straight, however, continued meeting with Michael Green and other Soviet agents for several more years. His wife, in turn, told Michael's story to her psychoanalyst, Dr. Jennie Wälder-Hall,[42] whose second husband was a British Embassy official. Dr. Wälder-Hall informed her husband of the situation, but apparently no action was taken.

Straight says that more contacts from Michael Green caused him to drop out of government service and shortly after the AYC Congress, he decided to work at his magazine, *The New Republic*. His first article was a 30,000 word argument for increased defense production and it was published as a supplement to the magazine, marking a sharp departure from its previously isolationist point of view.

Outwardly, Straight's active pro-interventionist writing and organizing separated him from the majority of the left, which, guided by the Communist Party, was cloaking itself in lofty platitudes of isolationism and pacifism. Yet, during this period, Straight continued his contact with Michael Green, whom he last saw in early 1942, after the United States entered the war overseas. The reasons for this were never made completely clear by Straight, but Yuri Modin, one of the Cambridge Five's Soviet controllers, shed some light on the question in 1994. In the first book to tell the "Atom Spy" story from the ex-Soviet side, Modin said that Straight's 1982 autobiography was "a completely truthful book," but that "the only thing he omitted to say was that his thirty-year silence was motivated above all by his fear of the KGB."

When the Americans finally entered the Second World War in December, 1941, Straight volunteered for military service, but not until writing *Make This the Last War*, dedicated to his late friend John Cornford, and in support of the Declaration of the United Nations which had been signed by twenty-six states. In this book, Straight argued for a synthesis of the communist and capitalist societies, the unification of Europe, and something very similar to the "new world order"[43] envisioned by his father's generation, when Woodrow Wilson enacted the League of Nations. In this, he was in public agreement with anti-Stalinist, leftist exile groups from Germany, such as Karl B. Frank's "New Beginning." Wilhelm Reich's old friend from the early days in Vienna, Frank was already advising the new Office of Strategic Services (OSS), as a representative of underground, anti-Nazi groups *within* Nazi Germany.

By November of 1943, Straight was in the Air Corps Reserve. He tried to pull strings with his old friend Tom Corcoran (with whom Straight had written some of President Roosevelt's speeches) to get himself into active combat, but this proved difficult, and before he finished training on B-17 "Flying Fortresses," the war was over.

After the war, Straight considered a run for Congress in New York, but balked when the Democratic Party bosses at Tammany Hall, Bert Stand and Clarence Laughlin, informed him that they knew about his Communist associations at Cambridge and asked him how he would handle the scandal when it surfaced. (They heard the story from Straight's friend, financial columnist Eliot Janeway.) He considered working for the United Nations but Alger Hiss talked him out of it. After briefly working with World Federalist organizations, and serving as secretary for Albert Einstein's Emergency Committee of Atomic Scientists, he returned to *The New Republic*.

On September 12, 1946, Secretary of Commerce Henry Wallace was forced to resign from his Cabinet position in the Truman administration as a result of the outcry caused by a speech Wallace had given denouncing the Marshall Plan, especially for its lack of support for the Soviet Union. Wallace, an eccentric man who used radioactive fertilizer[44] on his homegrown tomatoes, appeared to be the best hope of the nation's socialists who sought to regain the ground lost since the war.

Wilhelm Reich and the Cambridge Five

Straight took on Wallace as editor of *The New Republic*, he says, in an effort to increase the circulation of the magazine from twenty thousand to one hundred thousand, where it could once again play a leading force in the country's political life as it had in his parents' day.

Now we return to the events surrounding the Presidential Election of 1948 as described by Carroll Quigley. The Georgetown professor remarked, first of all, that many Communists had been hired by Straight upon his return to the magazine, and noted that Straight's parents had endowed it with a tax-exempt foundation. Quigley argued that this tactic, of providing the "left" with a vehicle to blow off steam, was a long-honored tradition by the "American Establishment"—in particular the finance capitalists in orbit around J. P. Morgan. Michael Straight's own account confirms many of the details upon which Quigley built his analysis.

In *After Long Silence,* Straight ridiculed Quigley's notion that there was any hidden agenda to his association with Wallace. Straight compared Quigley to novelist Evelyn Waugh's fictional character Colonel Marchpole, a paranoid British intelligence officer. Ignoring completely Quigley's main point, that the Morgan bank had a strategy in using philanthropic trusts such as the one that subsidized *The New Republic*, Straight zeroes in on a few highly speculative and parenthetic conjectures that Quigley made about Straight's inscrutable role in the election of 1948. Quigley had stated, "If Mike Straight planned to do what he did do to the Communists in 1946-1948, that is, to get them out of the progressive movements and unions, he pulled off the most skillful coup of twentieth century American politics. It is not clear that he did plan it or intend it. But as a very able and informed man, he must have had some motivation when he began, in 1947, the effort which he knew might defeat Truman in 1948."[45] This matches quite well with what Straight describes of his own behavior. Quigley had no way of knowing how intimate Straight had been with Communists. Quigley then goes a bit further and offers an alternate explanation that, as he plainly states, is based on inconclusive evidence.

Quigley noted that Michael Straight's brother, Whitney Straight, was a director of British Overseas Airways Corporation (BOAC) and that his cousin, Cornelius Vanderbilt "Sonny" Whitney, was a founding director of Pan American Airlines. The family, in other words, was heavily interested in airline regulation. Interesting, then, that President

123

Truman, upon being elected, appointed C.V. Whitney to head the Civil Aeronautics Board, "the most important post concerned with civil aviation in any Federal department," Quigley noted. "The connection, if any, between these appointments and Mike Straight's original support and later abandonment of Wallace has never been revealed." Certainly, this is legitimate conjecture on Quigley's part, and labeled as such. In fact, "Sonny" Whitney was involved in a business battle with Howard Hughes over international flights in 1948. Michael Straight, in his book, *After Long Silence*, merely debunks (without evidence but much dry humor) only one tenuous thread of Quigley's remarks.

Straight readily acknowledges that Wallace, while managing editor of his magazine, became the candidate who was actively endorsed by Moscow. Although the Iron Curtain had fallen by this time, many American leftists remained awash in the euphoria of the victorious wartime alliance with the Soviet Union. Even so, there was a bitter division between the democratic left and the authoritarian left, when it came to the issue of relations with the Soviet Union. Straight knew what was going on when the pro-Communist Progressive Citizens of America (PCA), which vied for preeminence among liberals with the more moderate Americans for Democratic Action (ADA), began sending its activists to Henry Wallace to urge him to mount a third-party candidacy. Many liberals were opposed to the President's "Truman Doctrine." In Greece the United States, through its intelligence agencies, was overtly and covertly intervening in the Greek civil war in a step to prevent the spread of Moscow-aligned regimes in Europe. Indeed, Straight would later point to the Truman Doctrine as an impetus for the Vietnam War. One week after Truman proposed his Doctrine, Wallace delivered a nationally broadcast radio address against it. It was written by Michael Straight, and marked the informal advent of Wallace's presidential candidacy.

Another Wallace speech, made in London, caused Winston Churchill to label Wallace a "crypto-communist," a charge he later withdrew. Straight would later write that "he sensed the danger" as Wallace drifted closer and closer to a pro-Stalinist position and as persons Straight knew to be Comintern agents gravitated toward the campaign, which was international in scope. Even then, as part of Wallace's entourage, Straight met privately with his old Comintern comrade, Guy Burgess, in a London men's club for drinks and a discussion of international nuclear arms control.

Wilhelm Reich and the Cambridge Five

Next, Wallace and Straight flew on to Norway, Sweden and, finally Paris. In France, Wallace's tour was handled by Pierre Cot, yet another Comintern agent. Cot had been brought into the mix by two Americans, Alfred and Martha Stern. The Sterns, upon their subsequent exposure as Soviet spies, would later flee to Czechoslovakia. Although Straight recounts that he no longer supported Stalinism, he found himself at this time "being drawn into the Soviet orbit." His whole relationship with the Wallace campaign was confused and ambivalent. "Should I have forced Wallace to choose between his new allies and me? It wasn't that easy." Ultimately, though the magazine's staff supported Wallace, *The New Republic* would endorse Truman, lending some credence to Carroll Quigley's assertion that there was more to the situation than met the eye.

While Lew Frank, a Communist, reorganized the Wallace campaign, Straight wrote many of Wallace's speeches. "At the national level, as I learned later, the Communist party was divided. The old-time officials were opposed to a third party, fearing they could not control it. The young, undercover leaders were confident of their ability to manipulate mass movements; they were ready to risk their forces in an aggressive political strategy." Straight's real intentions are exceedingly hard to divine from his own account; for instance, he tells of setting aside time to discuss the issue with Wallace on a plane trip, only to be distracted by a chess game.

In Denmark, the American ambassador refused to meet with Wallace. So, in Copenhagen Wallace and Straight were hosted by Ole Calving, the socialist publisher of *Extrabladet*, a Danish daily newspaper. Fifteen years earlier, Calving's newspaper had published numerous sensationalist articles[46] regarding Wilhelm Reich who had fled to Denmark from Germany in May of 1933. Reich's permission to stay in Denmark became a public issue and a referendum of sorts on psychoanalysis, which was new to the Danes. When Reich's permission to stay was refused, and Reich went across the channel to Sweden, Calving's *Extrabladet* crowed about the scandal: "Within a few days Reich goes to Mälmo, but then he has done so much for psychoanalysis in this country that it has gotten its martyrdom." The paper described Reich, who gave a final talk at a public meeting of the Danish Psychoanalytic Society, as a "Mephistopholes" whose brilliance as an orator had "his listeners nailed to their seats."

"In a way, " *Extrabladet* continued in its 1933 assessment of Reich, "it must be said that he is a 'dangerous' man, who with both hands plucks the fruits from the 'tree of knowledge.' Some of the fruits were so 'green' when he served them, that several young girls had to be helped from the auditorium in a hurry."

Obviously, Ole Calving was not sympathetic to Wilhelm Reich, and helped to push him out of Denmark in 1933—no matter that Reich was a refugee from Nazi Germany.

Wallace and Straight pushed themselves back from Ole Calving's hospitality table (Straight says they all got drunk) and returned from their European tour in April of 1947. One month later, they edited and published Mildred Edie Brady's "The Strange Case of Wilhelm Reich" in *The New Republic*.

A few months later, Wallace resigned as editor, when he accepted the nomination from the Progressive Party as its candidate in the Presidential election of 1948. Straight, reversing course, began informing on the Progressive Party campaign to the Truman people. He began stumping for Truman, delivering speeches in his favor across the country, dwelling on the Communists' control of the Progressive Party. Just as Carroll Quigley described, the pro-Stalinist left suddenly became uprooted from its base of support, both in the labor unions and the "united front" associations they had so carefully cultivated over the years, and the Wallace campaign failed. The only question was whether the Progressive Party would draw enough votes away from Truman's left flank to elect Republican Thomas E. Dewey.

It may well have been Straight's decision, to oppose Henry Wallace, that tipped the scales in Truman's direction.[47] In any case, Truman won the election. The election was very close and every American is familiar with the photograph of a toothy-grinned Truman holding an early-edition newspaper's front-page that had prematurely declared Dewey the winner. Truman's loss had been a foregone conclusion until election day.

After 1948, Straight would not see Wallace again until 1953. Throughout this period, Straight watched the rise of the House Un-American Activities Committee and secretly harbored anguish that his "past" would be exposed. In his 1983 book, he begins referring to his association with the Cambridge moles as his "past" even though he continued a relationship with them. It is difficult to gauge Straight's behavior. Did he hire Wallace as editor, and tour with Wallace in Europe, and write his speeches because he legitimately supported him

Wilhelm Reich and the Cambridge Five

as President? Did Straight support Wallace to force Truman toward the left? Did he want to soften Truman's stand against the Soviet Union, or did he really recoil from the Stalinist takeover of the Wallace campaign? Is it unthinkable that Straight's actions were motivated by the Soviets, using an old sexual blackmail? Above all there is a maddening ambivalence running through Straight's account. He's with them, and he's against them, over and over.

So it was that Straight met with Guy Burgess in 1949 in London at a gathering of the Apostles; and again in Virginia on March, 1951, just a few months before Burgess defected to the Soviet Union, along with Donald Maclean, on June 8, 1951. Burgess had been in Washington, D.C., since October, 1950, during which time he lodged in KGB master-mole Kim Philby's home. Philby was then the official liaison between the British Secret Service's counterintelligence and the US Government. Most historians agree that it was Philby who tipped off Burgess, and Maclean, about the impending exposure of their espionage activities and allowed them to escape to the Soviet Union. Philby had seen some of the NSA's early VENONA decrypts, and realized that they would eventually reveal Maclean as a key asset of Soviet intelligence.

During the 1951 meeting, Straight says he suddenly became aware that Burgess was actually passing valuable information to the Soviets. He wondered if Burgess might have been responsible for the October, 1950 rout of General MacArthur's forces in Korea by China's Red Army, by passing on the details of the American battle plans to the Soviets, who would have then bartered them to the Chinese. "If I was right, Guy could have caused the deaths of many American soldiers," Straight wrote. "That thought left me numb." Straight says he warned Guy Burgess that "if you aren't out of the government within a month from now, I swear to you, I'll turn you in." A month went by and nothing happened. Three months went by and Burgess defected along with Maclean.

The defection of the two highly-placed British diplomats sent the Cold War into high gear. Not only that, but the "special relationship" between the intelligence services of the Americans and the British was over. The repercussions of that day in 1951 lasted decades. Shake-ups ("purges"?) of career intelligence officers on both sides of the Atlantic were the order of the day. James Jesus Angleton began a decades-long molehunt that would ultimately paralyze the CIA, and the culture of international espionage took on a character to which the word "paranoia" does no justice.

Straight says he went to a British official he happened to know, shortly after the Burgess/Maclean defections, to report everything. He was told, he said, "to take a number"—as though dozens of people were attempting to confess. Somehow, nothing came of it.

In 1952 when he was called himself to testify before the US Congress' Reece Committee, established to investigate the activities of tax-exempt foundations, Straight testified, "I regard the Communist Party as a source of espionage and subversion."

And again, Straight reports meeting with a Soviet official named Striganov on several occasions in 1954, ostensibly in his capacity as an editor and journalist. Back and forth he went over the years, wanting nothing to do with the Soviets and yet, meeting them for consultations. The public knew nothing of Straight's dilemma. Earlier, in 1948, when Stalin crushed the governments of Poland and Czechoslovakia, Straight says he asked himself:

"Was it not plain by then that the Soviet state had forfeited its claim upon the allegiance of free men?

"It was plain to me. On three occasions between 1949 and 1951, I drove my car to the British embassy with the intention of walking in and asking to see an intelligence officer. On the fourth occasion, I took up with my cousin, a high official of our Central Intelligence Agency, my fears about the infiltration of the British intelligence services. I wanted to tell my story. I needed one beckoning word or gesture to lead me on. Without it, I lacked the resolution to carry my impulse through. [...] I still needed some external force to trigger into action my sense of what was right."[48] Straight does not name him in his book, but he refers here to his cousin, Tracy Barnes. During the war, Barnes had worked for the OSS in Switzerland on psychological warfare. He worked directly with Emmy Rado, a Swiss psychoanalyst, and a friend of Wilhelm Reich's in Berlin.

That force to speak the truth would not hit Michael Straight until the summer of 1963, when the Kennedy administration selected him as chairman of the National Endowment for the Arts, a position which required an FBI security check. Although Straight withdrew his name from consideration, he finally resolved to go to the FBI and tell his story.

Wilhelm Reich and the Cambridge Five

At the Justice Department, Straight was taken to William Sullivan, the head of counterintelligence for the FBI, to make a statement. (Another FBI Agent, James Hosty, said in a British documentary[49] that he got orders from Sullivan of the FBI to remove Lee Harvey Oswald's name from the FBI's security index after the assassination of President Kennedy, on the orders of Tracy Barnes of the CIA, Michael Straight's cousin. Sullivan was shot to death in a "hunting accident.") Over the hours and days and weeks, Straight was deposed on his past. He identified his KGB contact "Michael Green" from a book of mug shots, and the FBI agents who interviewed him for most of the month of June, 1963, were satisfied that he was telling the truth. "We spent forty of fifty hours in interviews, shook hands, and forgot each other. But many loose threads were left lying around, and so, for four years, the bureau continued to call me."

Wilhelm Reich's description of character armor resonates with these comments of Straight's:

"You bear a burden for many years; you say to yourself, *once I rid myself of this burden, everything will be different. I'll stand upright; I'll sleep well; I'll be considerate of other people because I'll respect myself; I'll be happy.*

"It did not work that way;" Straight lamented, "it did not for me. Our bodies and spirits become adjusted to the burdens we bear, and when they are lifted from us, we do not spring back into the ideal shapes that we picture for ourselves in our fantasies."

The information given to the FBI by Straight would not be made public for many years thereafter, but the private consequences were swift and enormous. Sullivan asked Straight to meet with Arthur Martin of British counterintelligence, who had been working on the still-smoking cases of the "Cambridge Five" espionage ring. When Martin heard the story, he confided, "You may not believe this but this is the first hard evidence that we've been able to obtain on Burgess and Blunt."

Burgess and Maclean and Philby had by that time all been exposed by their defections to the Soviet Union. Anthony Blunt, however, still remained in British government service although much suspicion had been cast his way. Straight agreed to confront Blunt personally, if necessary. In September of 1964, Straight flew to London and met with Arthur Martin once again, who informed him that Blunt had confessed. Discrepancies between Blunt's and Straight's stories remained, and

Martin suggested a face-to-face meeting. Walking to Blunt's flat in London, Straight thought to himself, "*This is the sort of thing that belongs in a movie, only it's been filmed too many times.*"

"Thank God you did what you did!" Blunt exclaimed upon seeing him. Blunt had bargained for immunity in return for a full confession, and since the charges would have been hard to prove even with Straight's information, the British establishment closed ranks behind one of its own. The entire matter was to be hushed up. Blunt insisted.

So it was that for the next ten years, the world would wait for an inkling of the real story behind the defections of Burgess, Maclean and Philby. For the first time, the British Government was forced to examine the old school-boy networks of covert operators that had thrived within its hidden culverts. Straight became an informer on his old Cambridge comrades, and over the next ten years met periodically with British intelligence officers and endeavored to identify other "moles" in sensitive positions. He salved his conscience with the notion that for every mole he identified, he cleared ten more innocent true believers.

It wasn't until November 15, 1979 that Michael Straight's story reached the world, when Prime Minister Margaret Thatcher announced that Anthony Blunt was the "fourth man" of the Cambridge ring—after Maclean, Burgess and Philby. Even then, Straight's identity as the source who exposed both Blunt and a fifth man, Leopold Henry Long, as operatives for Soviet intelligence was not revealed to the public until 1981, when the *Sunday Times* ran a story which was confirmed in Parliament. Straight objected to press reports of his own spying on behalf of Russia, but decided not to pursue the matter beyond a letter to the *Times,* because of bad conscience. He felt guilty about people he had informed upon.

Michael Straight's autobiographical account obliquely refers to the entire genre of spy novels, counterintelligence surveys, and press accounts engendered by the revelations of the case. He succeeds in ignoring completely the implications and fallout of the Cambridge Five's deep penetration of the British intelligence apparatus, and the consequent compromise of the CIA at the highest levels. He says nothing about the ring's role in the theft of atomic secrets.

Straight was never charged by the FBI with espionage, and he makes every effort to make the point that the information he gave to "Michael Green," though classified, was of no strategic importance. At first glance, it would appear that the most significant aid given by Straight to the Soviets was to remain silent about the covert work of Guy Burgess and Anthony Blunt, both of whom he knew to be spies. Straight's suppression of Wilhelm Reich was no small service, either. Through the publication of Mildred Edie Brady's "The Strange Case of Wilhelm Reich," Straight helped to suppress one of the Soviet Union's most telling critics.

However, Straight did even more. For example, Straight's FBI files[50] include reference to a meeting between Straight, "Green," and Green's wife (the niece of Earl Browder, leader of the Communist Party, USA) at a restaurant in New York. At Green's request, Straight provided a press credential for a Soviet agent named Mark Julius Gayn sometime in 1942.[51] Providing press credentials to a Soviet agent is a more serious event than Straight admits to. Three years later, in June of 1945, Gayn was arrested along with six others, including two State Department officials, Lauchlin Currie and John Stewart Service, and a young Office of Naval Intelligence officer named Andrew Roth, for possession of hundreds of highly classified documents in what would become known as "The Amerasia Case," the first inkling the American public gained as to the breadth of Soviet post-war espionage reaching into top levels of the government.

The overlapping territories of journalism and intelligence work can be demonstrated by a brief examination of the career of Mark Julius Gayn. Like Kim Philby, Guy Burgess, and so many others, Gayn used journalistic cover for his activities.

The Legend of Mark Gayn

Much of the information regarding Mark Julius Gayn's early life in China can be found in a disturbing book[52] written by his younger brother Sam, who remained there after the Maoist revolution of 1948. Sam was a translator and language teacher and later would suffer greatly during the Cultural Revolution.

Born Mark Ginsbourg in Siberia, Mark Gayn and his brother, Sam were raised in Harbin, China amongst a small but prosperous community of Russian exiles, many of whom were Jewish refugees from Czarist pogroms.

Sam's book never mentions his eldest brother by name, which is strange, since they were close. None of the names of family members are mentioned. Sam said his brother (Mark Gayn) worked in the Red Army's library in Vladivostok, and majored in "pedagogy"[53] at the Soviet "Institute for Teachers, Librarians and Propagandists" in 1926. A few years later, Mark came to America and went to Pomona College in Southern California. It's not at all clear how his family came up with the money to send Mark to California, but Sam says they did. Mark told the FBI that he worked his way through college, finishing at Columbia University in New York.

"In the mid-thirties he returned to China and did editorial work on some English-language newspapers published in Shanghai. He was violently anti-Japanese and did not try to conceal it. He finally made the Japanese military blacklist, on which he occupied a prominent place. He managed to escape, leaving Shanghai in the summer of 1939. Soon after arriving in the United States, he changed his name to make sure the Japanese did not take it out on me."[54]

"After victory, he was assigned to Japan as a war correspondent for one of the big American dailies. We exchanged letters. At the end of December 1946 he arrived in Shanghai. He stayed there for some weeks and then left for the Liberated Areas for some interviews with our Party leaders. (Recently he showed me a photograph of him and Liu Shaoqi taken in those days.) Not waiting for him to return to Shanghai, I went to the Shandong Liberated Area and left him a brief note informing him of what I had done.

"My departure made him unhappy. He wished I had talked things over with him before making such an important decision, of which he did not approve.

"We lost all touch with each other until the late seventies." Sam's daughter went to live with Mark Gayn while she attended a university in Canada, before returning to China.

The rest of the book describes Sam losing his post as a foreign languages teacher during the succession of upheavals in China starting with the Cultural Revolution, during which he was accused of being a "Russian spy." The brutality and humiliation during this period is clear enough, but Sam is still writing confessions with a Red Guard looking over his shoulder, so he sounds a bit like Winston Smith at the end of Orwell's "1984." He loves his country and his Party, despite its

Wilhelm Reich and the Cambridge Five

"excesses." Things are looking up.

After graduating from Pomona, Mark Gayn began a long career in journalism. Gayn wrote for national magazines such as *Collier's* and in 1934 he became the *Washington Post*'s special correspondent in China until his return to the US in 1939.

Michael Straight met with his Soviet handler ("Michael Green") at Longchamps restaurant in New York sometime after November 24, 1942. In later statements to the FBI, Straight recalled that at this meeting or the one previous, Straight supplied to "Green," at his request, *The New Republic*'s accreditation (i.e., official press passes) for a Swedish woman journalist and Mark Julius Gayn. Why would Gayn, already a widely published reporter, need additional accreditation from Straight? Why would "Green," rather than Gayn himself, ask Straight for these credentials?

Nobody knows. Mark Julius Gayn, subsequent to being accredited as a journalist for *The New Republic*, began to amass an impressive collection of classified documents from the US Departments of State, War, and the Office of Strategic Services (OSS). Many of the documents dealt with "psychological warfare" and propaganda, and nearly all of them dealt with the Chinese theatre in the war against Japan. When a classified report from the OSS turned up—almost verbatim—in the pages of *Amerasia,* US counter-intelligence agents broke into *Amerasia's* offices and began an intensive, 24-hour surveillance of everyone connected with it, including Mark Gayn. *Amerasia* was a small-circulation, pro-Maoist journal published by Philip Jaffee and partially funded by the Institute of Pacific Relations, which in turn was financially supported by Nelson Rockefeller, Michael Straight, and a host of "philanthropic" foundations that later became the subject of a congressional investigation called "The Reece Commission."

The FBI set up a massive, round-the-clock surveillance of the spy ring, capturing on tape, via wire-taps, conversations between Mark Gayn, Phil Jaffe, and the others. Jaffe referred to the secret documents he obtained as "cigarettes"—using an inept code; "I have some fine Japanese cigarettes," etc...

The FBI files leave no doubt that Gayn was working for the Soviets:

> Jaffe: It is just as you thought. She is talking all over Washington, especially in official circles, and plastering you with the good old 'red brush.'"

Gayn: Me personally?

Jaffe: Yeah, you personally. Very continuous. Continuous activity on her part and you should not turn your back on her, because if you do, she will stab you.

Gayn: My turning my back or not turning my back, I have nothing to do.

Jaffe: No, nothing you can do, but run the hell out of this country and over to the Soviet Union. You will be safer there. Joe will protect you.[55]

In their book *The Amerasia Spy Case, Prelude to McCarthyism*,[56] historians Harvey Klehr and Ronald Radosh mark the case as one of the most important in the Cold War. Coming as it did during the last years of WWII, it set the tone for future relations between the two post-war superpowers. The investigation uncovered a network of journalists, State Department officials, and servicemen who were caught red-handed spying for Russia. The evidence against those arrested was overwhelming. However, the charges were dropped.

Gayn, who was in as deep as the rest of them, was not even *indicted*. Klehr and Radosh suggest that a political fix crafted by none other than Michael Straight's old boss in the Roosevelt administration, Tommy "the Cork" Corcoran, worked the deal. There is also the possibility that Gayn, in particular, was let off in return for "turning" sides and becoming a double-agent. No other journalist, on either side of the Iron Curtain, could match Mark Julius Gayn's access to the highest echelons of power in Moscow, Peking and Washington, throughout the fifties.

Autumn, 1959: Mark Gayn's notebook records an exchange between CIA Director Allen Dulles and Soviet General Secretary, Nikita Khrushchev: "Dulles said to Khrushchev: 'You, Mr. Chairman, may have seen some of my intelligence reports from time to time.' Khrushchev said 'I believe we get the same reports. And probably from the same people.' Dulles said: 'Maybe we should pool our efforts.' Khrushchev: 'Yes, we should buy our intelligence data together and save money, we'd only have to pay the people once.'"[57]

So this is another bizarre twist, one of many in this investigation: Michael Straight gave credentials to a journalist-spy, who was subsequently cleared of treason in what should have been an open-and-shut case. While Straight was never implicated directly in the *Amerasia* case, it is hard to conceive what constructive purpose Straight had in

Wilhelm Reich and the Cambridge Five

mind when he provided his Soviet handler with an American press credential for Gayn, just three years prior to Gayn's arrest.

There is one final turning point in the Cold War saga of Mark Julius Gayn.

After the US Attorney General declined to file charges in the *Amerasia* case, Gayn resumed his work in journalism. There would be no further public scandals involving Gayn, until long after he died in 1981.

In 1992, journalist Dick Russell published one of the most well-documented and intriguing books ever written about the Kennedy assassination, *The Man Who Knew Too Much*. Russell unfolded the story of a Korean war-hero (two Purple Heart medals, with clusters) and former military intelligence officer in Japan, Army Captain Richard Case Nagell. About a month before the Kennedy assassination, Nagell sent a registered letter to J. Edgar Hoover at the FBI, warning him that Lee Harvey Oswald, along with two Cuban refugees, were planning to assassinate the President.

The FBI denied having received this letter when it was requested by Congressman Don Edwards in 1974. Captain Nagell replied that not only could he prove he sent that letter, but furthermore, he possessed a number of incriminating documents, such as photographs of himself and Oswald along with top-secret military intelligence reports, in a Swiss safe-deposit box. Nagell was a contract agent of the CIA at that time.

On September 20, 1963, Nagell walked into the State National Bank in El Paso, Texas, and asked a teller for one hundred dollars in American Express travelers checks. Nagell then stepped back from the window and fired two shots from a Colt .45 pistol into the bank's ceiling, put the gun back in his belt, and calmly left the bank. He got into his car and waited until the police arrived, raised his hands and said, "I guess you've got me, I surrender." When taken into custody by the FBI, Nagell would only say, "I would rather be arrested than commit murder and treason."

Nagell's arresting officer, Jim Brunden, would later recall during an interview by Russell, "When the assassination happened, I didn't think of it right away, because that was a pretty hectic time [...] Then a few days later, after Lee Harvey Oswald was killed, what Nagell said came back to me. I thought, this had to be what he was talking about. How the hell would he have previous knowledge of it? How would he know what was coming down in Dallas?"[58]

Nagell had hinted to Brunden about the coming assassination of John F. Kennedy during his arrest and trial. "I really don't want to be in Dallas," Nagell told him during the preliminary hearings.

To convince the FBI that his communication addressed to J. Edgar Hoover was not a crank letter, Nagell wrote it under an alias of a "known Communist agent" with whom he had been in contact in regard to the plot to assassinate Kennedy. This alias, "Joseph Kramer," was so well-known to the FBI that they could not have failed to realize that Nagell's warning should be taken seriously, because if "Joseph Kramer" had advance knowledge of the conspiracy, then so did the Soviets. Indeed, the warning letter was so carefully worded that it ought to have led Hoover to conclude that *a Soviet agent was planning to kill Kennedy*, even though Nagell (and, most likely, Hoover) knew otherwise.

"Joseph Kramer" was Mark Julius Gayn.

Nagell's account and Russell's investigative substantiation of it are far too complex to go into much detail here. In broad strokes, Captain Nagell was a lifer in the ultra-secret intelligence operations of the CIA and the Pentagon. In a game played out in Korea, Japan, China, and the Philippines, intelligence officers like Nagell tried to recruit their counterparts into becoming double-agents, while knowing all the while that their counterparts on the Communist side were trying to do just the same. As deceit covered falsehoods, as lies traded with truth, men like Nagell lost touch with reality, and no longer knew who it was, exactly, that they were working for. In despair and rootlessness, they tried to forge their own guiding lights of ethics and purpose, but after so many years of military service, they were unequipped to do so. Ideals of honor, duty, country became shattered as the operatives saw hellish tours of duty filled with torture, murder, and terrorism in service to shifting geopolitical goals crafted by craven ideologues on all sides.

Nagell had found himself in a situation where he might have been in Oswald's shoes. His contact with agents on the Soviet side, while sanctioned by US Intelligence, made him vulnerable to being cut loose, adrift, or framed. His principle contact with the Soviets, Nagell testified, was Mark Gayn, with whom he shared foreknowledge of *multiple* plots to kill the President of the United States. Nagell implied to Russell that he had been ordered to eliminate Oswald before he could pull the trigger. Nagell said this order was delivered by a Soviet agent, perhaps

Wilhelm Reich and the Cambridge Five

Mark Gayn, or, more startlingly, Tracy Barnes, of the CIA, Michael Straight's cousin. Recall that Straight's first approach to the US government to speak of his past had been made to his cousin Tracy Barnes.

The Amerasia Case was a cause celebré for the McCarthyites, yet Gayn somehow escaped prosecution even though he had been caught red-handed. All evidence points to the fact that he was a double agent found useful to both sides.

Was Michael Straight a Soviet spy? Yes. Straight himself deeply regrets his irresolute relationship with the Soviets, if for nothing more than his inaction when presented with clear choices. I think he said it best himself, when he told an interviewer for the Columbia Oral History Project that guilt weighed heavily on him:

Straight:

"I had not sought out any secrets or given away any confidential information, but I had accepted the shell, if not the reality, of that role. No, I felt deeply guilty about it all and I'd lived through some moments of terror."

The interviewer offered Straight an exculpatory "out":

"What you previously told me, by impression, would be that you were harassed and pursued in a subtle way, and that you bravely evaded doing anything distasteful."

Straight replied:

"Oh, no, I think it was highly distasteful, the whole thing. The association, the furtive surreptitious meetings with a man who was plainly a Soviet agent, the willingness to be placed in that role and to continue in that role, the inability to do anything about it. It all seemed to me highly reprehensible."

Straight's statement to the FBI, given a few weeks prior to the Kennedy assassination, led to the exposure of the chief of British intelligence in Germany as a Soviet spy. Straight would later recall:

"I remember on many occasions complaining to Isaiah Berlin and others in the postwar world about the shallowness of intellectuals such as himself who were complaining loudly about the excesses of McCarthyism in America, and yet were blinding themselves to the core of truth which McCarthy exploited and distorted, that there were problems of security created by Soviet infiltration of our governments."[59]

Above all, we know that Michael Straight possessed first-hand knowledge of that.

The Dolivet Memorandum

Another intriguing document is tucked away in Michael Straight's archive at Cornell University: a long memorandum regarding Straight's brother-in-law, Louis Dolivet. It was written by Ladislas Farago, who had served in American Naval Intelligence during the war, and who was later employed, as an editor, by Louis Dolivet, the publisher of *United Nations World*.

Louis Dolivet had been introduced to Straight, and to Straight's sister Beatrice ("Biddy"), by Pierre Cot, the French Comintern agent who would later surface as one of the organizers of Henry Wallace's European tour. Dolivet presented himself as a French refugee, a former officer of the French Air Force, and a representative of the Free French underground—none of it true, as it would turn out.

Biddy fell in love with him and they were married in 1942; the couple had one son. After the war, Dolivet, a minor consultant with the United Nations, was extremely successful in gathering the support of men such as oil scion Nelson Rockefeller, Dr. Max Ascoli from the New School for Social Research, and Pittsburgh's Richard Mellon, in Dolivet's ambitious plans to promote the U.N. However, Dolivet's close association with Henry Wallace later caused many of his benefactors to withdraw their support for Dolivet's *United Nations World*. Even so, Ladislas Farago estimated that Dolivet had been able to raise over $1 million in the years 1945-1946.

By then, Biddy and Louis Dolivet's marriage was failing. Biddy wanted to return to acting, and was to have a successful career in Hollywood.[60] Dolivet refused to grant her a divorce. A Republican lawyer at Straight's family law firm, Mudge Rose,[61] hired Ladislas

Wilhelm Reich and the Cambridge Five

Farago to investigate Louis Dolivet's past. Farago, who had resigned his job with *United Nations World*, had serious questions about Dolivet's associates at the UN. He believed that Dolivet was receiving instructions from people who, though formally underlings, were actually "political commissars."

In April, 1947, published accounts of Dolivet's true identity began to emerge. His birth-name was Ludovicu Brecher, he was a Romanian, and he had been an agent of the Comintern in France, Germany, and Switzerland during the 1930s. Dolivet denied these charges, but he declined to elaborate. In most cases, he was supported by the financiers who enabled his work. Attempts on the part of Straight's family to clear up the matter with the FBI and the State Department were met with official silence.

Biddy and Louis Dolivet were divorced in 1949, and Louis returned to France. The State Department, bowing to pressure from Congress, denied Louis Dolivet re-entry to the US; then, suddenly, the estranged couple's son, Willard, died tragically in a drowning accident. Michael Straight told the State Department that he "saw no useful purpose" in allowing Dolivet to return under an emergency visa to attend the funeral of his son. The visa was refused after an appeal went all the way to President Harry Truman.

Dolivet requested that the New York district attorney begin a murder investigation, demanding an autopsy. The autopsy confirmed the accident, but Dolivet pressed on.

Farago went to Europe in 1950, making a complete investigation of Louis Dolivet's past. He personally interviewed many of Dolivet's associates, including Arthur Koestler and Francois Mitterand. This unpublished document, "Memorandum on Louis Dolivet," is intriguing because its tone and tenor are so similar to cases brought up before the House Un-American Activities Committee.

Farago, in the employ of the Straight family, was apparently unable to prove any criminal or conspiratorial work on the part of Louis Dolivet while he was a resident of the US, but a familiar narrative does emerge. Dolivet had been active in the Communists' United Front organizations in France, taking leading roles in groups such as the Rassemblement pour le Paix and the World Federation of United Nations Associations.

Dolivet was born on March 26, 1908, in the Galician part of Poland. His parents were citizens of the Austro-Hungarian Empire. In 1918, his family moved to—of all places—Chernowitz, in the Bukovina, which is where Wilhelm Reich grew up and attended "gymnasium," or highschool. Reich had fled Bukovina in 1915 and joined the Austrian army; Dolivet's family moved into the area when he was ten years old. Like Reich, however, Dolivet left the Bukovina, at the age of seventeen, "never to return."

In 1925, Dolivet went to Geneva—most likely, Farago says, as a representative of a Romanian underground Communist student organization. In Geneva, he worked with Leon Nicole, Moscow's representative in Switzerland, and chief correspondent for *Pravda*. Nicole told Farago that Dolivet was an opportunist, probably a double-agent. In 1928, Dolivet went to Grenoble, France, on what he said was a scholarship from an American foundation, to study law. His university colleagues remembered him more for his "amorous escapades," Farago said. Though Dolivet received his law degree in 1931, he never practiced law. He became fully involved with the Comintern as a leader in the Popular Front organizations before the war.

Between 1926 and 1933, Farago recalled, the Soviet Union, through its agents in the Comintern, sought to promote peace organizations, to defend the Soviet Union from "Fascist aggression." "This was designed as the first practical step toward the creation of the Popular Front of all Communists, Socialists and independent progressives — a strictly political design," Farago wrote. "The second Stalinist version, still serving as a blueprint for the current peace propaganda campaign of the USSR and world communism, had an obvious conspiratorial orientation. It was designed to keep the Western democracies disarmed while *all* totalitarian countries would continue their armaments, in preparation for a joint Communist-Nazi-Fascist showdown with the West. In this later plan, Stalin envisaged an effective alliance between the USSR and Germany. He visualized the war between Nazi Germany and western democracies as inevitable, and expected it to end with the victory of Germany, however costly or difficult. He was ready to support Hitler's aspirations to power because Stalin regarded Hitler's seizure of power in Germany as an essential precondition for the implementation of the plan," Farago concluded.

In other words, the Comintern, whose agents influenced the world's peace organizations, was the cornerstone of Soviet foreign policy.

Dolivet's Comintern trail across Europe began in France, and after he graduated from law school, he returned to Geneva. He went to Berlin in 1932, and worked with Willi Münzenberg there, under the alias

Wilhelm Reich and the Cambridge Five

Ludovicu B. Udeanu. When Hitler came to power, the German sections of the Comintern relocated to Paris, if they were lucky. Re-entering France on a student visa, Dolivet enrolled at the Sorbonne. He arrived in 1933, Farago confirmed, "as a full-fledged agent of the Comintern, in charge of the student section of the Communist-dominated Amsterdam-Pleyel peace movement." Using his contacts among intellectuals, Dolivet secured a position with *Le Monde*, as a kind of editorial "commissar" who vetted the copy for ideological purity. This weekly journal had been founded in 1928 by Henri Barbusse, a famous man of letters, with 1.5 million francs supplied by the Soviet Commissariat of Foreign Affairs. "The idea behind *Le Monde* was Moscow's desire to have in Paris a journal of high intellectual standing for the propagation of Soviet achievements. The Soviet ownership and influence of the journal was to be kept secret," said Farago. Dolivet's Comintern status was completely unknown to the paper's editorial staff. Without even introducing himself to the editorial staff, Dolivet began working at the paper's printing facilities, quietly making small changes directly to the galleys being rushed to print. Soon he began to smuggle in whole articles of his own writing. Eventually, the editors resigned in protest. Though the paper lost readership and fell into decline under the covert editorship of Dolivet, he was a rising star in the Comintern. "The Bolshevik determinism his articles reflected, his Communist enthusiasm, and, above all, the ruthless efficiency with which he executed his orders and carried out the purge, assured him of a career within the Comintern hierarchy." In the summer of 1934, Dolivet traveled to Moscow to "reap his reward." He made direct contacts with important leaders in the Kremlin, and was subsequently appointed a leading role in the new Popular Front strategy. In 1935, his formal Party membership lapsed, but Dolivet continued his public support.

For a time, he received financial aid for his activities from Moscow, but charges of financial misdealing led to Dolivet's decline in influence within the Comintern. Farago's memorandum describes Dolivet as an opportunist who above all viewed his association with the Comintern as a means to power and influence: "an unscrupulous diplomatic adventurer, a double agent."

"He never tried to sit out anything on a fence. He was always intent on riding two horses simultaneously."

Still, Dolivet's activities continued in line with the larger Stalinist policy of infiltration of "peace groups," becoming what Farago called "a traveling salesman of the kind of peace Stalin tried to impose upon a genuinely peace-loving world." In 1937, Dolivet commissioned the construction of a "peace pavilion" at the Paris International Exhibition, "with funds whose origin remains obscure to these days."

"Hitler's invasion of Russia in the summer of 1941 brought Stalin's dreams to a close. But it is remarkable that this whole plan formed the chief motivation of Dolivet's activities already in 1936 when he outlined it in writing," Farago wrote. "In fact, the destruction of Britain and France was the actual assignment of Dolivet from the Comintern."

Some of the leaders in the Comintern had serious reservations about the policy of alliance with Germany against the western countries. Willi Münzenberg was one of these; he was found dead, hanging from a tree. Arthur Koestler recalled, in *The God that Failed*, "as usual in such cases, the murderers are unknown and there are only indirect clues, all pointing in one direction like magnetic needles to the pole." Farago recorded rumors that Louis Dolivet had a hand in the murder, but he was unable to substantiate them. Originally, General Walter Krivitsky, a Soviet foreign intelligence agent, had been assigned to assassinate Münzenberg, but he had spared Münzenburg's life because of his own doubts about the alliance between Russia and Germany. These doubts, and his failure to kill Münzenburg, precipitated Krivitsky's defection to the United States. Later, Krivitsky's own mysterious death, in a Washington, D.C. hotel room, terrified Michael Straight. Stalinism was not a game.

Thus, Farago allows, Dolivet may have had good reason to bury and never publicly discuss his work in the Comintern, for he must have known as well as anyone else the extreme penalty renegades earned for themselves from the Party. Once, Dolivet himself had been brought up on charges by the Communist Party in a special investigation. The Party charged Dolivet with being a "British spy." He was able to clear himself of these charges, a testimony to his flexibility. A co-worker from that time told Farago that he believed Dolivet had learned important lessons from that experience, which had reinforced his reticence concerning public disclosures of his past.

"Whatever it was," Farago assessed, "the fact is that young Ludwig Brecher [i.e., Dolivet] had no scruples when it came to the consideration of a Communist-Nazi conspiracy against the west. [...] As a result, Brecher is still alive today."

Wilhelm Reich and the Cambridge Five

After a brief stint in the French Air Force, Dolivet abandoned his unit when Hitler conquered France. He made his way to the US by misrepresenting himself to the American consul. Farago sums up the document by recounting Dolivet's activities involving the U.N., his supporters among the American financial elite, and his strange backroom coterie of apparent watchdogs from the Comintern, his frequent trips to Zurich and Vienna, "the purpose of which cannot be explained with the interests of *United Nations World.*"

"During his stay in the United States Dolivet, at various times, managed the political and diplomatic affairs of men ranging from the actor Orson Welles," whom, Farago says, was a regular guest along with Rita Hayworth at Dolivet's home, "to ex-Vice President Henry A. Wallace. The record of his work on behalf of Wallace, in the United States, in France, and in Sweden, is too well known to require recounting. According to Dolivet's own admission, he was responsible for Wallace's 'Century of the Common Man' speech[62] (which he says he wrote,) as well as for some of the speeches Wallace delivered during his ill-fated European trip."

Indeed. We might recall that Straight also was on this European tour, also wrote some of Wallace's speeches, but unaccountably reversed his course, when, he says suddenly he discovered that there were Communists directing the campaign.

"The Dolivet Memorandum" is a strange and melancholic document. It's ironic that, had this investigation been instigated by anyone other than Michael Straight, it might be construed as a "witch hunt."

Was Dolivet really a Soviet agent, or a true-believer in world federalism? Was he an opportunist, or a victim of Comintern terror, caught in some blackmail and rendered a tool of Moscow? Who he was really working for (aside from himself, obviously) remains obscure. Why did the State Department and the F.B.I. sit on the evidence until it became an issue in Congress? Did Dolivet's friends on Fifth Avenue lend a hand in quashing the investigation, or was he really a double agent, working both sides against the middle?

Dolivet had come a long way from Bukovina. He had been on hand in Berlin during Reich's own Comintern days, and was there in France among the German refugee community in Paris. Dolivet represented, at one time and in one man, the best hopes, of both the Rockefellers and the Communists, for world federalism. Was he such a flexible man, or did he offer something deeply and fundamentally shared by each?

Wilhelm Reich and the Cold War

Or was he ultimately a mere real estate agent? After his final return to France, after his divorce, Dolivet made a fortune in land dealings in Paris. He died in 1982.

Of course, Michael Straight says he was unaware of any Communist conspiracy using his periodical as a means to destroy the life and work of Wilhelm Reich. Yet on the other hand, he lived in an extremely ambiguous world, and -apparently- he did not even know his brother-in-law's real identity. He was surrounded by Communists, dupes, Republican lawyers, double spies, CIA cousins, KGB handlers, free-lance agents, and activists for a new world. Dry as the grass swaying on the dunes of Martha's vineyard, he was the perfect medium for intrigue. Too much money, and even more time.

He was the publisher of record for Mildred Edie Brady's public smear of Wilhelm Reich. She called for state sanction against Reich, but she did more than that. She carried weight with the government as a functionary of a "consumers" pressure-group. Armed with her own clippings from *The New Republic*, Brady marched on the Federal Trade Commission and the FDA, and demanded that the government protect the public from the "strange" Dr. Wilhelm Reich. Straight says he had reason to believe that Brady was a Communist, and says he had felt manipulated by Mildred and Robert Brady in a separate incident, when they pressured him to publicly attack a government official over low-cost housing.

To be sure, Straight had other fish to fry in the election-campaign year of 1947. Yet he was certainly surrounded by people who had the motive and opportunity to make sure that Reich's critique of "Red Fascism," as embodied by the Soviet Union, never gained a full hearing in the United States. Reich had published *The Mass Psychology of Fascism* in English in 1946, and was immediately denounced in the pages of *The New Republic*. Reich was viewed as a renegade Communist, and he knew the secrets of the Comintern. He could have exposed many covert Stalinists active in the US and elsewhere, but never did. But more essential than that, the Stalinist agents of "emotional plague" simply could not tolerate open, self-regulated life, autonomous movements, nor a scientific discovery of Life Energy (orgone biophysics). The Comintern had degenerated into a support-group for the impotent mass, a training-ground for conspiracy, and a haven for the perverse.

Wilhelm Reich and the Cambridge Five

Was it just a coincidence that Dolivet married Straight's sister? Not likely, if he was, as the "Memorandum on Louis Dolivet" by Ladislas Farago suggests, an active, or semi-active agent of the Comintern. Throughout Farago's account runs a theme of Dolivet acting as a "honey pot" manipulator of women. Details of Straight's life and family were filtered back to Moscow through Arnold Deutsch's written character-analyses. Moscow Center knew all about Straight, his father, his mother, and his sister. Dolivet had first arrived in America penniless, with a broken hip from a shipboard accident, not knowing anybody; yet he quickly ensconced himself at Straight's well-appointed home.

Dolivet married Straight's sister, and the family trust invested $250,000 in his *United Nations World*. One either must possess an extremely charitable view of humankind, or criminal ignorance, to suggest that Dolivet had married Biddy, Straight's sister, as a "coincidence."

When a guy starts rolling straight sevens and elevens all in a row, in fairness to everyone at the table, it's time to check the dice.

My interviews with Michael Straight

The evidence presented in previously published accounts, including Michael Straight's own autobiography, leave us with a murky portrait. We are confronted with the fact that the publisher of "The Strange Case of Wilhelm Reich" was a self-admitted recruit of Soviet intelligence. Although there is evidence that Straight did commit acts of treason, no charges were brought against him by the US government.

I began to read a portion of the vast literature that dealt with the Cold War's so-called "atom-spies" who worked not for money but for organizational commitment. My first stop was John Costello's *Mask of Treachery*, an exhaustive tome that traces the career of art historian and double agent, Anthony Blunt. Costello documents the sexual economics of homosexuality that was a kind of unifying glue for the spy ring.

I had formed an opinion about Straight, based on what I had read in the literature, but it was not a positive one. I had no evidence that orgonomy meant anything to him, or that he had any direct role in orchestrating the show-trial against Reich. Even so, I knew that I needed to ask him directly about it, and interview him for this book. It was easy for me to categorize Michael Straight as a dilettante dupe, historically, and it would be just as easy to hook into my basic envy of the well-born. Straight was in his eighties when the first edition of this book was published and died in 2004.

Through some diligence and blind luck, I obtained his telephone number and was startled to hear him answer the telephone himself, with a gruff, no-nonsense greeting: "Yes?"

What, I thought to myself, no Jeeves the Butler?

Straight had been interviewed by dozens of authors and journalists since he publicly disclosed his youthful interlude with the Comintern. It's evident from the transcript of our interviews that Straight has a very practiced narrative, a story he has retold in every detail. Yet I would be asking him a different set of questions, beginning with my central question: did he remember Wilhelm Reich and Mildred Edie Brady's article in *The New Republic*, and was that article inspired, as Reich believed, by Stalinist "red fascists?"

"Wilhelm Reich?" he answered with a question quickly, pronouncing Reich's name a little too perfectly, I thought. I tend to pronounce it in the American style, like "rike" but Straight said "reish." That's as close as I can come, but the "ch" sound in German is very difficult sound for Americans to make; it's a soft "sch" coming through the back of the throat. My suspicions were raised by Straight's easy pronunciation. "I didn't know a thing about him. The name rings a very faint bell with me. I had absolutely zilch interest in him at all," Straight replied.

He remembers Robert Brady, whom he knew at the Office of Price Controls. He said Helen Fuller "brought him in," and that both Bradys wrote some articles for *The New Republic*. Straight remembers Robert Brady as a "hard liner" because he refused to translate a German text, because it was anti-Bolshevik, or too sympathetic to the Fascists.

Straight confirmed that the Bradys were Stalinists. The couple had worked in Washington, D.C., in the early forties, and were both removed from government positions when Robert Brady's book on German economics was read on the floor of Congress. "Young economist Helen Fuller (later an editor for *The New Republic*) had worked for the National Youth Administration. Perhaps she had met the Bradys there. She and I, along with Alfred Sherrard and William Salant, of the Federal Reserve Board, wrote a 30,000 word report published in *The New Republic* insisting on 1) support of the Allies during the Hitler-Stalin Pact and 2) a 'democratic defense.'" It turned *The New Republic* 180 degrees from its isolationist policy. It was very well-written."[63]

Wilhelm Reich and the Cambridge Five

"Helen was a liberal. She had some fellow-traveling friends, who thought it wouldn't be so bad if England was defeated. Her work was good, very sound, but she brought along with her some associations."

Did Straight mean "fellow-traveling" as in "Stalinists"?

"Yes," he replied firmly. "Robert and Mildred Brady were friends of Helen's. I was very clear about where I stood on foreign policy. On domestic policy, there was no disagreement."

I informed him that Kim Philby, when he went to Vienna in 1935, became involved with Reich's political organization, Sexpol. "It's all total news to me," Straight said.

Another crossover I found was when Straight went to Copenhagen, and met with Ole Calving at the *Extrabladet* newspaper. That meeting was set up by Pierre Cot, I proposed.

"Probably. Perhaps," Straight answered. "I can't answer that. You're looking back fifty years, and I don't really know. I think it was set up by *The New Statesman*. Cot handled the French end of it."

Ole was the publisher of Extrabladet *and they had published some of the first attack-articles on Reich in Denmark when Reich lived there.*

"Really? I didn't know a thing about him. I think Kingsley Martin had arranged this trip. I remember the French part of the trip got out of control because Pierre Cot got into it. He was brought in by two friends of Henry Wallace's, Alfred and Martha Stern. She was the daughter of the American ambassador to Germany. They turned out to be Comintern agents. They couldn't get back to this country. My friend who used to work for the FBI, a Congressman from San Francisco, Don Edwards tried to get them back, and the IRS hit them with back taxes when they returned. I think Alfred Stern died in Mexico. He was from a wealthy family in Connecticut, and somehow he and his wife got mixed up with the Comintern. He was turned in by another former Russian agent, I think his name was Boris... Anyway, they had arranged the French trip [taken by Henry Wallace and Michael Straight in 1947]."

A. S. Neill spoke at Dartington Hall. He was a friend of Reich's.

"Really? He was a friend of Reich's? Well, that makes sense. Neill was kind of a crazy nut, you know, and I guess Reich was too. Both eccentrics. Neill had the alternate school, well-to-the-left of Dartington Hall, in terms of stuff like nude bathing and living together with both sexes and so forth. Neill was to the left of us being really far out there, taking kids who were totally unmanageable. Dartington took kids who public schools wouldn't take, that is to say the private schools, but Neill

Wilhelm Reich and the Cold War

was way to the left of Dartington. Probably more successful than Dartington. In his own way. I mean, more well-known for his so-called 'miracle cures' for kids who wouldn't speak, that he could do wonderful things with them, you know. His school was not endowed at all; Dartington was, of course, endowed by my parents."

Did you know Lewis W. Douglas from Tucson, Arizona?

"I knew who he was, I don't know whether I ever met him."

He met your father at Plattsburg, NY.[64]

"That may be. Teddy Roosevelt was a good friend of my father's. He went up there to Plattsburg where my father was in training. It was a place you went if you believed in 'Preparedness.'"

It was kind of like a militia...

"Yeah. Well, no! They were volunteers training for service in the US Army."

It was the same situation as today because the US was not at war. Also, they weren't in the army.

"*Militia* carries this connotation of nuts..."

I wanted to ask you about Charles Rycroft, who you went with to the Soviet Union.

"Yes. He was a contemporary of mine, I think at Trinity College, Cambridge. He was a member of that cell. Became a psychiatrist. Went to Russia in that student group with me, organized by John Madge in August of 1935. They talked about expelling him from the cell, because he was showing signs of independence. They felt a little discipline would be in order after John Cornford died and James Klugman had gone to Paris. I had nothing to do with all that. He became a well-known psychiatrist, that's the last I heard of him."

Charles Rycroft wrote a book on Reich.[65]

"Did he really?"

Yeah.

"Well, he would have left the movement when he left Cambridge. It was not the Party, there were no dues paid. In order to belong to the thing you would have had to join the Young Communist League, which most people laughed at, you know. But he belonged to that student movement which had college cells. He was a perfectly decent guy, with no great fervor."

He worked for the Tavistock Institute...

"Did he really?"

Yeah.

Wilhelm Reich and the Cambridge Five

"That's very possible. Tavistock of course was down in Devon, and established ties much later on with my family and Dartington, because Irene Champernowne, a psychoanalyst at Tavistock, became a very close friend of my mother's. She had this vision at the moment my mother died, of my mother standing before the fireplace in front of her, which was very strange.

"The Champernowne family had owned Dartington. They couldn't keep it up, and my step-father and my mother bought it. David Champernowne was a friend of mine and he got me into the Apostles and all of that. Irene married a local Champernowne and was the head of Tavistock."

Moving up to 1940, when the American Youth Congress occurred in Washington, you were involved...

"I was only involved to the extent that I was helping Mrs. Roosevelt."

Was there a relationship between what was going on in that situation and the Stalin-Hitler Pact, and what was going on with the "People's Convention" for a "People's Peace" demonstrations in the U.K.?

"Probably. I don't know."

There was a People's Convention in 1940 and Claud Cockburn organized it.

"I may have forgotten about that. Cockburn? He was a terrible organizer, It's hard to believe him organizing anything like that."

You knew him?

"No, I didn't, I read *The Week,* as did Senator Borah. Brilliant guy, obviously. Wrote a very funny book about Spain. His son, as you know, is on *The Nation* now. No, I never met Claud but John Cornford knew him well, so I knew about him."

Was the American Youth Congress a part of that Comintern policy to support the Stalin-Hitler Pact?

"No. There was no effort, within the American Youth Congress, to support the Pact. They were promoting isolationism, saying America had to keep out of the war. They don't want to defend the British Empire. They brought in a lot of Indians like Rajni Patel, who came in from Cambridge. It was an in-gathering, without any question at all, of some very shady characters who appeared in behalf of the Comintern one way or another. Behind the scenes, along with the American Communist Party, pulling the strings, the guise being 'Christian Youth' and 'Black Friends' – you know, all the front groups of the period. I never knew who they all were, but I knew instinctively what the hell was going on. My self-assigned job was to try to protect Mrs. Roosevelt from their influence and control."

Was she naive that way?

"Yes. She knew nothing about the inner workings. She was being advised by Joe Lash,[66] who apparently, according to my friend Gus Tyler, was an undercover member of the Party itself. He was in the course of slowly breaking with it, but he was still closely tied at that time with the CP. He came into see me, and he and I discussed, after the war began, what to do about it. He thought that I was a member of the Comintern, because I asked so many probing questions about the Youth Congress, he told me that shortly before he died. I said, 'Are you crazy? Do you think I was being that dishonest? Of course not.' He still believed I was involved with the CP, some thirty years later."

"He was there [at the A.Y.C] but he couldn't really help her much because he was too tied into it to tell her the real truth about it all. But I kept on spotting very odd people around that meeting, such as Betty Shields-Collins, she was some kind of underground agent from England who slept with everyone in sight including Philip Noel-Baker, who was the shadow foreign minister. I said to myself, 'what the hell is going on here.' This was obviously an in-gathering of the clan.

"Their line was very confused, because the smarter among them knew that the Nazi-Soviet Pact wasn't going to last very long, and therefore they couldn't support the Pact. They had a very mixed line, saying, 'Keep America Out of War.'"

Around that same time, Guy Burgess visited you?

"He arrived in Washington with Isaiah Berlin. He was going to Russia to work for the British ambassador, Sir Stafford Cripps. That had nothing to do with the AYC. He just called me up and said he wanted to go to supper. So I took him out to supper, and he asked, 'Could you put me back in touch with my friends?' I said, 'I can't, and I wouldn't if I could.' I don't know what Berlin did, but they both had to go back to London, since Cripps was smart enough to say no [to Burgess and Berlin going to Moscow with the British Embassy]. On both counts, too, because I think Isaiah Berlin was an asshole, too."

At that point, you could have turned Burgess in, but you didn't. Was it loyalty to old friends, or what?

"That's a much tougher question. Um. [pause] I don't know if that was the period of the Nazi-Soviet Pact... yeah, it was.

"It was as if you'd been raped. That's the only thing I could think of. As to whether you'd tell anybody, I'd been raped by Burgess and Blunt. This is the closest emotional parallel I can give to you. You're so appalled and horrified by what has happened, that you just can't communicate it even to your closest friends. As to the question, why didn't I tell Gustavo Duran about it, why didn't I tell Roger Baldwin about it? It's as if I'd been raped.

"It wasn't loyalty, no. It was a horror of being an informer.

"I said to Guy, later on, 'if you are going to go on with this, I'm going to turn you in.' That wasn't at that time, that wasn't until just after the war. At that point, the war was on, I was about to go and so forth and so on."

Did you know George Soule?

"Yes."

He worked for The New Republic?

"Yes. He was an editor."

He worked with the Bradys at Consumers Union.[67]

"I didn't know that. He was a totally boring academic. A total washout, just wanted to sit back and opine after having read the *Wall Street Journal* or something. Lazy, superficial academic, with no conviction."

Do you think he was a CPer?

"No way, in my opinion."

Were there any other contacts with the FDA? Was there any crossover between Consumers' Research, Inc., Consumers Union and the FDA?

"No, I wasn't aware of that. I was persuaded by Helen to make an attack on a government official who had made a proposal on low-cost housing, at the instigation of the Bradys. And I did it, feeling very uneasy about it. A housing conference was held in Washington D.C. and I identified the guy who made the proposal, and said this guy is selling everybody out. I did not want to make the speech and I made it under protest, but Helen really wanted me to do that. I think that was the Bradys pressing her, that's all I knew about it."

What was the point of that? Why would they be against low-cost housing?

"This guy was not doing it the way they wanted him to do in that area. But I don't think that had anything to do with Russian foreign policy. They just didn't think he was being aggressive enough."

Wilhelm Reich and the Cold War

It may not have had anything to do with "Russian foreign policy," I thought to myself, but it certainly was a clear indication of the way the Bradys operated in a manipulative, back-channel political mode.

I questioned Straight about the raid on the office of *Amerasia* magazine. Straight had given press credentials, for *Amerasia*'s Mark Gayn, to his Soviet contact, "Michael Green."

"To Green? Given him press credentials? If so, I've forgotten it."

I told Straight about Richard Case Nagell, an associate of Oswald's, went into a federal bank and shot a couple of bullets into to ceiling to get himself arrested in advance of the Kennedy Assassination. He was trying to get himself arrested. Nagell had sent a letter to J. Edgar Hoover prior to all of this, informing Hoover of a conspiracy to kill JFK, well before November 22nd, 1963. The letter had a return address with Mark Gayn's pseudonym, "Joseph Kramer."

"That's weird. Doesn't have anything to do with me," Straight said.

But, I thought to myself, how did he know?

I was once again impressed with the perils of the covert life; Straight gave little thought as to how his assistance might have served "Green's" interests.

I said, *in studying these intelligence-counterintelligence operations, it's as if it's a world unto itself.*

"Yeah, you get the files and they don't tell you much, they black out the interesting things, particularly the names of people."

Right, but also, to be a counterintelligence agent, you must pretend to help the other side. So you have this whole other world of people who you don't know where they stand, and sometimes they don't even know where they stand, or who they're working for. It seems as though this may have happened to you, when you were recruited, you weren't sure if it was Blunt who was pulling the strings, or whether it was Stalin? Did you really think Stalin had given direct orders for you to go to America to find a Wall St. banking job?

"No, I think that was Burgess' clever idea. Their organization was called 'The International' and in those days, as a kid of 20 years old, who was working ten hours a day on his own stuff, I didn't think through the relationship between the so-called International and Stalin, who had obviously taken over and was using this framework they inherited during the international phase for purely nationalist ends. The reason why Stalin put down roots in Russia was precisely because he was a nationalist responding against the Jewish-dominated International of the Comintern. None of this occurred to a 20 year-old living in Cambridge, England."

When Reich was brought up on charges by the FDA, he felt that there was some type of conspiracy against him. He went on the basis of the general drift of The New Republic *and also some personal information about where the Bradys sympathies lay. This was back in 1947, when this article appeared. I wonder what your reaction would have been had he brought libel charges on the basis of this article? Later in his own trial he tried to address the issue of conspiracy unsuccessfully. What would have happened if, in 1947, you had been put on the stand and asked about your connections to Soviet intelligence?*

[pause]

"Well, that wouldn't have been relevant. I didn't belong to either the British Communist Party, strictly speaking, or the American Communist Party. I never advocated revolution or anything else. My FBI files of the period consisted largely of Hoover's extreme annoyance with me, based on one reference in *The New Republic* about the FBI. Hoover didn't like anybody criticizing the FBI so he marked me down as a typical intellectual-romantic liberal."

Reich had a strong suspicion that the Soviet Union was involved...

"In *The New Republic*?"

No, in the general persecution of his ideas.

"I see. No... heaven knows how I would have reacted to that. A little later on, if anybody had asked me, I would have been glad to volunteer the truth. At that point, since we were in the war, I didn't know what I would have done."

No, it wasn't during the war, it was 1947. What would you have done, had Wilhelm Reich sued you, Mildred Edie Brady, and The New Republic *for libel in 1948, right in the middle of Henry Wallace's Presidential campaign? He had asked his lawyer to do just that, and he most certainly would have put you on the stand and questioned you about your relationship to the Communist conspiracy.*

"Well, Jim, I can't answer that. Everything you told the FBI went straight to McCarthy.[68] McCarthy would have identified it not only with my family, with five children, but with my sister-in-law's family. I would have had to think about a lot of things besides myself."

Well, it wouldn't be through the FBI, it would have been through the context of a libel trial.

" I don't know the answer to that. You're just probing the impossible."

Right. After that point, the exposure of the atomic spies occurred, and Maclean—who you didn't know, right?

"No."

And Philby, you didn't know him?

"No."

So you didn't know these guys. But what I'm thinking, maybe if your relationship to Blunt had come out then [in 1948], what would have changed, would they have been able to track down these guys?

"No, I don't think so. Everybody would have denied it, and my stuff was only second-hand. That was one of the reasons they gave Blunt immunity, they knew it wouldn't stand up in court. I said I would go to court."

But you wrote in your book that this was the first hard evidence they had on Blunt?

"It was, but it wouldn't stand up in court, which was one of the reasons they decided not to prosecute him legally."

At the same time, Maclean was essential in passing atomic secrets to the Soviets. [At that time, in 1947, Maclean was not only the supervisor to the code-room in the British Embassy, he was the official liaison between the British and the Americans for everything related to the atomic bomb program. He even had a no-escort pass to the A.E.C. facility in Oak Ridge. J. Edgar Hoover himself was required to have an escort at Oak Ridge.] *Mightn't that have changed things, had Maclean been exposed in 1948?*

"It's possible. I just don't know."

As I sit here, writing up this interchange, a year later, I know. Of course it would have changed things, had Maclean been exposed. If Straight had told the true story of his attempted recruitment, and named those who had recruited him, the entire social network of the Cambridge Five would have been under investigation in a heartbeat. But then, Straight would have perjured himself, rather than expose his relationship with Soviet intelligence, not because he remained a committed Communist, not because he feared repercussions against his family from "McCarthyism-hysteria," but rather because, as Yuri Modin recently said, he was terrified of Soviet assassins. He was in fear for his own life. Some might call that "paranoia," but Straight knew

Wilhelm Reich and the Cambridge Five

better. It was an entirely rational fear.

At this point, I felt a tinge of pity for Michael Straight, but not much. It took me another month to let it all sink in; it's not often one converses with one of FDR.'s speech-writers about Cold War espionage.

About a month later, I interviewed Straight once more, and again, I found him seemingly open to my questions, while tight-lipped with details. I wanted to ask him more specifically concerning the questions raised by Professor Carroll Quigley, about Quigley's assessment that the movement toward "world federalism" was founded in the English ruling classes and their wealthy, Anglophile American counterparts, and that Straight himself offered a singularly instructive example.

I asked Straight about the chapter of his own book that dealt with Quigley, but he answered with a digression about his former brother-in-law, Louis Dolivet.

"When I put it in, I gave absolute assurance to the British publisher, Collins, that they would be protected against libel. So then they decided that if they got sued, my corporation could go bankrupt. So then I said, fine, you can hold me personally responsible. This related to the coincidence that my sister married a man named Louis Dolivet, who was at one time during his weird life, an assistant to Willi Münzenberg. And that my wife's sister married a fine man, veteran of the Spanish Civil War, who was named Gustavo Duran. I thought it was obvious at some point that somebody was going to try to put this all together. And indeed, my mad half-brother did. He said that this was all a giant conspiracy and that this was all linked together. It was nonsense."[69]

Your brother wrote this? What was his name?

"My little half-brother, William Knight Elmhirst... it doesn't matter, he was a total inconsequence to me."

"Now, as to Quigley, I used him as an example of Evelyn Waugh's point about somebody keeping little notes, you know, putting things together in a kind of counter-intelligence book, which form no patterns at all. Mostly peoples lives are made up of patterns. So I quoted Quigley and put these things together and said somebody like Quigley would make a pattern out of it, when there's absolutely no pattern here at all."

Were you aware that Carroll Quigley was President Clinton's professor at Georgetown?

"No, I didn't know that. I guess if you had an hour to spend listening to somebody, it might as well be [Quigley].

Instead of continuing on the topic, Straight diverted again into a discussion of why his British publisher pulled all the copies of *After Long Silence* off the shelves in England, because of Dolivet's lawsuit, and the curious fact that the publisher's lawyer, who agreed to the recall, was Hilary Rubinstein—Blunt's lawyer, as well.

I began to ask again: *the Louis Dolivet part of Quigley's conspiracy theory was only one part of a larger theme...* but I was interrupted.

"He talks about this marvelous conspiracy in international aviation involving my brother and myself and 'Sonny' Whitney," Straight interjected, remarking on a very tangential conjecture of Quigley's, regarding a "possible" explanation for what appeared to be a conscious manipulation of the election of 1948 on Straight's part: Quigley had noted that, subsequent to Straight's reversal of support from Wallace to Truman, President Truman appointed 'Sonny' Whitney to the most important post in civilian aeronautics and that the Straight-Whitney family was heavily invested in airlines.[70]

Once again, I attempted to point out that *Quigley went back even further to Straight's father's generation and argued that Michael Straight was following in the tradition of Wall Street...* but I was interrupted again.

"This is so much more interesting than the truth. I only wish it were true."

I found myself unable to complete a sentence. I persevered.

So you weren't playing "left" off "right," in the interests of Wall Street...

"Too bad it wasn't true. Quigley shows a very fertile imagination. It's a wonderful concept, you know. You can play around with these things that are coincidences and have a lot of fun, but they just doesn't happen to relate to reality."

Quigley also said that if you intended to do what you did in the election of 1948, it would have been "the biggest coup in 20th century American politics."

"Right," Straight said dryly, "I appreciated that."

I had to laugh, as much admiration for his diligence in dodging as his pat answer. I wasn't going to get anywhere.

I tried another tack, moving on to another famous, wealthy supporter of world federalism, and a prominent figure at the formation of the United Nations: *Nelson Rockefeller. Was he an investor in* The New Republic?

"No. He began as an investor, then pulled out of the *UN World*, set

Wilhelm Reich and the Cambridge Five

up by my brother-in-law [Louis Dolivet], along with Jock Whitney and Max Ascoli. They pulled out when Henry Wallace became editor of *The New Republic*. They were ready to pull out anyway. I think that was their excuse.

Did you know Nelson pretty well?

"Not very well, no. We went to school together at Lincoln School, but he was four years ahead of me."

While I was name-dropping, I had another nagging question:

Did you know Timothy Leary?

"No."

Really? Well. Listen to this, and maybe you can tell me if this is a bunch of garbage. This is from Timothy Leary's book, Flashbacks:

> Mary [Pinchot Meyer][71] sipped her glass and shook her head.
>
> "I'd hate to be the one to break the news to you. Do you remember the American Veterans Committee [AVC], that liberal GI group you belonged to after the war? The CIA started that. Just like Teddy Roosevelt started the American Legion after the first World War. Remember your liberal friend Gilbert Harrison? He ran the radicals out of AVC. And later he bought *The New Republic*— the so-called progressive magazine—from Michael Straight, your hero. Do you know why Michael Straight backed Henry Wallace for president in 1948? To siphon liberal votes away from Truman."
>
> "How do you know all this? " [Leary asked.] "How did you know I knew Michael Straight?"
>
> "I knocked you with those facts to get your attention. It's a standard intelligence trick. I could tell you hundreds of little stories like that."[72]

"I don't know how I became Leary's hero..." Straight reacted with a touch of amusement.

Now, Mary Pinchot Meyer was in Washington and it sounds like she picked this up from...

"Cord Meyer," Straight interjected.[73]

Cord? Or Quigley?

"Him? Oh, I don't think so."

That was strange. Cord Meyer and Carroll Quigley were of the same opinion, that Straight had manipulated the national election of 1948.

Did you know Cord Meyer?
"Very well, yes."
But you didn't know Timothy Leary?
"No."
Huh. Why would Leary write this?
"I have no idea."
And Cord Meyer?
"Cord was a member of the American Planning board, or whatever it was called, of the American Veterans Committee. And it's a lot of nonsense to say the CIA started it, I mean, Jesus, that's totally misinformed."
So Cord Meyer wasn't CIA back then?
"No, much later."
This was his wife talking here.
"Ex-wife. Cord was a right-wing member of the AVC board, as I was. He then left, and under the tutelage of Tom Finletter,[74] he went into the CIA. He spent three or four years in the interim as head of the World Federalists."[75]
Who supported that group, the World Federalists? Was that Nelson's[76] group?
"No, no, no. That was just a way-out thing, saying we had to have world government and so forth. Very ill-advised and abstract. A lot of people were taken in by it."
So you knew Tom Finletter also?
"Yes."
After Mildred Brady left the Office of Price Administration, she worked for D. S. Gillmor, the publisher of Friday *magazine. Do you remember* Friday *or Gillmor?*
"I knew Gillmor from cadet training school in the Air Force, we met briefly somewhere down South. Very eccentric. I didn't know much about *Friday* other than he did a long review of my life in the magazine at one point."
Who funded Wallace's salary at The New Republic*?*
"The parent company trust funded it."

Summary

Mildred Edie Brady's article, "The Strange Case of Wilhelm Reich," was published by Michael Straight in 1947. Straight lived at the

Wilhelm Reich and the Cambridge Five

epicenter of an apparent attempt to manipulate the results of the US Presidential election of 1948; he was surrounded by covert agents of the Comintern at the same time he lived in mortal fear of the KGB. His cooperation, his silence, in the post-war years, allowed the Cambridge Five spy ring to work uninterrupted until 1951. If it is true that Straight himself played no active role in the theft of atomic bombs, it remains apparent that his inaction assisted the Russians. According to his FBI files, he passed press credentials for journalist-spy Mark Gayn to a man he knew to be a Soviet intelligence official, and Gayn subsequently was arrested in the Amerasia affair. Later on, Gayn was implicated as to having advance knowledge of several plots to assassinate President John F. Kennedy. I believe that Michael Straight, to this day, has not revealed what services he performed for the Soviet Union.

Kim Philby, " the greatest spy of the century," was recruited and trained by Arnold Deutsch, a psychologist who had known Wilhelm Reich in Austria for ten years, who had been active (along with Philby's wife, Litzi) in Reich's Sex-Pol organizations, and who incorporated Reich's "character analysis" in his agent-profiles. Philby would come within a heartbeat of assuming command of the entire British counter-intelligence apparatus. Philby knew Burgess and Blunt, who were Apostles secret society members in the mid-30s at Cambridge. Donald Maclean (diplomat from Britain's Foreign office, liaison to Washington) and Guy Burgess (SIS officer, MI6) defected to the Soviet Union in 1951, just as they were about to be exposed as spies. Blunt and Philby remained in the West to be interrogated—eventually, they were cleared. Michael Straight failed to identify Blunt until June 1963, and somehow Philby was also able to defect that month, escaping from the CIA and the SIS. The Philby-Deutsch relationship offers a corroborating instance of a malignant interest in Wilhelm Reich on the part of covert Communists.

The central point here is that Reich earned the animus not only of "left-wingers," "liberals," or "fellow-traveling" Communist sympathizers, but active agents of Soviet espionage. Reich's enraged blast about "red fascist spies" has been dismissed as a symptom of an "hysterical" anti-communism in the 1950s—even by those who claim sympathy for much of his scientific work. We have now established ample evidence for a direct conduit between Reich's successful persecutors and important elements of the Soviet Union's espionage apparatus.

Wilhelm Reich and the Cold War

It's hard to think of two more dissimilar persons than Wilhelm Reich and Michael Straight. Yet they were in complete agreement about one fact: that if they told the truth they would be murdered by Stalin's henchmen. Reich told the truth, Straight did not.

EPILOGUE

Since I wrote this chapter, a great deal of new information has been published about Michael Straight, who apparently has never told the full story of his involvement with the KGB.

Straight supplied the *Daily Worker* with 1,500 pounds sterling as his Party dues annually while at Cambridge, and later while he was working for the US State Department.[77]

In their book, *The Haunted Wood,* Allen Weinstein and Alexander Vassiliev had access to the KGB's file on Straight (#58380) but the documents were released selectively with the approval of the Russian government, so they may have been limited or censored.

These files reveal that when Straight went to the US in 1937, he reported to his Soviet contacts about his "contacts, employment prospects, and operational potential."[78]

We know today that Michael Straight, in 1937, actively sought a position with Henry Morgenthau in the Treasury Department, or the Federal Reserve Board. "In those places," Straight told Deutsch, "possibilities are great because of the influence on Roosevelt[...]" Straight also proposed that he try to get himself appointed as one of FDR's personal secretaries. "My parents know Roosevelt well," Straight wrote to Moscow. He continued:

"All my relatives treat me negatively, since they think I was in Spain. Now I try to dispel this impression by the following means:

a. I use brilliantine and keep my nails clean;

b. by ardent speeches against the Reds in some places, in other places, I present myself as a radical."[79]

Straight met with Arnold Deutsch several times in June and July of 1937.

Here is Deutsch's report on Straight (August, 1937):

"He differs very much from the people we have dealt with

Wilhelm Reich and the Cambridge Five

before. He is a typical American, a man of wide-ranging experience who thinks he can do everything himself....He is full of enthusiasm, well-read, very intelligent, and a perfect student. He wants to do much for us and, of course, has all the possibilities for this... but he also gives the impression of being a dilettante, a young guy who has everything he wants, more money than he can spend, and therefore in part who has a restless conscience.... I think, under experienced guidance, he could achieve a lot. However, he needs to be educated and to have control over his personal life. It is precisely contact with people in his future profession which may turn out dangerously for him. So far, he has been an active member of the Party and constantly surrounded by his friends. ...

Tomorrow he is leaving for America.... He has very little experience and sometimes behaves like a child in his romanticism. He thinks he is working for the Comintern, and he must be left in this delusion for a while."[80]

Straight's income that year was $50,000. He started paying his party dues directly to Deutsch, beginning with 500 pounds for the *Daily Worker*.

In America, "Michael Green" made contact with Straight, but the connection was botched by the failure of the NKVD to provide Green with the proper handshake.

Fall 1937: Straight offered Green ten or twelve thousand dollars, "spare pocket money," in addition to his regular party dues ($8000/year). Green told him to keep it. The revolutionaries at Moscow Center told Green to go back and get that $12,000. "Receive this money and send it to us."

Green convinced Straight to take a job at the State Department, and also that Straight should avoid contact with Duggan, thus Straight knew that Duggan was also a mole in 1937. The CPUSA's leader, Earl Browder, was ordered by Moscow to keep Party members away from Straight to protect his cover. Moscow wanted to use Straight as a talent-spotter of potential Soviet assets in the US government. Recently released KGB files indicate that Straight suggested early on that Alger Hiss was a likely recruit, and this set off considerable unease in Moscow, since Hiss was already working for GRU military-intelligence, and Straight was to be kept in the dark about this.

Wilhelm Reich and the Cold War

Only Straight's lowly position at the State Department kept him from access to highly classified material. Straight temporarily withdrew from Green and his espionage apparatus after August 23, 1939, when the Stalin-Hitler Pact was signed.[81]

From the 1999 book, *VENONA* by John Earl Haynes and Harvey Klehr we now know that even after Straight's "final break" with the Communist underground in 1941, Moscow did not abandon hopes of gaining his assistance. They "dispatched Arnold Deutsch to renew contact with him and several other sources [...]"[82] However, Arnold Deutsch died in 1942 when the Soviet freighter bringing him to the US was torpedoed by a German U-boat in the Atlantic.

Solomon Adler, another mole in the State Department, told Straight to "lay low" after the defection of Whittaker Chambers in 1938. Adler was part of the Silvermaster network in Washington, DC. Straight himself had a connection in New York to the Silvermaster ring, as one of his liaisons to Soviet intelligence was Alexander Koral, who was a contact person for the Silvermasters also. Haynes and Klehr describe Silvermaster as "a man who headed one of the largest and most productive spy rings that Stalin's espionage agencies maintained inside the US government."[83] The next chapter, about Mildred Edie Brady, demonstrates that she was a close associate of the Silvermasters.

Notes

1. Frederic Wertham, M.D., German psychiatrist, was born in 1895 and immigrated to the US in 1921 and would later forge a career in addressing the problems of juvenile delinquency, which he attributed to images of sex and violence in children's comic books. See *Seduction of the Innocent*, Reinhart, NY, 1953.

2. Testimony of James Lilly. Hearing of the commission on the roles and capabilities of the United States intelligence community. Room SD-106, Dirksen Senate Office Building, Washington, D.C. Friday, January 19, 1996. Lilly was Ambassador both to China and to the Republic of Korea, having served in Defense Department positions, including Assistant Secretary of Defense for International Security Affairs. National Intelligence Officer for China, also on the National Security Council staff.

3. Borovik, Genrikh and Knightley, Phillip. *The Philby Files*. New York: Little and Brown, 1994, p. 611.

4. Yergin, Daniel. *The Prize; The Epic Quest For Oil, Money And Power*. Simon & Schuster, NY, 1991, p. 290.

Wilhelm Reich and the Cambridge Five

5. Brown, Anthony Cave. *Treason in the Blood; H. St. John Philby, Kim Philby, and the Spy Case of the Century.* New York: Houghton Mifflin, 1994, p. 142.

6 Rabinbach, Anson Gilbert, 1945-: *Ernst Fischer And The Left Opposition In Austrian Social Democracy; The Crisis In Austrian Socialism 1927-1934.* University of Wisconsin Ph.D.. (Modern History) Thesis 1973, Xerox University Microfilms, Ann Arbor Michigan, p. 101.

7. This group influenced only a small section of the young people in the SPÖ and of Schutzbund. Reich anticipated many points of a book later that later became very important for the left wing of the SPÖ, written by Ernst Fischer: *Die Krise der Jugend* (The Crisis of Youth, Wien 1931.) One of the editors of the official Social Democratic newspaper, *Arbeiter-Zeitung*, Ernst Fischer, organized this "youth opposition." The *Revolutionäre Sozialisten* developed directly out of this leftist opposition within the SPÖ.

8. Teddy Kollek would also become one of the first people to remark on Philby's past, when he informed US counterintelligence official James Jesus Angleton, of the CIA. Angleton, who had been trained in spycraft by none other than Philby himself, did nothing with the news that Kollek had been present at the wedding of Philby and Friedman, an open Communist.

9. West and Tsarev, *The Crown Jewels.* Yale 1999 p. 105. This book contains the most definitive biography of Arnold Deutsch to date.

10. Fritz Keller, *Gegen den Strom. Fraktionskämpfe in the KPÖ - Trotzkisten und andere Gruppen 1919-1945.* Wien, Europaverlag, 1978.

11. West and Tsarev, *The Crown Jewels.* Yale 1999 p. 108.

12. ibid.

13. Andrew, Christopher, and Mitrokhin, Vasili, *The Sword and Shield; the Mitrokhin Archive and the Secret History of the KGB.* Basic Books, New York, 1999, p. 1.

14. See also, *KGB: The Inside Story*, by Christopher Andrew and Oleg Gordievsky, HarperCollins, New York, 1990, p. 202 .

15. Wilhelm Reich: *Sexualerregung und Sexualbefriedigung.* Schriften der Sozialistischen Gesellschaft für Sexualberatung und Sexualforschung in Wien, Nr.1, Wien, [Copyright: Dr. Arnold Deutsch,] Münster-Verlag, 1929.

16. Reviews: *Berlin Am Morgen* 9/8/29; *Kommunistische Arbeiter Zeitung* 11/2/29; *ZSS* XVI, 3/4 1930 (Fenichel); *Naplo,* Hungary Vol. 33, No. 220, p. 2. 1929-1930; *Arbeiterpolitik*, Leipzig 2/27/30; *ZSS* Dec 1931; *Vi Gymnasiaster* (Scandinavia) 12/1932.

17. Wilhelm Reich: *Geschlechtsreife, Enthaltsamkeit, Ehemoral. Eine Kritik der bürgerlichen Sexualreform,* Wien: Münster-Verlag, 1930, 182pp.

18. Marie Frischauf/Annie Reich: *Ist Abtreibung schädlich?* - Schriften der Sozialistischen Gesellschaft für Sexualberatung und Sexualforschung in Wien, Nr. 2, Wien: Münster-Verlag, 1930.

19. Reich, Wilhelm, *Biographical Material; History of the Discovery of Life Energy, Documentary Volume A-IX-B: Bibliography on Orgonomy*, Orgone Institute Press, Rangeley, 1953.

20. *The Sword and Shield; the Mitrokhin Archive and the Secret History of the KGB*. Basic Books, New York, 1999, p. 273.

21. ibid., p. 57.

22. ibid., p. 59.

23. Brown, Anthony Cave. *Treason in the Blood; H. St. John Philby, Kim Philby, and the Spy Case of the Century*. New York: Houghton Mifflin, 1994, p. 169.

24. Borovik, Genrikh and Knightley, Phillip. *The Philby Files*. New York: Little and Brown, 1994. p. 33

25. For a fascinating character analysis of *Philby*, see "Communism and Character: The Case of Kim Philby," by Douglas F. Lewinson, M.D., *Journal of Orgonomy*, Vol. 17, No. 1, May, 1983.

26. "Kim Philby and the Age of Paranoia," by Ron Rosenbaum. *NYT Magazine* 7/10/94.

27. Quigley, Carroll, *Tragedy and Hope*. MacMillan, 1966, p. 938.

28. Wolfe, Theodore P. *Emotional Plague versus Orgone Biophysics*. New York, Orgone Institute Press, 1948.

29. Quigley, Carroll, *Tragedy and Hope*. MacMillan, 1966, page 938.

30. See, *Record of a Friendship; The Correspondence of Wilhelm Reich and A.S. Neill*. New York. Farrar, Straus and Giroux, 1981.

31. Straight, Michael. *After Long Silence*, Thorndike Press, Maine, 1983, p. 227.

32. Quigley, Carroll, *Tragedy and Hope*. MacMillan, 1966, page 939.

33. "The Wallace Plan vs. the Marshall Plan," testimony by Henry A. Wallace before the House Committee on Foreign Affairs, February 24, 1948.

34. Charles Rycroft would later become a psychoanalyst associated with the Tavistock Institute for Human Relations, founded in London, England in 1947, with a grant from the Rockefeller Foundation. Originally set up as a psychological study center, focusing on shell-shock and battle fatigue, Tavistock continues today, providing management techniques based on psychology and sociology. Rycroft would ultimately write a dismissive book about the life and work of Wilhelm Reich. See Roycroft, Charles. *Wilhelm Reich*. Viking Modern Masters Series, New York, 1972.

35. Straight, Michael. *After Long Silence*, Thorndike Press, Maine, 1983, p. 125.

36. Straight, *op cit,* p. 190.

37. Straight, p. 209.

38. Straight, p. 217.

Wilhelm Reich and the Cambridge Five

39. "Covername ALBERT (earlier covername MER), is found in VENONA more than 50 times, sometimes as signatory to messages sent by the New York Residency. ALBERT was Iskhak Abdulovich Akhmerov, a veteran KGB officer who had two tours of duty in the U.S. as an "illegal," that is, an officer using a false identity and background (*legenda* in KGB parlance) and without diplomatic cover or immunity. Aliases (1937-45): William Greinke, Michael Green, Michael Adamic, and more. Street names: Michael, Bill. For example, Elizabeth Bentley knew him only as 'Bill'—no last name, not even the alias, much less his truename. She knew his wife, also an 'illegal,' as 'Catherine.' She actually was VENONA covername EL'ZA, truename Helen Lowry, a niece of KGB agent and Communist Party leader Earl Browder. True name: Iskhak Abdulovich Akhmerov. (In some cases so-called true names of KGB officers, the names that appeared on their passports and the diplomatic lists, were false. For example, Vassili Zubilin, the sometimes KGB Resident in New York and Washington, was actually named Zarubin.) To add to the difficulty in understanding the names, other KGB officers used truenames that were not names at all, Russian words, but not traditional Russian names. Some of these nommes de guerre included senior KGB officers Vladimir Pravdin and Jacob Golos, whose last names mean, respectively, 'truth' and 'voice.' In 1946, ALBERT reportedly moved to Baltimore. Actually, he and EL'ZA secretly left the U.S., probably because the KGB had learned about Elizabeth Bentley's statements to the FBI. KGB veterans consider him to have been one of their service's most successful officers."-from Robert Louis Benson, VENONA Historical Monograph #3: The 1944-1945 New York and Washington-Moscow KGB Messages.

40. Eva Reich recalls that her stepfather, Thomas Rubinstein, lived with a similar fear. As a former colleague of Lenin, Rubinstein had opposed Stalin and was living under an assumed identity in the United States. Eva reports that Rubinstein "always expected 'them' to come for him, literally every time the doorbell rang in our apartment on the sixteenth floor at 27 W 96th St., New York City."

41. Ironically, Eva Reich attended these demonstrations!

42. Straight, in his autobiography, incorrectly identifies his wife's analyst as "Jennie Welderhall." As a child, Eva Reich knew Dr. Jennie Wälder-Hall in Vienna, when she was married to Robert Wälder, a psychoanalyst. The Wälders belonged to the circle of younger Viennese psychoanalysts, along with Eva's mother and father, Annie and Wilhelm Reich. Eva recalls playing with the Wälders' two daughters in Vienna. In 1960, the executrix of Wilhelm Reich's estate, Mary Boyd Higgins, sued Aurora Karrer, Reich's common-law wife at the time of his death, to obtain allegedly missing papers of Reich's. The case was tried in a Bethesda, MD, court, and Eva Reich stayed with Jennie Wälder-Hall while testifying as a witness.

43. Straight's words in 1983.

44. Wallace discussed his theory in an editorial in *The New Republic;* ironically he had stumbled onto the same phenomena Wilhelm Reich identified as the Oranur effect in 1952.

45. Quigley does not mention that the election of 1948 was made even more complex by the fourth-party candidacy of Strom Thurmond and the segregationist Dixiecrat Party. Although Thurmond and Wallace each drew less than three percent of the popular vote, Thurmond's concentrated support in the Deep South allowed him to capture 39 electoral votes, nearly throwing the election into the House of Representatives.

46. See "Wilhelm Reich in Denmark (1933-1939)" by Ellen Siersted, *Pulse of the Planet* #4, Orgone Biophysical Research Laboratory, 1993.

47. A famous document, drawn up by a Democratic Party strategist for the election, advised Truman on "The insulation of Henry Wallace. Wallace should be put under attack whenever the moment is psychologically correct. If it is clear that organizational work is being undertaken by his men in the West either for a third party or for delegates to the Democratic Convention—and that work seems to be taking effect—the Administration must persuade prominent liberals and progressives—and no one else—to move publicly into the fray. They must point out that the core of the Wallace backing is made up of Communists and the fellow-travelers. At the same time some lines should be kept out so that if the unpredictable Henry finally sees the light and can be talked into supporting the Administration, he will have handy rope to climb back on the bandwagon—if he is wanted." - "The Rowe/Webb Memo," September 18, 1947.

48. Straight, p. 649.

49. "The Men Who Killed Kennedy." (1988) Nigel Turner, ITV.

50. Costello, John. *Mask of Treachery.* William Collins, London, 1988, p. 479.

51. Ibid. Costello cites four sources for this fact, but mention of Gayn, if any, was missing from Straight's FBI file, as he received it as a result of his late-seventies FOIA request. Straight's FBI file, which I personally examined, is part of his collection archived at Cornell University.

52. Ginsbourg, Sam. *My First Sixty Years in China.* New World Press, Beijing, 1982.

53. At the time, psychoanalysis had received the patronage of Leon Trotsky in the Soviet Union, where Russian intellectuals blended Pavlovian behaviorism with Freud's new science and the mixture was termed, "pedagogy." See *Eros of the Impossible.*

54. Ginsbourg, p. 110.

55. FBI Bufile 100-267360, sec. 11, serial 10.

56. Klehr, Harvey, and Radosh, Ronald. *The Amerasia Spy Case, Prelude to McCarthyism.* UNC Press, Chapel Hill, 1996.

Wilhelm Reich and the Cambridge Five

57. Russell, Dick. *The Man Who Knew Too Much*. Carroll & Graf, New York, 1992. p. 117.

58. Ibid. p 45.

59. Columbia Oral History Project, interviewer Peter Jessup.

60. She appeared in *King Lear* (1953) and *Poltergeist* (1982).

61. Disgraced former President Richard Nixon was hired by this firm after his resignation.

62. Wallace's pandering to the powerless/impotent "common man" in this speech inspired much of Wilhelm Reich's book, *Listen Little Man,* and certainly Wallace's editorship of Brady's slander in *The New Republic* gave Reich the impulse to publish that book in 1948.

63. Author's interview with Michael Straight, 10/1/96.

64. Bird, Kai. *The Chairman; John J. McCloy and the making of the American establishment.* Simon & Schuster, New York, 1992.

65. Rycroft, Charles. *Wilhelm Reich. Modern Masters Series.* Viking, NY, 1969.

66. Joe Lash was allegedly Mrs. Roosevelt's lover.

67. George Henry Soule, I later found, married Helen Flanders Dunbar on July 13, 1940. Dunbar had been married until December 1939 to Theodore Wolfe, Reich's close associate and translator. Helen Dunbar was known as a pioneer in psychosomatic medicine (*Emotions and Bodily Changes*). Thanks to Bernd Laska for this information from the *Dictionary of American Biography*, Supplement 6 (1956-1960).

68. NB: McCarthy was not elected to the Senate until late 1948; there was no "McCarthyism" in 1948.

69. Author's interview with Michael Straight (11/1/96).

70. A suggested source for the conscious manipulation of the Pan Am/TWA struggle over international flights is *The Chosen Instrument*, by Marylin Bender and Selig Altschul.

71. JFK's paramour, ex-wife of Cord Meyer.

72. Leary, Timothy, *Flashbacks,* Tarcher, 1990, p. 154.

73. Cord Meyer, Mary Pinchot Meyer's ex-husband, was a CIA officer.

74. Thomas K. Finletter was born Nov. 11, 1893, in Philadelphia. Minister in charge of the Economic Cooperation Administration Mission to the United Kingdom, 1948-49; Secretary of the Air Force, 1950-53. Oral history at the Harry Truman Library, 84 pages. [Finletter: Air Force, President's Advisory Committee on Weather Control Committee, documented WR contactee during Desert OROP Ea].

75. Six community-based groups joined their 18,000 active members and 70,000 adherents into one group called the United World Federalists on February 22, 1947. Led by Thomas H. Mahoney and T.K. Finletter, the group's manifesto announced: "We believe that peace is not merely the absence of war but the presence of justice, of law, and order—in short, of

government and the institutions of government; that world peace can be created and maintained only under world law, universal and strong enough to prevent armed conflict between nations...Therefore, while endorsing the efforts of the United Nations to bring about a world community favorable to peace, we will work primarily to strengthen the United Nations into a world government of limited powers adequate to prevent war and having direct jurisdiction over the individual." In April of 1948, Albert Einstein and other famous atomic scientists began lobbying for the establishment of world government.

76. "At the old Inter-American Office in the Commerce Building here in Roosevelt's time, as Assistant Secretary of State for Latin American Affairs under President Truman, as chief whip with Adlai Stevenson and Tom Finletter at the founding of the United Nations in San Francisco, Nelson Rockefeller was in the forefront of the struggle to establish not only an American system of political and economic security but a new world order." -*New York Times* , November, 1975.

77. *The Haunted Wood*, p. 72.
78. ibid, p. 74.
79. ibid, p. 74-5.
80. ibid, p. 75.
81. ibid, p. 79.
82. Haynes, Klehr: *VENONA*. Yale, 1999. p. 156.
83. ibid, p. 132.

Photo Section

Young Wilhelm Reich in the Wienerwald near Vienna, c.1920.

Reich in Sletten, Denmark, 1934

Photo Section

Ruins of the farmhouse where Wilhelm Reich was born,
(at point of black triangle) in Dobrzanica, Bukovina, then
part of the Austro-Hungarian Empire,
now a part of Ukraine.
(1997 Photo by Oksana Melnyk.)

Above: Typical German colonial home,
built 100 years ago, in Dobrzanica.
Below: Parsonage built from the bricks from
Reich's birthplace. (Photos by Oksana Melnik)

Photo Section

Above: Reich with his arm resting on Sergei Feitelberg.
Below: Reich (tossing ball) Relaxes at the beach, his wife
Annie standing next to him, around 1928.

Wilhelm Reich and the Cold War

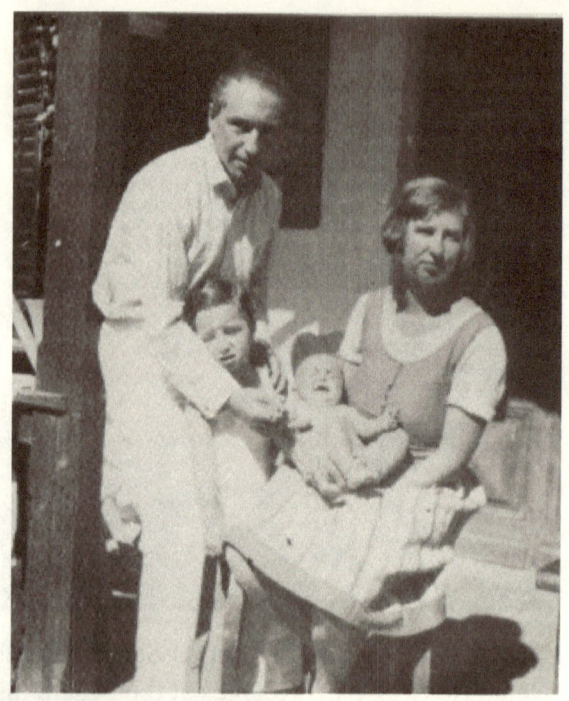

Top: Reich and Annie with daughters Eva and Lore.
Bottom: Reich, after his divorce, enjoys a rare and
cherished visit with his children.
Both photos around 1928.

Photo Section

Reich holding the baby of A.E. Hamilton, 1946

Working in his Laboratory

Top: Reich's Observatory building at Orgonon, Rangeley, Maine. It provided a study, a laboratory and observation platforms. Bottom: The Student's Laboratory at Orgonon. Orgonon is today open to the public as the Wilhelm Reich Museum, with a bookstore and exhibits.
(www.wilhelmreichtrust.org)

(Orgonon photos by James DeMeo)

Photo Section

The Cloudbuster, c.1954.
Photo courtesy of the
Wilhelm Reich Museum & Trust

Under Arrest, 1957.

Photo Section

Karl Frank

Above Left: Kim Philby, photograph taken by Edith Tudor Hart (Above Right) in Vienna, 1934. Right: Alice "Litzi" Friedman, first wife of Kim Philby.

Photo Section

Soviet postage stamp honors Kim Philby.

James Jesus Angleton, in London, during WWII, where he learned spycraft from Kim Philby.

Left, Max Eitingon; Right, Naum "Leonid" Eitingon.

Left: Arnold Deutsch's portrait hangs in the "Hall of the Great Illegals" in Moscow today. Right: Clockwise from Top Left: Arnold Deutsch and the Cambridge Five: Kim Philby, Donald Maclean, Guy Burgess, Anthony Blunt and John Cairncross.

Photo Section

Left: Arnold Deutsch and his daughter in London. he had been recruited by the OGPU while he was Reich's publisher; he used character-analysis to recruit the Cambridge Five.

Below: Cambridge students bound for Leningrad in 1935, (from Left) Michael Straight, Brian Simon, Charles Fletcher-Cooke, and Anthony Blunt.

Michael Whitney Straight at Cambridge. His parents founded the *New Republic* and he assumed control of the magazine in 1937. In 1947 he published Mildred Brady's "The Strange Case of Wilhelm Reich."

Cord Meyer, a moving force for the World Federalists, with his wife, Mary Pinchot Meyer. He and Michael Straight worked together in the American Veterans Committee. Meyer became a CIA officer and wrote of his suspicions about Straight.

Photo Section

Iskak Akhmerov, aka "Michael Green" – Michael Straight's NKVD controller in Washington and New York.

Professor Carroll Quigley, Georgetown University, was an important influence on President Clinton. His magnum opus, *Tragedy and Hope*, deals extensively with Michael Straight's manipulation of the 1948 presidential election.

Wilhelm Reich and the Cold War

Mark Gayn: born in Harbin, China, he was arrested in the Amerasia Spy Case after receiving press credentials from Michael Straight.

Vice-President Henry Wallace

Photo Section

Left: Sandor Rado at the Lucerne Congress of the International Psychoanalytic Association, 1934. Right: Emmy Rado, his wife, from her OSS identity card.

Left: Ruth Oesterreich Jensen, second wife of Arnold Thomas Rubinstein. Right: Allen Dulles, shown here as Director of the CIA, worked closely with Emmy Rado.

1935—*just before CU's founding; in the organization's early years Mrs. Brady ran CU's West Coast division*

Top Left: Mildred Edie in 1935.
Top Right: Mildred Edie Brady c.1965, prior to her death.
Bottom Left: Reich's lawyer, Arthur Garfield Hays (center) in Germany for the trial of Georg Dimitri for the Reichstag Fire. Bottom Right: Robert Brady.

Photo Section

Left: Nathan Gregory Silvermaster denies the charges in Congress; he was never prosecuted.

Right: J. Robert Oppenheimer was socially and politically active with Mildred and Robert Brady at UC Berkeley. He told Eleanor Roosevelt that Reich's Oranur Experiment was a "hoax."

Left: J.B. Matthews, who coined the term "Fellow Traveler," was a former socialist who educated the American public as well as the US Congress about the CP-USA. He exposed Mildred and Robert Brady in 1940 as likely Soviet moles in the Federal Government.

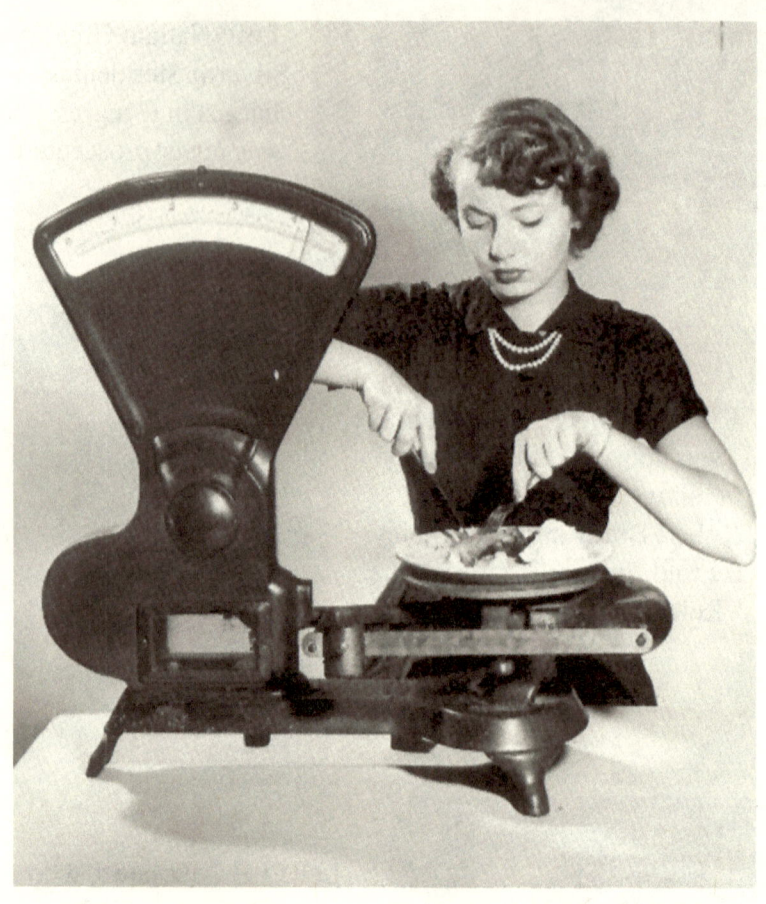

Consumers Union (CU) conducts testing on plastic tableware, 1951. CU was a splinter group, having formed in 1936 after a violent strike in 1935 against its parent group, Consumers' Research. The strike was supported by the CP-USA and many leaders of CU were covert party members.

Photo Section

Nelson Rockefeller (left) and his family appear with Secretary of Health, Education and Welfare (HEW), Oveta Culp Hobby (seated), and President Eisenhower at a White House ceremony. At the time of Wilhelm Reich's Trial, the FDA answered to Rockefeller.

Above: Little Orgonon, Reich's base of operations in Arizona, was an isolated area in the 1950s. Mount Catalina, where Reich observed multiple UFOs and conducted cloudbusting operations, stands in the background. Today, the area has been engulfed by Tucson's sprawl. Below: A typical rural Tucson ranch home of the 1950s, probably similar to what Reich lived in at Little Orgonon.

Photo Section

Eisenhower Makes a campaign stop in Tucson

Lewis W. Douglas greets Eisenhower. Although a registered Democrat, Douglas was the Chairman of the Eisenhower campaign in Arizona.

Photo Section

After his election, Eisenhower supported Douglas' interest in weather modification.

John J. McCloy (center), "The Chairman" of the American establishment, CEO of Chase Manhattan Bank, hits the ground running with his brother-in-law, Lew Douglas.

Wilhelm Reich and the Cold War

James E. Mcdonald (Above) worked closely with Lew Douglas (Left) as Director of Research at the Institute of Atmospheric Physics in Tucson, Arizona after 1954.

Photo Section

Institute of Atmospheric Physics at the
University of Arizona. (1996)

File cabinets at the Institute of Atmospheric
Physics, filled with time lapse films from weather
modification studies. (1954-1974)

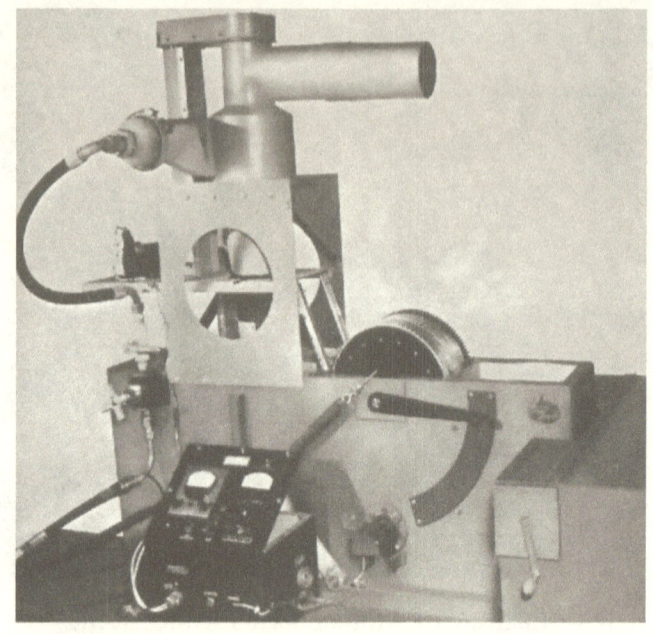

Top: In 1966, William W. Vickers and James F. Church, working for the US Air Force, developed a dry ice seeding machine called the "Cloudbuster."

Left: Roscoe Hillenkoetter, First Director of the CIA.

Photo Section

Vincent Schaefer (left) and Irving Langmuir (right) developed cloudseeding at General Electric. Below, GE's logo was burned into a layer of clouds.

"Exactly! There's A Plot To Make Us Look Foolish"

Harvey Matusow (Left) was in Lewisburg Prison with Reich, and he confirms he did indeed plot to discredit McCarthy.

Photo Section

Trevor James Constable, Left, with Air Force General Curtis LeMay in 1990. LeMay assisted Constable's smog removal experiment and provided his home as a base of operations.

Long-time Tucson resident Joe Blankenship with his old Orgone Accumulator.

Above-ground atomic bomb test, Nevada, 1952.

Photo Section

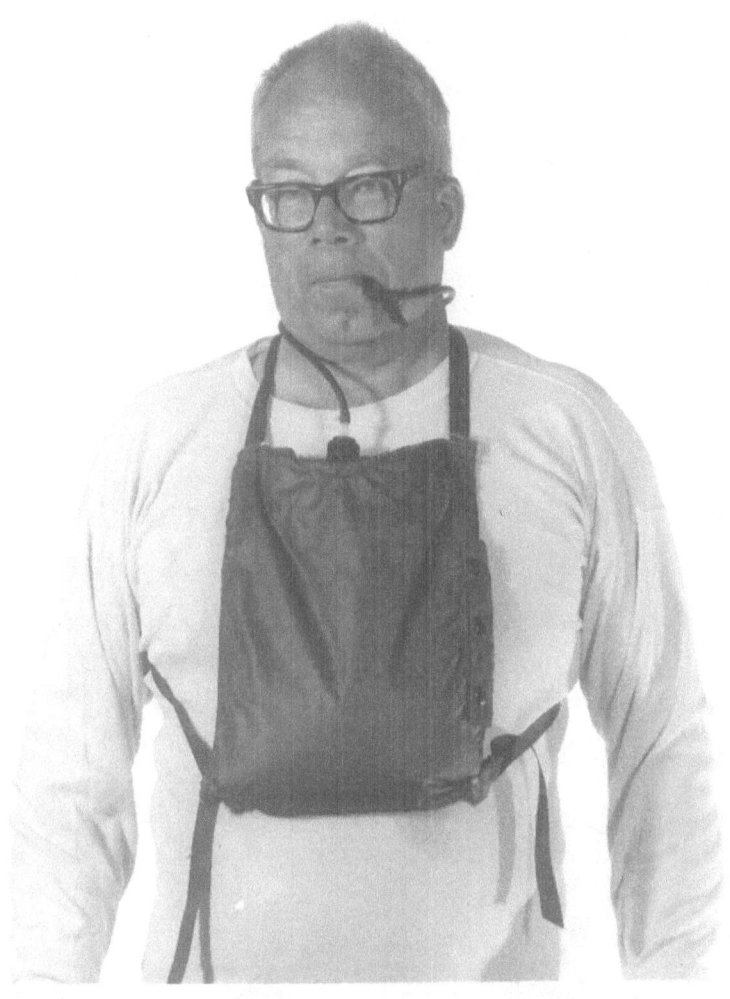

Robert C. McCullough, Reich's research assistant in Tucson, was devastated by Reich's imprisonment and death in 1957. Here, Mccullough demonstrates a drinking device in 1974, while working for the Army at Dugway Proving Grounds, Utah.

Robert C. McCullough cloudbusting in 1978

Reporter Dorothy Kilgallen

Photo Section

Eva Reich and author James
Martin (Photo by Marilyn Bacon)

Wilhelm Reich and the Cold War

Chapter Five: The Secret Life of Mildred Brady

"Political interests of a foreign power in combination with national USA commercial interests entered the scene of orgonomic research when, in 1948, the orgonomic motor function was discovered in my laboratory and documents relating to this discovery were executed in the law office of Mills and Mills, the later prosecutor of the Discovery."
– Wilhelm Reich, from "Outlook at Crossroads Ahead," a statement included in his Brief to the US Court of Appeals for the First Circuit, October, 1956.

"What happened was that, sometime in the autumn of 1946, a woman telephoned Reich at his Long Island house, and told him that she was bringing him greetings from a West Coast friend. Reich invited her out to Long Island. She was a forty-year old writer and journalist named Mildred Edie Brady, the wife of a professor of Economics at the University of California. It was while working as an industrial analyst, studying the beer industry, that Mrs. Brady became interested in the subject of alcoholism, and read various books on psychoanalysis, including some of Reich's. A colleague who was suffering from terminal cancer renewed her interest in Reich's cancer therapy (which she regarded as 'crack-pot nonsense'), and this explains why she decided to get an interview with Reich."
– Colin Wilson, The Quest for Wilhelm Reich.

"Thus, the sole cause for alarm that was ever mentioned by the investigators was their reading of a lying, slanderous article by a Red Fascist, Mildred Brady. It seems incredible that a single

Wilhelm Reich and the Cold War

Red Fascist smear article, without a word of truth, used as bait to lure a government agency of the USA into blindly fighting Red Fascism's opponent, could be so effective, could take the FDA for such a ride, such a long, costly and dishonoring ride.....unless there were those in that agency who had deliberately manipulated the whole thing, baited the hook, perhaps, and confused the other members of the agency into swallowing it."

– from the draft "The Strange Case of the USA Food and Drug Administration" by Wilhelm Reich, September, 1952.[1]

The decade 1947-1957 marked a very dangerous first phase of the Cold War between West and East. In order to end World War II, the United States had secretly spent $20 billion and produced the atomic bomb. The taxpaying public, at first surprised to find out what it had bought, was shocked to learn that this costly secret had been stolen by spies working for the Soviet Union and that some of these spies were American Communists occupying military and government positions. This was the decade of the FDA's investigation of Wilhelm Reich. These were also the ten years that saw the national trajectory of Senator Joe McCarthy. It was a time of fear-based conformity and curtailment of individual liberty. The sprawl of New Deal Federalism hardened into the national security state, and Americans were uncertain just who was dancing behind the seven veils of official secrecy. Along with most Americans, Wilhelm Reich was to become a victim of veiled, underhanded Communist tactics.

Recently, when the Archives of the Comintern[2] were opened to scholars, it was proven once and for all that Soviet intelligence in the United States depended upon the voluntary, ideologically committed participation of American Communists, and that in turn, the Communist Party of the USA (CPUSA) depended upon the influx of cash from the Soviet Union.[3] The spies were rarely open members of the CPUSA. Some spies considered themselves members or agents of the Comintern, rather than the Party, but most understood that the Soviet Union directed the Comintern. The national Communist Parties were merely local chapters of the international Communist Party: the Comintern.

Here I use the word *Communist,* with a capital C, to describe the people of the 30s and 40s who accepted Comintern violence, conspiracy and "party discipline." By Communist I mean both Party members and "fellow travelers." The Bradys were not Party members, but "only" fellow travelers, yet they submitted to Party discipline nonetheless. All Communists were spies. Their diligence in pursuing their convictions under adverse conditions was equaled by their pathology and self-delusion. They were trained in ciphers, infiltration, camera-work and propaganda. All the tools of conspiracy were built into Party work.

The Communists examined here were not "proletarians" or industrial workers, but rather highly educated, well-heeled and guilt-stricken individuals. Rank-and-file, working class members of the Party viewed intellectuals with utter contempt. Among Communists, Reich described a particularly malignant strain, which he called *Red Fascists*, intellectuals and Party functionaries who functioned much like the *Black Fascists*, such as the Nazis.

> "March, March, till impure blood
> Overflows our ditches."
> – "The Marseillaise"

We can now confirm that the Americans who betrayed secrets to Russia were ideological Communists, and we are obliged to understand how they came of age during the Great Depression. As the world Depression grew longer and destroyed even the most basic of social relations and simple personal dignities of people, and one reformist movement after another failed to address their immediate needs, the niceties of ethics and of philosophical detachment lost real-world relevance. There was a sense of immanence, a need for action, *now*, and without much patience for the discussion of what or how. As Reich put it at that time, ethics are for those who have bread in their bellies. Rest assured, however, that the key figures in American treason all had full bellies.

Reich clearly had some interesting things to say about the nature of this conspiracy, including the idea of an "emotional conspiracy" between individuals who were conscious only of a deep animosity to Reich's open discussions of sexuality, his life-affirmation. Yet there is a vague, mushy quality to this idea as a final explanation of the characters and, more importantly, the sequence of events that led up to Reich's death inside Lewisburg Federal Penitentiary. Although he was aware of the deepest motivations of all participating parties – the religious moralists, the professional competitors and so on – who wanted to squelch his efforts, Reich identified a very particular conspiracy involving faithful, hard-line Communists. He left volumes of documentary evidence behind to substantiate this conspiracy.

In the May 26, 1947 issue of *The New Republic,* "The Strange Case of Wilhelm Reich" appeared. The byline of the article described its author, Mildred Edie Brady (June 3, 1906-July 27, 1965) as a "freelancer"—although she would never write more than a half-dozen freelance articles. Reich interpreted Brady's attack in historic terms by comparing it with the deeds of Mocenigo, the tormentor of Giordano Bruno, and Stalin, whose real name was Joseph Djugashvilli. He combined the first parts of their names to coin the term *Modju* to describe the killer of truth: "As indicated in the very title 'The Strange Case of Wilhelm Reich,' Modju is very clever. Modju knew, in throwing the match, that one could arouse the deepest *fears* in *armored* people through the 'strange' ('bizarre,' 'weird') quality of the cosmic energy, helped by the international projection of the 'psychotic' slander (Neuropsychiatrists, Psychoanalysts), implying WR is a 'case' of paranoia with delusions, as well as creating a *legal* case by claiming jurisdiction in the cosmic energy realm (FDA in conjunction with Chemists, AMA, APA, Classical Cancer-Research Societies, Geneticists, the law system, etc.) in which case WR is a 'charlatan, quack, fraud, promoter,' in other words, a *criminal.*"[4]

As Reich's biographer, Myron Sharaf, put it, "Brady correctly plays Reich's individual notes, but she totally distorts his melody." However, Brady libeled Reich on an important point: she flatly states that Reich offered the orgone accumulator as a cure-all.

Reich was understandably puzzled later as to how this single article became the driving force behind the investigation by the US Food and Drug Administration (FDA). He needn't have been. As we shall see, Mildred Brady was a Godmother of government bureaucracy of drug regulations and best knew how to manipulate its machinery. With Charles Crawford, the FDA's commissioner when Reich was prosecuted, she crafted some of the FDA's governing legislation back in 1938. Mildred Brady was specifically responsible for the legal language regarding "mislabeling" of products, as well as some of the agency's strongest enforcement tools, under which Wilhelm Reich's books would be condemned as false "labeling." As a citizen-lobbyist in the 1930s for Consumers Union, Brady endowed the 1938 FDA with sharp enforcement teeth, and she could make that watchdog bite.

> "Wilhelm Reich, immediately upon reading the articles, said that they came from Communist headquarters, that they showed unmistakably the stamp of Red Fascism. Since the Communists had wanted to kill his work since 1931, Reich had learned the 'smell' of the typical Red Fascist smear, as unmistakable an odor to the experienced sociologist as is the odor of a smear from infected tissue to the nose of an experienced medical doctor."[5]

The first FDA agent who appeared at Orgonon in August 26, 1947 was Charles A. Wood.[6] He openly informed Reich that the Brady article had alerted the FDA to the problem of the orgone accumulators. Only six years earlier, Mildred Edie Brady and her husband had been forced from their jobs in the Federal Government, under public Congressional scrutiny for their adherence to Communist doctrines. To date not one of Reich's biographers has pointed this out, although it has always been a matter of public record. In fact, since his death, few writers on Reich's life have seriously given consideration to a factual basis for Reich's claims of a conspiracy against his work.[7] It's significant that, until now, no one has ever requested Mildred Edie Brady's FBI files under the Freedom of Information Act. These files, newly released to the author in 1998, confirm Reich's suspicions. They show a long-term working relationship between Mildred Brady and a little-known, unindicted mole for the Soviets: Nathan Gregory Silvermaster. The evidence strongly suggests that Mildred Edie Brady herself was an active agent of the Comintern throughout the thirties and forties.

Mildred Edie Brady left marks on this world. Pick up many packaged foods in the USA and you'll find the wrapper labeled "packed by weight, not by volume." *Contents may settle during shipping.* "This ham contains less than 10% water." *Truth in Lending.* *"Orgone Energy does not exist,"* said the Court. These are Mildred Brady's empty leavings.

Born in 1906, a native of Little Rock, Arkansas, Mildred Edie grew up in Kansas City, Missouri. Her father, Stewart Carson Edie, was a pharmacist for the Katz drug chain. Her mother, Maude (*nee* White) was the thirteenth child in a joyless family. Mildred attended the University of Missouri, but never completed her degree. She was expelled from another college as a result of her early journalistic endeavor with a campus periodical called *The Dove*. She later claimed she was expelled for being too radical. She left home in 1929, seeking distance from her family, and, being a strikingly attractive young woman, worked as a model in Chicago. She had a lean, edgy look, perfect for the times. Moving on to New York, she married an old boyfriend named Gerald Fling, but it didn't last; they were divorced after a few months. Mildred cashiered at Brentano's Bookstore until she met Dexter Masters, who worked at *Tide*, a marketing trade journal that was an offshoot of Henry Luce's *Time*. Masters became her lover and hired her to report on the American advertising trade, focusing particularly on the impact of New Deal programs on this business.

Masters and Edie had a five-year love affair, but Masters did not wish to have children. Then she met Robert A. Brady, an up-and-coming, Marxist economist at the University of California at Berkeley. Mildred, according to one account, met Robert when she interviewed him for a major article she was writing for *Tide* on the 1935 strike at Consumers' Research, the first national, non-profit consumers organization in the USA. [8]

Although Professor Brady already had a wife and young baby at the time, Mildred turned his head and he ran away with her. Colston Warne, long-time director of the Consumers Union, offered this soft-focus bio of Mildred Edie Brady:

> "In 1936 Arthur Kallet and others split from Consumers' Research to found Consumers Union (CU), a competing organization. Dexter Masters became editor of CU publications and Mildred Edie and Robert Brady joined the new group... Robert Brady... became vice president of the organization; Mildred Edie became manager of the short-lived western branch of the organization, which was established at her suggestion...In 1940 she went to New York as managing editor of the reform publication Friday and then helped launch a new weekly publication for CU. Called Bread & Butter, it dealt with inflationary tendencies of a defense economy. She worked briefly during World War II in Washington, DC, for the Consumer Division of the Office of Price Administration, and she edited a Washington consumer newsletter for *McCall's*. In April 1947 she wrote a famous article for Harper's Magazine based on an investigation of therapist

Wilhelm Reich and some of his associates. Her work also appeared in *Collier's* and *The New Republic*.

"In 1950 Mildred Brady rejoined CU as a feature writer for *Consumers Reports*...In 1958 Dexter Masters, who had become executive director of CU four years earlier, brought Brady to Mount Vernon, N.Y., as editorial director and senior editor of Consumers Reports...In 1961 Brady's story 'The Great Ham Robbery' focused public concern on the practice of injecting hams with water to increase their weight...Though she was burdened by Robert Brady's illness (he suffered a stroke in 1952 and was an invalid until his death in 1963), Mildred Brady was through the 1950s and early 1960s an influential and creative force in the consumer movement...She participated in the International Organization of Consumers Unions, founded in 1960, and lectured on the role of consumer testing before technical groups in the United States, Europe, and Asia...her investigations laid the groundwork for the rise of world consumerism. When she died of heart failure in 1965, while in the midst of a conference with CU leaders about product standardization, *Consumers Reports* paid tribute to her 'intelligence and drive' in guiding the publication and noted the widespread impact of her work."[9]

Mildred Brady's public life began in the winter of 1935 on the picket line in support of striking office workers at a non-profit organization called Consumers' Research, Inc., (CR).

Here, she found her voice. CR had been founded by Frederick J. Schlink, an engineer who had been employed by the National Bureau of Standards, and by Stuart Chase, Schlink's coauthor of a two-part article published in 1925-1926 by *The New Republic* called "A Few Billion Consumers."

The New Republic and its founder, Mrs. Dorothy Payne Whitney Straight Elmhirst, were central to the formation of the CR, with a generous promotion in the pages of the magazine as well as Mrs. Elmhirst's donation of $10,000 in seed money for clerical and publishing expenses for 1927-8, and a mass-mailing to solicit members for Schlink's "Consumers' Club." The effort was sufficient to found Consumers' Research, Inc. on December 9th, 1929. There was an element of secrecy built into the early CR; subscribers to its "Confidential Bulletin Service" were required to sign a confidentiality clause to preclude libel suits from maligned businesses.

By 1935, the year of the strike, CR had over fifty thousand members. Exposing deception in advertising, shoddy goods, and the generally raw deal buyers get in an unregulated market economy, the group attracted housewives and engineers, leftists, idealists and tightwads.

J.B. Matthews and Consumers' Research

J.B. Matthews was on the board of Consumers' Research at the time of the 1935 strike by the office and editorial workers. During this strike, Matthews' son was pulled off a bus and severely beaten by the strikers, while Mildred stood and watched along the picket line.

Matthews, a bespectacled middle-aged former preacher who had a long history of supporting a variety of socialist, pacifist, and communist organizations, firmly believed that the strike was an effort by CPUSA to take over Consumers' Research, and the historical record bears this out. Mildred Brady, along with the strike's CP supporters and the strike's leaders, saw the consumers' movement as a revolutionary weapon as long as it could be "unionized" under CP control. The strike occurred at the beginning of the Popular Front period (1935)—when the Comintern adopted a policy of infiltration and conversion of a wide spectrum of political groups, non-profits, and governmental agencies. It was at this time Moscow ordered its more reliable supporters to stage a "break" by severing overt and personal ties to the Party; this was a direct order obeyed by Michael Straight, Kim Philby, and countless other recruits.

1935 was a crucible year; Mussolini's fascist centurions sought to begin a new Roman empire in Ethiopia; and in the Soviet empire, show trials and purges were just beginning.

The Consumers' Research strike of 1935 marked a turning point in the life of board member J.B. Matthews. Formerly, Matthews had been a radical himself. "I moved from religious fundamentalism to the social gospel... I went, finally, from socialism to communism." Scarred, embittered and obsessed, Matthews emerged from the CR strike of 1935 as a canny anti-communist. He regularly fed data to the House Committee on Un-American Activities and he testified before the Dies Committee, as well as at the McCarthy hearings later on. Because Matthews had life experience with American Communists, he was an effective witness and investigator. While the conservative southern

Secret Life of Mildred Brady

congressmen on the Dies Committee were prone to laughable gaffs, which served to highlight their ignorance of left-wing politics, Matthews was able to guide them toward the factual data. He coined the term "fellow traveler." As Mildred Brady's smear attacks in *The New Republic,* ten years later, caused Reich's anti-communism to ratchet up a few notches, so J.B. Matthews had a similar response to the Consumers' Research strike of 1935, when the CP beat up his son to within an inch of his life.

During the strike, and long afterwards, Consumers' Research compiled an enormous archive on the communist support and organization of the strike, as well as its supporters—including Mildred Edie Brady. Today, this collection is archived at Rutgers State University in New Jersey. Even Brady's *New Republic* article on Wilhelm Reich was duly filed here. Most of this documentation is comprised of the public statements and articles of the early leaders of the Consumers Union, which was subsequently founded by Brady and the other losers in the power struggle for control of Consumers' Research.

Using their own words and statements, rather than the seedy, "informer-rat" hearsay often associated with the McCarthy era, CR archivists gathered powerful evidence of the Stalinist sympathies of the Bradys. One of these incriminating documents in Mildred Brady's file leaps out: in 1940, she was managing editor of a newsweekly called *Friday*. Included in the CR archives is a copy of the masthead page with an editorial by *Friday*'s president, D.S. Gillmor. Addressing the issue of intervention in 1940, Gillmor, an American Jew, wrote a strangely detached argument *against* US intervention against the Nazis in Europe in 1940. Gillmor concluded:

"Many readers will interpret this to mean that we [*Friday*'s editors, Gillmor and Brady] favor neither side in the war. This is true. I cannot see how a victory of either side will make the world any better a place to live in. A Nazi victory would be worse than an Allied victory to be sure. Would it be so much worse as to make it worth the spilling of so much as one drop of American blood on European soil? I do not think so. I do not think the possible difference is worth so much as taking even the slightest risk of our becoming involved."

Gillmor goes on to say the real "Fifth Columnists[10] are Hunger, Unemployment, and Insecurity." How could Gillmor oppose American intervention in 1940, and then enlist in the Army Air Corps in 1942?[11] His statements against intervention in 1940 make more sense in light of the Non-Aggression Pact of 1939 between Germany and the Soviet Union. Influenced by Moscow's propaganda in favor of the Moscow-Berlin alliance, many leftists suddenly joined the "peace movement" after many years of raising money for arms to fight the rising tide of fascism in Europe. Some of the more ethical, soul-searching Communists chose this historical moment, with the "Non-Aggression Pact" in effect, to fully break with the Party. Only by the most contorted mental gymnastics and emotional self-suppression, were Stalin's faithful allies able to publicly toe this line.[12]

The 1939 Hitler-Stalin pact exposed the hardened party-liners who advocated non-intervention during this critical phase of WWII. Hitler, having already carved up Poland with Stalin, having absorbed Austria and Czechoslovakia, and who was about to defeat France, faced only Britain as the remaining obstacle to fascist imperialism in Europe. Close observation of shifting attitudes before, during and after the Hitler-Stalin Pact serve as a litmus test for true allegiances. In 1940, well-disciplined Communists invariably shifted from an anti-fascist stance to a "keep America out of the war" stand, which was functionally pro-fascist. Clearly, the editorial direction of *Friday* took its cue from Moscow, and Stalin's managing agent on the scene was Mildred Edie Brady.

D.S. Gillmor's wife, Noelle, was a film translator and dubber with significant credits.[13] One of the Gillmors' sons, Steve, remembered "the late Daniel S. Gillmor, amidst many philanthropic efforts, published a magazine (1938-1942) called *Friday*. The magazine, sporting fascinating photos of the WWII era, was an interesting blend of politics and Hollywood." D.S. Gillmor also wrote a book about the McCarthy era, called *Fear the Accuser*.

Mary, D.S. Gillmor's first wife, remembered the days before WWII as "exciting times." She remembered Mildred Edie as "blond, tall," and she said with a meaningful giggle, "outspoken." She described the outlook of *Friday* as "anti-fascist, pro-union, pro-Spanish revolution, and in the format of *Look* magazine."

Dan Gillmor, another son, who currently works as science editor for the *San Jose Mercury News*, said his father's magazine was a "basically left-wing, muck-raking, investigative, entertainment, you-name-it journal." He also described it as "populist" and as a "friend of every political persuasion." Significantly, he confirmed that his father wrote the official government debunking of Unidentified Flying Objects, popularly known as the "Condon Report." Dan said his father "did a lot of science writing."

"He was not isolationist," Dan insisted without my having asked, "he served in the Air Force during WWII, and was very anti-fascist." When I read him the passage in his father's editorial in 1940 where he questioned whether the cause was worth "one drop of American blood," and asked whether this might not be an indication of close sympathy with the Party Line, he replied, "I can't imagine him taking orders from anyone." Gillmor's son understood that it was not so much a Communist Party, as a Communist Army, and in an army, one takes orders, period.

Son of a Slave

"My father... said sex had the power to blot out everything else." – Joan Brady, daughter of Robert and Mildred Edie Brady.[14]

Mildred's second husband, Robert A. Brady, was born in Marysville, Washington, in 1901. His father, a backwoods, "healing" minister, was an escaped slave, a white boy sold into servitude after the Civil War when he was just four years old.

Working his way through school, Brady received a B.A. from Reed College in 1923. He worked with Research Staff of the National Industrial Council Board (1926-7) before receiving his Ph.D. from Columbia in 1929. He was appointed Assistant Professor of Economics at the University of California at Berkeley (1929) and published his first book *Industrial Standardization* that same year.

Brady had been associated with the Consumers' Research (CR) organization since his days at Columbia University and was a friend of its founder, Fred Schlink. In accepting membership in CR's corporation, Brady wrote, "Be assured of my desire to cooperate. I would go a long distance out of my way to give the Consumers' Club a hand even though I have quakings re your ultimate success in kicking against 'the pricks' – unless, as Montague Norman suggests, the capitalistic system is going to pot before too many moons have rolled around." (December 17, 1931).

During the Depression, Brady, like many young academics, was drawn to government service by the New Deal. As an advocate of consumers' rights, Brady in 1934 became chief of the Standards

Section at Consumers Division, National Recovery Administration (NRA), a central New Deal agency. At that time, Schlink complained to Brady that the bureaucrats on Brady's Consumers Advisory Board (CAB) were using Consumers' Research publications and research without properly crediting the source. "You are not saying the truth," Schlink wrote to Brady when the latter protested his innocence, "...If you mean yourself and your small personal group, I take no objection, but again you are not independent of your small group of conspirators against the public welfare..." This is an interesting statement on Schlink's part, especially as he and Brady were still on relatively good terms and corresponding frequently.

While Brady and Schlink were in accord for the need to develop standards and regulations in industry, an interesting double standard appears in a letter from Brady to Schlink, April 26, 1934:

"I am reliably informed that you have nothing to fear about the Code for Commercial Testing Laboratories. Apparently the entire Code is going to be thrown out and most probably no code will be drawn for this so-called industry at all. Even as the matter stands, however, specific exemptions have been made of the following: All government testing bureaus; all college and university laboratories, and *specifically Consumers' Research*." [Emphasis in the original]

Regulate *them* folks, not us, and specifically not Consumers' Research.

By April of 1934, Schlink was openly aggravated with Brady, when he again sent an emissary to CR's headquarters in New Jersey in order to pick Schlink's brain about testing methods. Given Brady's position within a government agency, funded to do precisely the work he was asking CR to perform, Schlink's frustration is understandable. He wrote, "Frankly and personally, I think your crowd ought to start scratching for a living for awhile and not just soak things up."

"We don't mind having our stuff soaked up, but we decidedly do object to being ordered to prepare it immediately and in convenient form for the sponge."

From that point on, communications became strained, and "Fred" and "Bob" began addressing their letters more formally, "Mr."

Secret Life of Mildred Brady

Robert Brady made numerous trips to Europe; he visited Germany in 1930, 1931, 1934, and 1935 as well as Russia in 1931 and 1935.[15] Brady spent several years working on a book on the economics of German Nazism under a grant from the Carnegie Corporation (1935-1937).

In Germany, Brady and his first wife stayed at the home of fellow economist Jürgen Kuczynski (1904-1997), at his family estate at Schlachtensee near Berlin. Brady lived with Kuczynski for part of the year 1930. Kuczynski joined the German Communist Party in that year, and would rise in the party's hierarchy thereafter. He knew Albert Einstein quite well from the 1920s. The director of the Marxistische Arbeiterschule (or MASCH, where Wilhelm Reich was involved in the 1930s) had the idea to invite Einstein to speak to the group. But the MASCH office was in the center of the red-light district of Berlin so the director asked Kuczynski to personally escort Einstein to the building.[16]

Kuczynski kept up contacts with Einstein up until May, 1939, just prior to the beginning of the Manhattan Project. In his 1997 book, Kuczynski wrote that the Party asked him to contact Einstein in 1938 to inquire about raising funds for a secret radio station in Europe. He wrote of their last meeting at Princeton in 1938, "They were beautiful hours; sadly-wisely he [Einstein] judged the world situation. Not bitterly, but also without the cheerful pugnacity which he showed before from time to time. Six years had passed since our last meeting. With great interest he listened when I told him about the radio station [called "29.8"] and generally about our [Comintern] work. A day after my visit, he sent me via his secretary [Helen Dukas] the address of someone, who a mutual friend of ours knew, to get more money [for Communist underground work in Europe.]" It was Jürgen Kuczynski that physicist Karl Fuchs first approached with the idea of providing information about Britain's atomic bomb project to the Soviets.[17] Kuczynski's sister, Ursula, became Fuch's primary contact with Soviet intelligence.

Comrade Kuczynski remembered "Bob" Brady as a committed communist, "with heart and mind completely on our side, but forced to teach all kind of nonsense at the University of Berkeley." Dorothy Brady wrote to Kuczinski's wife in 1931: "We always will look back at our year in Germany as one of the happiest years of our life. How clearly I realize it when I see how Bob's susceptibility has returned here. He is a bag of nerves... maybe because he has the feeling to sell his soul to mammon. I wish I could somehow operate on his head and put some good knowledge of Latin and Greek into it, so that he could become a historian of the Ancient World and could speak his mind without fear of the consequences. He is not the man to teach embryonic businessmen to find their way to wealth by fraud—no more than your Jürgen." Kuczynski responded, "Bob shouldn't have concern about his soul, since he wrote such great books against German fascism, and monopoly capitalism in general!" Kuczynski noted that Dorothy was not in the Party but like Bob, "socially active on our side."[18] In 1935, Kuczynski worked in the underground for the Party in Germany. His parents were in England. Kuczynski reports about his correspondence with his parents: "On June 22 I wrote on the occasion of the visit of American friends—the Bradys: 'Bob asked me whether I would come next summer to California. Then I would be nearer to Ursula (who was in China) but farther away from Europe."[19]

In 1935 Brady lived in Switzerland while gathering data first-hand. Brady was married to Dorothy, but apparently brought Mildred Edie along with him to Switzerland. Dorothy had recently given birth to a baby boy.

While in Switzerland, Mildred Edie and Robert Brady reportedly[20] attended a lecture, given by Wilhelm Reich on August 30, 1934 on "Psychic Contact and Vegetative Current." Not only was this his last appearance before the International Psychoanalytic Association Congress,[21] but the speech marked Reich's first use of the term "vegetative energy." (See Sharaf, *Fury on Earth*, p. 188, for a detailed discussion of this paper, which was later published in the 1945 edition of *Character Analysis*.) Reich would soon begin experiments recording the bioelectric charge on the surface of the skin. He had not yet discovered orgone energy, but he was already taking a wholly new direction in his scientific work.

By September, 1935, Robert Brady was writing to the CR strike committee in support of their actions. Judging from the strikers' demands, this was not an economic strike, but rather a politically motivated strike to restore Arthur Kallet, a former board member, to his position. "Unless Consumers' Research allies itself with labor it will continue to bark like a Pekinese dog until it is ready for capture as a perfect instrument for Fascist propaganda purposes." Unless CR's "efforts are bent to serve the masses that produce... then it is better that Consumers' Research be destroyed."[22]

In 1935, from Switzerland, Robert Brady wrote to Fred Schlink regarding his research: "I am working on the job on bureaucracy, to name the title, and I have got some of the hottest material you ever laid eyes on, so hot in fact, one of these days you may be having to supply me with a little storm cellar in your nice comfortable bourgeois house, set up on the side of your middleclass hill." Brady asked Schlink to forward a selection of material on the New Deal to Brady's next destination: "c/o Intourist, Hotel Metropole, Moscow, U.S.S.R."[23] Brady was living in Switzerland and collecting information for a study funded by the Carnegie Foundation with a grant of $17,000—a huge sum in the Depression, and the largest given to the University of California in 1934. Perhaps Brady's 1935 visit to Moscow was part of his research on the bureaucracies of fascist, communist and capitalist economies. However, the book he produced, *The Spirit and Structure of German Fascism*,[24] contained nothing about Soviet bureaucracy. Brady's central thesis was that fascism was capitalism in its utmost extremity.

It will be instructive for our purposes to examine Robert Brady's 1937 study of German fascism, since it followed one of Reich's most important works, *Massenpsychologie des Faschismus (Mass Psychology of Fascism)* by four years. The perspective of Brady's book is diametrically opposed to Reich's *Massenpsychologie des Faschismus*. Brady argues that gross material forces, embodied by monopoly capitalism, are the foundation of fascism. Brady never misses a chance to point out the similarities of Nazi Germany and New Deal America. In this, Brady was in line with the Dimitrov Doctrine, as promulgated at the 1935 Seventh Comintern Congress by Giorgi Dimitrov. At the same time, Dimitrov was busy overseeing the rounding up of suspected Trotskyists in Russia including one Otto Knobel (Brandt) whose prime offense was his association with Wilhelm Reich. Knobel was sentenced to five years in the Gulag. Brady offers scant mention of Russia beyond gratuitous lines such as this: "further successes of socialist planning in the Soviet Union" during the Weimar era could have strengthened the position of a "united, left-oriented front" in Germany.

Brady noted: "American readers, who may be shocked at this cynical combination of [the Nazis'] simulated concern for the public well-being, maintenance of the husks of representative government, and complete subjection of labor to the power of employers, may well be interested to note that the entire Nazi labor program is identical in effect with one of the earliest arguments on behalf of company unions—the Industrial Creed announced by John D. Rockefeller..."[25]

Brady follows this logic only up to a point. When discussing Nazi "borrowings" of Soviet Union's trade union structure, Brady is insistent that there is no underlying functional identity: "The reader should, for sake of clarity, be very sure to distinguish here between *form* and *intent*. [Emphasis in the original] The Nazis have borrowed Soviet *forms*; the *intent* of the borrowing is exactly the reverse of that which dominates the use of those forms in the Soviet Union."[26]

Brady's book has foldout flowcharts depicting the organizational structure of the German Labor Front (with its "joy division"—*Kraft durch Freude*—literally "Strength Through Joy"); National Food Estate; National Chamber of Industry. Most of it is dry-as-dust economic analysis. Brady broke out in florid style for the conclusion of his book. In the last two pages, Brady calls up imagery of capitalist Argonauts:

> "Here we have the modern parallel to the ancient story of the hero Cadmus, who, in the quest for the Golden Fleece, slew the fiery dragon. Following the voice of an unseen oracle, he sowed the dragon's teeth in the soil where the battle took place, and from there sprang a race of warriors who rose and fought each other to death. When the business forces of the world openly sow dragon's teeth of discord among otherwise peaceful peoples, they prepare, if not themselves, coming generations for the wasteland shambles; for ten to fifty or a hundred years of sanguinary war and terror-ridden struggle in what may be the growing darkness of a fading and broken civilization.
>
> "They condemn, likewise, the arts and sciences. In the stifling atmosphere of fascism, no originality of mind, no creative power, can flower. The writer must drop his novel or his poem for the bitter propaganda tract. The natural scientist must search for that which promotes the ends of those who command what he shall know, think and believe. The social scientist must, like a high-school debater, supply proof for arguments advanced by those who have no interest in the truth.
>
> "There is no hope against all this, except that of sweeping away the very foundation on which fascism constructs its brittle

edifice. If the world-wide fascist trend is to be reversed, what is left of democracy must be reinforced, and the sweep of the tide turned. But this means an erosion of fascist foundations, not by humanizing business enterprise, but by arraying together all those forces whose face is turned the other way.

"Against an opponent who believes singly and solely in force and guile, force must be massed. The hope of the people of the United States is to be found, not in giving free reign to monopoly-oriented and fascist-inclined capitalism, but in turning back its fields, factories and workshops to those who fought its war of freedom against a tyrannical power, and who built, with their muscles and their brains, all the real wealth and all that there is in America that deserves the name of culture. But it will not come to them as a 'gift'; they must learn that the only solution to recovery of their heritage is within themselves.

"And the solution is no different for any other people or nation in the world."[27]

There concluded Robert Brady's *Spirit and Structure of German Fascism*.

"Charge of the Letterhead Brigade"

After his return to the United States, Brady gave a speech at Wheeler Auditorium at Berkeley, and predicted that revolution in Fascist Italy and Germany would be "inevitable" in 1937. This was the Stalinist political line that year. A student-produced handbook from the University of California at Berkeley that assessed the school's teachers commented that Prof. Brady "talks through his beard." After several years back at Berkeley, Brady was appointed Chief Senior Economist (Standards) and Economic Advisor to the Director, Consumers Division, in the Office of Price Administration (where he would serve between 1941-42).

In 1939, Robert Brady and Haakon Chevalier served as associate editors of the periodical *Black & White*. Under murky circumstances, in 1942, former Berkeley physics professor J. Robert Oppenheimer would later denounce his close personal friend, Chevalier, as a Soviet agent. Oppenheimer told his military security officer that Chevalier had approached him about passing scientific information onto the Soviet

Union. On the basis of Oppenheimer's denunciation, Chevalier was fired from his position at the University of California and found it impossible to obtain work throughout the fifties. Later, Oppenheimer reversed himself on the "Chevalier Affair" at a 1954 security hearing in which he described his earlier story as "a tissue of lies." Both men would later write books on the subject, although their respective versions of events never would match up.

Certainly the Bradys were part of the small, and somewhat isolated, community of leftists among the faculty at UC Berkeley, which at that time was comprised mainly of conservative Republicans, as it is today. According to recently released files from the KGB Archives, Oppenheimer was a secret Party member at Berkeley, and I believe he belonged to the same covert cell as the Bradys. Chevalier described an "innocent time" before the outbreak of WWII, when Oppenheimer threw fundraising parties, complete with "B-girls" hired for the occasion to sell drinks and cigarettes in support of Russian War Relief.[28]

Along with Mildred Edie and Robert A. Brady, J. Robert Oppenheimer was a founding Board Member of the Western Branch of Consumers Union in 1938. We will consider Oppenheimer's relationship to Reich separately, but for now, let's note that Oppenheimer supported the Bradys' political work. The Western Branch of CU opened up offices at 251 Kearny St. in San Francisco that spring.

Picture the social set: earnest intellectuals, talented artists, university professors and professional revolutionaries. There were house parties in Berkeley, set in the summer evening's fog of 1939, where the Bradys hosted the Oppenheimers, the Haakon Chevaliers, Otto Fenichel, Steve Nelson (a Party official closely associated with atomic espionage). They were joined by a spymaster in San Francisco named Gregory Heifetz from the Soviet Embassy.[29]

Oppenheimer was questioned about his relationship with the Bradys on May 6, 1954 by the Personnel Security Board of the Atomic Energy Commission:

> Q: You say on page 6: "I also became involved in other organizations. For perhaps a year I was a member of the Western Council of the Consumers Union." Who composed the Western Council of the Consumers Union?
> A: Chairman and the man I knew best was Robert Bradley [sic], a professor of economics at the university.

Secret Life of Mildred Brady

Q: Was he a communist sympathizer?
A: No, I don't think so. His wife was Mildred Eddy [sic] and the two of them were what made this. They had an enthusiasm for this.
Q: Did they recruit to it?
A: Yes, they asked me to come. It was a very inappropriate thing for me to do. I know nothing about the business. [30]

Prominent people, such as the lawyer, Arthur Garfield Hays, and writers such as Dashiell Hammett and Dorothy Parker, were among those listed as Sponsors of Consumers Union, along with Oppenheimer and the Bradys. Arthur Garfield Hays (1881-1954) was a professional colleague of Mildred Edie Brady, prior to Hays' handling of Wilhelm Reich's legal affairs in the 1940s. Reich was completely unaware of Hays' professional relationship with Consumers Union, or with Mildred Brady. Over the six years Hays represented Reich, Hays consistently failed to serve Reich as professionally as he might have.

For example, Hays dissuaded Reich when his client asked him to initiate a libel suit against Mildred Edie Brady and her publisher, *The New Republic*. In November, 1947, as officials of the FDA began making sidelong inquiries into Reich's alleged "sex racket," Reich wrote to his lawyer, "Dear Mr. Hays: I wish to stress the fact that *the slandering article by Miss Brady was the beginning of a chain reaction set into motion, beyond any doubt, by communist quarters*. The husband of Miss Brady is a communist." [Emphasis in original]

Reich went on to note twelve specific slanderous statements in the article; "I think a libel suit is necessary in order to put matters straight and to put an end to this pestilent chain reaction."[31] As he composed a private memo, not sent, titled "Warning Against Communist Snipers" that December, Reich certainly would have been interested in learning that his lawyer, Hays, was the author of the *What I Saw in Russia, Informal Report of an American's Tour of the Country Ruled by the Soviet*—one of the "Little Blue Books" published by E. Haldeman-Julius and Henry J. Haldeman.

To Reich's reasoned request that Hays initiate a libel action against *The New Republic*, Hays replied, "If you could sue a person for what he *thinks* of you, you could do it; but the writer of this article was clever enough to avoid a libel suit."[32] Yet Brady's article clearly libeled Reich as a money-grubbing charlatan. And, as long as there are lawyers, no writer will be "clever enough" to avoid a lawsuit. Yet Hays did not act. Certainly Hays was aware of the circumstances under which Consumers Union was founded, after the failed strike against Consumers' Research in 1935.

In his 1938 book, *Odyssey of a Fellow Traveler*, J.B. Matthews identified Arthur Garfield Hays as a supporter of the Communist Party through its "united front" organizations in the thirties. Matthews said that the CP, acting on the presupposition that the average American worker took his cues from observing his "betters," used intellectuals and professionals like Matthews and Hays as "window dressing" for front groups that appealed to a broad range of liberals and progressives, but in fact these groups were bureaucratically managed by the Party.

Matthews, a lively writer of no small wit, coined the term "fellow-traveler" to describe non-Party members' relationship with Stalinism, and he characterized the proliferation of "united front" groups in the thirties as the "charge of the Letterhead Brigade." These front-groups, consisting of little more than a masthead of prominent persons and a mailing list, represented the fruits of the abrupt 1935 turnabout for the Soviet Union, which had originally seen the rise of fascism as a welcomed precursor to world revolution. Nazism, according to the Party line, "accelerated the rate of Germany's development towards proletarian dictatorship" by "destroying all democratic illusions among the masses and liberating them from the influence of social democracy." In 1935, however, Stalin encouraged a "unified front" with Social Democrats and other leftists, while using a small fraction of disciplined Communists within each front-group to coordinate their overall movement and directions. Matthews himself had been on the masthead of dozens of such groups, and had shared the speaking dais with Arthur Garfield Hays on many occasions.

Much has been made of the fact that another of Reich's private attorneys, Peter Mills, became the prosecuting attorney in Reich's criminal trial. Yet this earlier incident may have been even more damaging to Reich's work. Hays had been specifically asked by his client to file a libel suit against the *New Republic* and Mildred Brady, but Hays advised Reich against the idea. Reich was never satisfied with Hays' advice, and his medical colleagues, including Theodore Wolfe, also wrote to Hays to urge him to initiate a libel action. In this context, given his experiences first with Hays and then with Mills, Reich chose to represent himself in later court proceedings.

Secret Life of Mildred Brady

At no time did Hays advise Reich that he knew Mildred Brady, or that Mildred Brady had been employed as a writer for the League of Women Shoppers, which in turn was directed by Mrs. Arthur Garfield Hays. (*The Daily Worker*—the official organ of the CPUSA—identified the League of Women Shoppers as an authorized united front organization, suitable for housewives who were not politically advanced enough for full Party membership.) J.B. Matthews, who had originally consulted with the League of Women Shoppers as to its organization program, called Mrs. Hays a "stooge" for the front-groups in his book. Hays himself had been a sponsor of the Consumers Union. To say the least, Hays had a conflict of interest while representing Reich in the 1940s and Hays had a legal obligation to advise Reich of this conflict. He never did.[33]

In May, 1947, when Brady's article first appeared, the Soviet Union had not yet completed its first atomic bomb. However, Michael Straight, Brady's *New Republic* publisher, had first-hand knowledge that his old comrades from Cambridge had become Soviet moles operating deep inside British intelligence. (See chapter on Wilhelm Reich and the Cambridge Five). It was Straight's fellow Cambridge student, Donald Maclean, who relayed some of the first details about the Manhattan Project to the Soviets.

If Reich had been adequately defended, his lawyer would have sued Mildred Brady, Michael Straight and the *New Republic* for libel. He would have put Straight on the witness stand under oath. Asked directly about his connections to Soviet espionage, Straight may well have told the truth in 1947, when it still made a difference, and identified Anthony Blunt and Guy Burgess as Soviet agents. (Straight has since claimed that he would have liked to confess and that he was waiting only for someone to ask.) Straight's testimony in such a trial, if truthful, would have exposed Soviet penetration of the highest echelons of the British secret services, just as Straight's testimony ultimately did in 1965. Reich's life and work could have been saved. It's a moot point, in legal terms.

Mr. Bursler was a Spy:
"Norman will try to find out in detail about S."

In 1941, F.J. Schlink wrote to J.B. Matthews at the House Committee on Un-American Activities, providing him with a file of letters he had received from Brady in the thirties. "Your attention is especially called to the letter of April 10, 1935, in which a Mr. Bursler is referred to. At the time, we took him to be a graduate student of some sort. Since then, however, I have wondered if he were not a political commissar, since in conversation with us he boasted of having been one of three (?) who voted against Lovestone at the time of his election as Secretary of the Communist Party." Norman Chandler Bursler, a research assistant for Dr. Brady's book, was a research expert in the Anti-Trust Division of the Justice Department in 1944. Bursler's name appears in declassified decrypts of cable traffic between Moscow and its illegal agents in the United States. The documents identify Bursler as a conscious agent of the Soviet Union, a link in Nathan Gregory Silvermaster's espionage ring. Another scholar whom Brady thanked in his book's acknowledgments was William Henry Taylor, a professor at the University of Hawaii. Taylor was named as a covert Party member in sworn Congressional testimony in 1950.

In 1996, the National Security Agency (NSA) released decrypted cable traffic between Moscow and its agents in New York, Washington, San Francisco, Mexico City, and elsewhere. The NSA, colloquially known as "The Puzzle Palace," houses the code-breaking work of the US government. The VENONA documents were released at a widely publicized press conference in 1997 attended by Russian and American intelligence officials (significantly, the British were not invited, since many of the documents relate to the Cambridge Five.)

The transmissions are fragmentary and only parts of each cable have been decoded. Left and right partisans among historians and journalists can still take issue with differing interpretations of the true meaning of the individual cables. For instance, the NSA identified one agent with a code name "ALES" as "probably being Alger Hiss." Today's partisans of the Party argue that this qualification is too vague, but Hiss has been positively identified as an active Soviet spy, and he was given a medal for his work immediately after the Yalta Summit, where Stalin was given control of Eastern Europe. The wartime documents from VENONA offer a fascinating look into the massive Soviet intelligence effort that utilized a broad spectrum of American Communists in government service.

While some of the agents identified in the VENONA decrypts are open to subjective interpretations, the document naming Brady's associate Norman Bursler is clear. A cable, dated November 20, 1944,

Secret Life of Mildred Brady

from the Soviet Vice-Consul in New York to Lt. Gen. P.M. Fitin in Moscow reported on Silvermaster's efforts to obtain information about Felix Somary, an Austrian businessman. "ROBERT [Silvermaster] received this information from the FELLOW-COUNTRYMAN [member of the CPUSA] Norman BURSLER. [unrecoverable text] Later Norman will try to find out in detail about S. [Somary] and his activities."

Thus, Robert Brady's primary research assistant was an agent of the Soviet Union. Norman Bursler had been with the Bradys in Lucerne, Switzerland, in 1934 when Reich gave his farewell address to the International Psychoanalytic Association. Reich's lecture would later be expanded into a paper called "Psychic Contact and Vegetative Current," so he was breaking new ground and leaving the old behind. Reich still considered himself a communist at that time, albeit a dissenter from Stalinism. The Comintern had by then reversed its earlier support of Reich's Sex-Pol organizing, and as Reich made plans in Scandanavia for conducting bioelectric investigations of sexuality and anxiety, the Party apparatus began treating him as an enemy of Socialism.

Alger Hiss' conduit to Moscow, Nathan Gregory Silvermaster, had been Robert Brady's student, and had been recommended for government service by Brady. In formerly classified documents newly released for this book, we now know that the FBI had taped several phone conversations between Mildred Edie Brady and Helen Silvermaster in the spring of 1947, and that Robert Brady had met with the Silvermasters in Washington, D.C in 1946. Today, the evidence records over ten years of direct contact between Moscow's espionage networks and Mildred Edie Brady, who had the ear of the FDA as a top functionary of the Consumers Union—and the Communist Party considered this "Union" as a front group of its own. As federal proponents of "consumerism," the natural constituency of the FDA was the Consumers Union, which had lobbied for, and got, strict enforcement powers for the FDA. Mildred Brady had written parts of the FDA's 1938 enhanced regulations.

Perhaps even more startling than the direct connection between the Bradys and Soviet espionage agents is the indication that Mildred Brady was in attendance at the 1934 IPA Congress in Lucerne, over a decade before she aimed her poison pen at Reich. In 1948, Dexter Masters, Mildred's boss at *Consumers Reports*, told a colleague at

Wilhelm Reich and the Cold War

Columbia University "he knew Reich was psychotic." Dexter Masters had been a member of the Communist faction that split the original Consumers' Research organization and later he served as an editor at *Consumers Reports*. Masters was also Mildred Edie's lover for the five years previous to her meeting Robert Brady. Joan Brady, Robert and Mildred's daughter, would subsequently marry Dexter Masters, crushing her mother, who had hoped to renew her own relationship with him after Robert Brady's death. An affidavit[34] by Angelica M. Haymes reported to Wilhelm Reich that she had asked "on which grounds Miss Brady and Mr. Masters declared Reich psychotic. The answer was that Miss Brady had attended a Psychoanalytic Congress sometime in 1934, where, Miss Brady says, Dr. Reich made the soundest speeches, but the signs of the psychosis were already in Dr. Reich's behavior: He was living in a tent on the hotel lawn, with a dancer, and wore a dagger."

Reich referred to this rumor in an official 1952 interview at Orgonon with Dr. Kurt Eissler, head of the Freud Archives. "It is quite amusing to think back on that today. But to give you a picture of some analysts at that time: They lived in hotels, sat around in smoky lobbies, and so on. I didn't. I lived with my wife in a tent at the Lucerne Lake. I had a dagger, you know, as you have when camping. Today, nobody would find anything peculiar in it."[35] Yet there was sufficient detail in the rumor spread by Masters and Brady to indicate that Mildred Brady was in the audience at Lucerne, with an early interest in the "strange case of Wilhelm Reich."

"Think of that, my colleagues!"

In 1941, Robert Brady was appointed Chief Senior Economist at the Office of Price Administration (OPA). Brady brought along with him his second wife, Mildred Edie Brady, as well as Dewey H. Palmer, both of whom had been involved with the strike at Consumers' Research in 1935. All three of them had been founders of Consumers Union in 1936.

Conservatives in Congress were already alarmed with the prospect of the Roosevelt Administration setting retail prices, but they became even more alarmed when Rep. Martin Dies, of Texas, Chairman of the House Un-American Activities Committee, began reading from Brady's book on Fascism on the House floor:

"As all the parties realize," Brady had written, "the end result will be either the triumph of reaction and a new lease on life for capitalism, or else a victory for socialism and the extermination of the rich and powerful of the bank, factory, bivouac, and cloister."

"Think of that, my colleagues!" blasted Congressman Dies.

As Walter Goodman, a reporter from *The New York Times* who followed the entire career of the House Committee on Un-American Activities from an even-handed perspective, put it, "this was heady stuff for a government consultant." Professor Brady suffered infinitely bad timing in having signed a proclamation in support of the Soviet Union in August, 1939, on the very eve of the Soviet alliance with Nazi Germany.[36] He was shifted to a less sensitive "consultancy" within the OPA in March of 1942, while the Civil Service Commission considered the evidence. Mildred, on the other hand, resigned as soon as Dies leveled his charges. Leon Henderson, the OPA chief, defended his colleagues with a typical defense tactic of ridiculing the charges as "red-baiting" and he jokingly offered to "eat any subversive organization [that he or his aides had belonged to] on the Treasury steps."

Much of the data provided to the House Committee on Un-American Activities had come out of the offices of Consumers' Research, where Fred Schlink still smoldered in anger over the 1935 strike by many of the same people who now served at the OPA. In a letter to Thomas M. Leary, Special Agent in the Treasury Department's New York Intelligence unit, Schlink recalled his impression of Mildred Edie Brady. "She expressed herself as wholly incapable to understand the political value or utility of the consumer movement unless it fitted in with Marxist theory by proving or tending to the inevitableness of the breakdown of capitalism; i.e., she was interested in the consumer problem only in so far as it represented revolutionary aspects which fitted in with Marxist predictions of the crumbling of our economic order." Schlink went on to describe the 1935 strike, which was given broad coverage in the "left press, including specifically the *New Masses* and the *Daily Worker*." Schlink claimed that Mildred Edie "took an active part in advising and counseling the leaders and on one occasion, appeared with the pickets."

Schlink concluded, "Miss Edie is an extremely bright and hardworking person, whose loyalties are wholeheartedly for the left-wing cause she believes in. Her loyalty to this government will, in my judgment, be wholly dependent on its foreign policy with respect to Russia and its domestic policy in relation to left-wing labor."

Some of the ideas coming out of the Bradys' department at the OPA bred the kind of nuisance regulations that have become the bane of American existence. There was talk of standardizing underwear, eliminating trademarks, and a rigid system of grading products. Reading the contemporary press accounts of the OPA's mission, both pro and con, it's hard to escape the utopian vision of social engineers like the Bradys. Walk into the store and choose between an array of generic products in clearly marked labels: Good Cheese, Pretty Good Cheese, and Okay Cheese.

In the light of the FDA's use of the statute against "mislabeling" of products to force Reich and his associates to burn his own books, inventions and research journals, it's important to remember that Consumers Union, Inc., and the Bradys in particular, were instrumental in writing these codes when the Food and Drug Act was beefed up in 1938. In 1941, the FDA seized 400 cases of "mislabeled" cans of corn, based on standards that had been set at Robert Brady's division of the OPA. When questioned about the case, Brady responded that there would be no need to set up additional staff at the OPA for enforcement since the FDA already existed for that purpose.[37]

James L. Goddard, MD, then Commissioner of Food and Drugs, addressed a convention of Consumers Union commemorating that organization's 30th anniversary on September 30, 1966. Full text of this speech appears on FDA stationery. "Rarely have I felt as 'at home' with an audience as I do tonight," he began. He noted the close cooperation between the FDA and the young Consumers Union of the late thirties, particularly in drafting food and drug legislation:

> "During that period, when the Consumers Union was begun, the Congress was wrestling with some form of full-scale food and drug legislation. Five years of hearings and draft bills passed before a final version took meaningful shape.
>
> "It is interesting to note," Goddard continued, "that one of the provisions of the early draft that was thrown out forbade labeling or advertising 'if in any particular it [was]...untrue, or by ambiguity or inference create[d] a misleading impression...' This kind of thing was just too much for industry to swallow in 1936. But by 1938, the time had come for major legislation. President Roosevelt, who had been re-elected in 1936 by an electoral landslide—every State except Maine and Vermont—signed the Food, Drug and Cosmetic Act into law on June 25, 1938."

Secret Life of Mildred Brady

Goddard emphasized how Consumers Union set much of the enforcement agenda at the FDA throughout the years:

> "Especially important has been the coverage in your magazine, *Consumer Reports*, on matters of health and economy. [...] Investigative reporting by *Consumer Reports* has repeatedly contributed to our own work to track down and prosecute those who threaten the marketplace and the consumer. We have drawn much from your close attention of your work. This collaboration of ideals and of effort has produced a fruitful 30 years thus far."

Goddard singled out Mrs. Brady for her contributions.

> "I would like to pay a special tribute at this time to a woman whom all of us knew as one of the greatest champions of consumer rights: the late Mildred Edie Brady, whose life was devoted to the hard job of communicating these facts in print. Mildred Brady was a special kind of person, although she would probably be the first person to blue-pencil that statement out my speech, were she with us today. She was special because she was not willing to compromise the right of the consumer, as she saw them...I did not feel, ladies and gentlemen, that I could speak to you tonight without taking a moment out to pay tribute to Mrs. Brady. Her work as an expert witness was only one aspect of a busy, even turbulent, life as the consumer's best friend."

Mildred Brady, Wife and Mother

So far we have traced the public record regarding Mildred and Robert Brady's connections to Soviet espionage, as well as their connection to the FDA. What kind of people were they? This is a question the stale FBI reports and the dusty archives can't answer. What we do have, however, are the deftly heart-rending writings crafted by their daughter, Joan Brady, who published two works of fiction that drew heavily from her childhood in Berkeley, California. Joan Brady painted shattering images of her parents, Robert and Mildred. Their relationship was filled with spite, violence, and sadism.

On Mildred Edie Brady, Joan wrote, "She had the easy sureness of manner good-looking people have, and she was intelligent, responsive, forceful, perservering, but she was also a changeable woman, possessive, easily stung, self-righteous, and given to sexual jealousy as profound in her as the faith in humankind to which she clung more and more tightly in the face of my father's increasing hatreds."[38]

Mildred and Bob maintained "The Open Door Policy" with their daughters, Joan and Judy. Joan, in her book, describes the Open Door Policy as a rigid invasion of privacy, what Joan called an "uncomfortable mysticism that prevailed at home." It's more than ironic that these Marxists would choose the name for the forced-entry of China by Western powers, to describe sexual "openness."

"And she didn't like the thoughts she sensed I thought about myself, either. Mine was, she said, an important age for a girl brought up under the Open Door Policy. An exciting age. Pubic hair. Boys, perhaps. The menarche. Breasts. Sex emergent. Judy had responded properly to all these beginnings only three years before."[39]

Two vignettes that recur in Joan Brady's novels—*Theory of War* and *The Unmaking of a Dancer*—stand out it my memory. The most gripping scene occurs when Mildred and Bob, at the conclusion of one of their chronic evening shouting matches, break new territory. Mildred has discovered an incriminating, "raspberry-scented condom" under the bed in Bob's study. For the first time, Bob strikes her. He then locks himself in the bathroom, and swallows 500 Nembutals. (Although this is a novel, Robert Brady did in fact, make a suicide attempt in 1951.) "God help us," Joan wrote of the same scene in her other novel, *Theory of War*, "what a thing to die for: a raspberry-scented condom."[40] Robert Brady temporarily recovered for a time, but soon became convinced he was dying of cancer, although his doctors declared him free of the disease. He believed they were lying to protect him. Mildred eventually had him committed to a psychiatric hospital, taken away in a straitjacket.

The other memorable scene occurs just before the debut of Joan Brady's ballet performance in New York City. Throughout her childhood, she has lived and prepared for this day; her sole goal in life is to become an accomplished ballerina. She is an adult now, and living on her own, but with her parents' financial support. Bob is still alive, but his suicide attempt has damaged him, and he's had a stroke and is now a vegetable. Mildred has wheeled him back to New York, following Joan. "...Her eyes came, in the years after our arrival in New York, to have the aged, marooned resignation of a turtle's eyes."[41] Mildred

Secret Life of Mildred Brady

hopes to renew her relationship with Dexter Masters, her lover during the Thirties. But Joan has begun sleeping with Masters, an old family friend twice her age. Mildred shows up just as her daughter is trying to learn her part for the performance, and she lashes out at Joan in a vulgar and sadistic fashion, leaving Joan in a state of deathly emotional paralysis (Wilhelm Reich would have called it "anorgonotic shock.") Joan passes out; when she awakes, her mother is gone, but her sister, Judy, is there, comforting her. Joan has lost her voice, she cannot speak or explain what is wrong. She is unable to perform, unable to achieve her dream.

I was able to contact Joan Brady and asked her a few questions about her books and her parents. After a few pleasantries, I wrote, *"I assume you know about a famous article Mildred Edie Brady wrote about Dr. Reich in* The New Republic *in 1947. A few months later, the US Food and Drug Administration began an investigation of Reich. His books were destroyed by court order and Reich died in Lewisburg Federal Penitentiary in 1957."*

Joan leapt to her mother's defense. "But you must not take what I say about my mother as a full portrait; she didn't like me, which must have been as hard for her as it was for me, but I have great respect for her in her field. She had a long and distinguished career as a journalist; she exposed frauds of almost every variety from local rackets to national fiddles, testified before several Congressional hearings, and is almost single-handedly responsible for the Truth in Lending Bill that means you aren't being ripped off by your mortgage company right now. You'll find Senator Hart's tribute to her in the Congressional Record of July 29, 1965."

"As for The Strange Case of Wilhelm Reich, of course I know about it, but I doubt I can give you what you want," she warned. "The article is as cleanly written and as clearly researched as any I've read. She makes Reich's quackery manifest without resorting to histrionics, and what she quotes directly from him exposes the physics behind his theories as too silly even to be comic." I was beginning to feel a little of the Brady heat myself. "If, given my endorsement of her position, you still want to ask some questions, do feel free to write me. I would prefer to keep the record written, but I promise you I'll do my best."[42]

I was game. She wasn't the first person to tell me Reich was both crazy and a crook. I offered to provide Joan with the documents I had found; she had no interest. A pity, I think; she might be interested to learn that sections of her mother's FBI files remain highly classified in 1999, under the National Security Act of 1947 and the CIA Act of 1949.

Hello, Joan

I assume the biographical details about your parents in your novels are accurate and true, rather than fictionalization. Is that fair?

"I don't mean to sound stuffy; but when the word 'novel' appears on the cover of a book, you cannot make the assumption that *anything* in it is fact. When I write fiction I play with what's true, like any other illusionist."

Did your mother ever discuss Reich with you? If so, what did she say? How did she become interested with him?

"Reich was a subject of dinner table discussion from time to time along with other fraudsters my mother helped expose. I don't know how she got interested in him."

Did your mother have a professional relationship with federal officials in the FDA and Federal Trade Commission in connection with her work at Consumers Union?

"I assume you would call her relationship with various government officials professional. Information certainly got exchanged, although I never heard of any contractual arrangement. Consumers Union could clarify the position for you."

When did your parents' relationship begin? In Reich's court papers, there was a reference to your mother attending the 1934 Congress of the International Psychoanalytic Association in Lucerne, Switzerland. (Dexter Masters was given as the source of this information.) Reich spoke at this Congress, and—apparently—both your parents were in attendance. Can you verify this, or recall any of their reminiscences of this trip? Did they ever mention a trip to Moscow in 1935?

"I can't be precise about when my mother and father got together, but I'd guess it was 1936—too late for Switzerland in 1934 or Moscow in 1935. I didn't know anything about Switzerland; I don't know that my mother ever went there (I wouldn't rely on a quote of a quote if I were you), but I do know my father went to Moscow—and without her. Much to his surprise, he was not impressed."

In his book, The Spirit and Structure of German Fascism, *your father thanked his research assistant, Norman Chandler Bursler. Do you*

Secret Life of Mildred Brady

recall meeting Mr. Bursler? Did he accompany your father on his research trip to Switzerland in 1934-1935?*

"Bursler is just a name to me."

Many of the articles in Consumers Reports *are without bylines, what proportion of them were written by your mother?*

"Again I'd refer you to Consumers Union. My mother's style was fairly distinctive; if you're good at styles, perhaps you can identify her work that way. If not, I think you can assume that a high proportion of the articles on consumer economics were hers."

There were a limited number of articles credited to Mildred Edie Brady in the Reader's Guide to Periodicals; *did she work as a freelancer much or did she primarily work under contract with trade journals? Did she ever use a pseudonym?*

"My mother did do some free-lance work, but I think you're probably right in assuming she worked mainly under contract. I don't know that she ever used a pseudonym—unless you consider the work she did as a ghost. She wrote many of Colston Warne's speeches. Other people's too—probably some of Kallet's."

In your books, you briefly discussed some of the difficulties your father experienced during the McCarthy era. Could you discuss that? How did this pressure filter down to the children?

"*The Unmaking of a Dancer* tells most of what I remember of the effects of McCarthyism on my father—and on me."

Your parents knew J. Robert Oppenheimer, and also your father co-edited Black and White *with Haakon Chevalier. Oppenheimer denounced Chevalier as a Soviet agent, then recanted; their stories never did match up. Did your parents ever discuss this affair?*

"I remember both Haakon and Oppenheimer, Oppenheimer more fondly than Haakon; and I remember discussion of both of them—but I'm sorry to say I can't remember its content. Most of the controversy, though, came about after my father's stroke; which is to say there wasn't any discussion to listen to."

How did Dexter Masters come to meet and become friends with Alger Hiss?

"When Alger Hiss got out of prison and that paper supplier gave him a job as a salesman, Dexter offered him the contract for Consumer Reports. Unfortunately, the supplier ran too small an operation to take on so large a customer; but Dexter and Alger became friends."

Did your parents know Alger Hiss, as well, and what kind of relationship did they have, if any?

"Alger had somewhat romantic ideas about my father's struggle with his stroke; beyond that, there was nothing that I know about in the way of a friendship. There might have been acquaintance – but probably not even that."

A good deal of archival material on your parents political associations resides at Rutgers University's Special Collections, in New Jersey, in the files of Consumers' Research, Inc., from which a number of workers broke away in 1935 to form Consumers Union. (I would be more than willing to share any of these research materials with you.) The leaders of CR, especially Fred Schlink and J.B. Matthews, became obsessed with proving that this "strike of 1935" was Communist-inspired and that your parents, along with the strikers, were Communists. Were they?

"So far as I know, neither of my parents was a member of the Communist party. Dexter certainly wasn't. As to CR, I hate to admit it, but I've forgotten what the strike issues were. I do know, as you indicate, that Schlink and Matthews were reds-under-the-bed obsessed. Most of the strikers were idealistic young people who believed that if they tried hard enough they could create a better world. I imagine they were also self-righteous and thoroughly irritating—people with causes often are—but their aim was to disseminate information, not aid Communist takeover."

The Cold War between the Americans and the Russians is over, and a host of new books have come out recently to settle old scores. Looking back, from a child's-eye view, what was the role of your parents (and, perhaps, people like them) in the 50's drama of McCarthyism, the Cold War, and left-wing politics in the USA? As an example: was there ever any sense of your being an outcast, because of suspicions about your parents? How did they deal with the pressure?

"My parents' motivations were the same as the CR strikers'. I would say that whatever else McCarthy achieved or didn't achieve, he helped quash any such optimism in me; I grew up—and remain—a political cynic. But I can't blame my parents' response to him for any trouble at school or in the neighborhood. As to how they dealt with the pressure themselves, I can only guess. Some people killed themselves because of it. Some had nervous breakdowns. Some collapsed physically. Many collapsed financially and professionally. Marriages broke up. I don't see how such a profound threat could fail to contribute to the tension in a household. It may well have been a factor in my father's suicide. I couldn't say for sure."

Without a doubt, your mother was a true crusader involved in some righteous causes. In some ways, she and Reich were on the same page; for example, she was publicly campaigning, through CU, for freer access to birth control. To what do you ascribe your mother's drive, determination, and willingness to be a vocal minority for progress?

"I'd argue that my mother's genetic makeup accounted for her drive and determination and that the times supplied her with an outlet."

A recurring word in your mother's (and father's) writing and work is "standards." What does the word standards mean to you?

"Standards? Uniformity of technical terms for specifications: dimensions, equations, formulae, graphical symbols, etc. Benchmarking of types, sizes, grades of manufactured products. Benchmarking of and rules of operation for machinery and apparatus. Safety provisions."

Your parents had an Open Door Policy, yet your mother lied about some significant things such as her relationship with Dexter Masters, and her "bigamous" marriage with your father. Is that correct? How did that conflict affect you? She also used subterfuge to establish contact with Reich for an interview, telling him she was bringing "greetings from a friend from California." Then, there is the way she sabotaged your efforts to be a dancer, while paying for your lessons. Was this typical of her?

"What parents don't lie to their children? When my sister and I found out about the late marriage, we were—both of us—amused and pleased. It's always fun to catch your parents out. As to ballet, it is as I wrote you. My mother didn't like me. Because of this, she did me a few injustices and caused both of us some unnecessary pain. A pity. But these flaws had nothing to do with her as a journalist. You show me a journalist who doesn't use subterfuge, and I'll show you a hack out of a job. I don't want to sound stuffy again, but people are complex. It's true that she was nastier to me than she should have been; but she was also a courageous, highly-principled crusader who got her facts—and prided herself on getting them right—by whatever means she could."

You mentioned a time, in the 50's, when your father became convinced that he had cancer after his suicide attempt. What was your mother's attitude about cancer, vs. your father's? Did your father's fears have anything to do with your mother's suspicions about "alternative" and/or "quack" cancer cures?

"My father took a massive overdose of Nembutal; his blood pressure was over the roof and his kidney function was closing down. Even at the best of times—which this plainly wasn't—he was a hypochondriac and a depressive. What I'd guess is that he was scared; he'd just discovered that nature didn't give a damn whether he lived or died, and he was flailing around for something solid to pin his fears on. But despite the personal friction, my parents agreed on fundamentals; he was every bit as contemptuous as she of quacks, panaceas, untested remedies, exploitive fads."

It's been reported that your mother's interest in Reich's cancer therapy was provoked by a friend of hers who sought help from Reich in the forties. Who was that?

"It's entirely possible that my mother heard about Reich through a friend. She often picked up ideas from friends or associates who'd been bilked. Dorothea Lange, the photographer, is a possibility. If so, the 'greetings from a friend from California' would have come from Dorothea—and the subterfuge that worries you would have been nonexistent. But... I really don't know."

$64,000 question: The raspberry-scented condom? A moralist might say that your mother "stole" your father from another woman; Dorothy Brady had a small child when the marriage failed. So: why did your mother have such a problem with jealousy? Guilt?

"The raspberry-scented condom? See my answer to your first question. As for that moralist, he might also say that my father was falling all over himself to be stolen. It does take two, you know. Jealousy? Some people are jealous. Some aren't. My mother was one of the jealous ones: just another aspect of her complexity. I don't think she suffered overly from guilt."

"That it?"

As I re-read this interview with Joan Brady, it makes for great reading, but the whole thing sounds false to me. It was not that I thought that Joan Brady was being consciously untruthful, but that the format of a written interview allows for no real give-and-take. It sounds removed from human contact, because Joan Brady refused to speak with me. She refused me her voice. I feel more comfortable talking to people in person, but the expense of traveling limits what I can do. In a telephone interview, I can hear the tone and quality of the human voice, which speak more than words.

Secret Life of Mildred Brady

I realized that, of course, this is not the first time Joan Brady has had to answer for her mother's role in the persecution of Wilhelm Reich. Her answers were tight and defensive. She wrote a number of outrageous statements, charges of quackery and fraud, without offering an opportunity to follow up immediately, and to see where it leads. Just like her mother.

How could Joan Brady maintain the illusion that her parents were just liberal folks, not Communists? Because her parents lied to her, like they did to everybody else. Joan chuckled to herself in discovering a few of her mother's lies, but she did not discover them all.

New Revelations from the Bradys' FBI files

On September 7th, 1941, Representative Martin Dies (D-Texas) wrote an open letter to President Roosevelt, demanding that Leon Henderson, head of the Office of Price Administration (OPA) and four members of his staff be removed from their positions on the grounds of past Communist affiliations. The staff members were Robert A. Brady, Mildred Edie Brady, Dewey Palmer (also associated, along with the Bradys, with Consumers Union), and Tom Tippett.

Congressman Dies was, as has been mentioned, acting on information given to him by Fred Schlink and J.B. Matthews of Consumers' Research, Inc. Having spent the last five years collecting data regarding the Bradys and other members of the Communist fraction of the strike, they had voluminous documentation, and the whole affair had caught the Federal Bureau of Investigation somewhat flat-footed.

In the summer of 1941, the FBI learned the first details of the brewing scandal in Congress, and opened an investigation of the Bradys. These files, released to the public for the first time for this book, reveal that J. Edgar Hoover took a personal interest in the investigation. The Director sent testy messages to FBI Field Offices in New York and San Francisco, demanding a timely delivery of the completed investigation. FBI agents interviewed Professor Brady's associates at UC Berkeley, and culled through the reports of "Confidential Informants" in the field.

Robert Brady's university colleagues described him to the FBI agents as decidedly "left-wing," though not a member of the Communist Party. Confidential informants from within Party circles in California said that it "was common knowledge" that Professor Brady was in close sympathy with the Party, though not a Party member. One informant "described Brady as one of the most capable men in the Communist Party work in the Bay Area and stated that he was greatly relied upon by the Party." This informant told the FBI that Brady belonged to a group of professionals affiliated with the Party, who, in view of their positions, "in order not to be labeled Communists, stayed more or less on the outside in their activities."

"He stated BRADY was not foolish enough to put his true name to any Communist petition."[43]

Professor Brady supported a number of groups—including the Harry Bridges Defense Committee, the American League for Peace and Democracy, Friends of the Chinese People, The Tom Mooney Defense Committee, the Simon J. Lubin Society and the Medical Bureau to Aid Spanish Democracy—that were alleged by the FBI to be either "subversive" organizations or "Communist Front groups." No one at the FBI had taken the trouble to read Dr. Brady's book, *The Spirit and Structure of German Fascism,* but agents were reliably informed, that "this book adheres to the Communist Party line with respect to the subject Fascism."

Some of the "confidential sources of information"—as the F.B.I terms the people it interviews—were professional colleagues who emphatically stated that Brady was definitely *not* a Communist. One, whose name was redacted in the FOIA release along with the names of all other informants, said he "felt BRADY had done nothing to warrant such a hearing [i.e., investigation by the Civil Service Commission] and 'it looks as if he is being framed.' [Deleted] stated that he wanted to make it clear that he did not like Brady personally and that he was not giving a prejudiced statement in saying that BRADY was never a member of the Communist Party, to his knowledge. He said BRADY was very strongly anti-Fascist and very much opposed to capitalism and big business. Informant stated that BRADY was very active in the Consumers Union but, at the same time, informant stated that during his contacts with BRADY, BRADY had never spoken anything un-American and informant felt that BRADY was 100% American with liberal views, which are possessed by many economics professors."[44]

These comments appear in the FBI's report on the investigation of Brady's ties to the Communists. Statements such as this, and many others like it, show a surprising professionalism on the part of the Bureau. It's true that J. Edgar Hoover created a national political police

Secret Life of Mildred Brady

with the FBI, and that this investigation was started on the basis of Congressional politics. Under law, the FBI was required to determine whether or not Robert Brady, a government employee, had ever been a member of the Party, and were unable to discover any proof of that. The negative material (and Brady's own writings were not the least of these) was given balance with positive reports, including a colleague who said, "to state that he is a Communist, is, in my opinion, pure 'bunk.' He is no more of a Communist than you or me. I would state my life on the fact that he is not a Communist. I would do the same in regard to his wife."[45]

Obviously, avowed Communist Party members were rare birds in the academic aviary of the 1940s. If Brady was a Party member, he must have played his hand very close to his chest with his professional colleagues. Some of those who gave statements to the FBI for its investigation were very clearly concerned that an innocent man might have been "tarred with the Communist brush."

On April 14, 1942, Brady himself made a statement to the FBI, under oath, regarding the charges. He emphatically denied that he had ever been a member of the Communist Party, nor any organization controlled by the Communist Party. As he explained to the FBI: "The last paragraph of the book was said to be an appeal to revolution because one word in it, 'force' was interpreted to mean force and violence."[46] Brady said he didn't believe in force and violence. "There are," Brady continued, "a number of those remarks which I now believe were unnecessarily loose in a sense that they did not make my meaning clear." He denied that he was a Communist, or that he ever worked with Communist-dominated organizations. "I never had anything to hide and I probably said what I had to say too openly."

However, there are some troubling indications in the FBI's report that Robert Brady was something more than a "broadly Marxist" professor with left-wing New Deal ideals. A month after making his statement to the FBI, after hiring an attorney to fight the Commission's recommendation of dismissal, Brady resigned his post, effective July 15, 1942. Why? Brady's FBI file, though full of suspicious associations, had no proof of Party membership. His university colleagues, for the most part, had supported him admirably, and the FBI's snitches in the Party could produce no documentary proof that Brady had ever been a Party member. Mildred had resigned immediately, but Robert Brady characteristically spoiled for a fight, and demanded a hearing. For the most part, Brady is defined in the FBI investigation as a leftwing blowhard, without portfolio.

The files document Professor Brady's close and continuing contact with his old student from Berkeley, Nathan Gregory Silvermaster. In 1942, the FBI did not understand the significance of this relationship.

Nathan Gregory Silvermaster was a Russian-born economist who served at the Farm Security Administration. Silvermaster had been a student of Robert A. Brady's in the early 30s, and Brady helped him write his thesis, titled "Lenin's Economic Thought Prior to the October Revolution." Silvermaster provided Brady's name as a reference in his application to the Farm Security Commission. When Brady took his post at the OPA., Silvermaster's wife, Helen, accompanied Brady while apartment-hunting in Washington, DC.[47]

Elizabeth Bentley identified Silvermaster as a Soviet espionage agent in 1948. He collected information, she said, from a red cell in Washington, DC, which included a top White House assistant, Lauchlin Currie, and Harry Dexter White, who was Assistant Secretary of the Treasury under Henry Morgenthau; White helped write the Bretton Woods Monetary Plan, which created the International Monetary Fund and the World Bank. Silvermaster collected information from a variety of pro-Communist New Dealers in a broad array of Federal agencies, most of whom were aware of the ultimate destination of the information. Edward Condon's association with Silvermaster became the object of Congressional scrutiny; he would later sign his name to a whitewash report on Unidentified Flying Objects, also known as the *Condon Report*:

"Miss Bentley named a score of lesser known persons as members of two rings for which she acted as courier during the war, and whatever doubts an observer might have been left with regarding specific incidents or the value of the information that had been collected and passed along to the Russians, he could hardly help concluding that US security precautions regarding Communists during the war years were less than stringent. The central figure in the major Washington cell was Nathan Gregory Silvermaster, an economist who had worked in various government agencies for a period of twelve years, from August 1935 to November 1947. His record was thick with evidence of Commu-

Secret Life of Mildred Brady

nist sympathies and activities—A Civil Service Commission report concluded that 'Nathan Gregory Silvermaster is now and has for years been a member and a leader of the Communist Party and very probably a secret agent of the OGPU'—and yet he was able not only to hold onto his jobs but to use his house as a photographic lab and transfer point for documents. Even granting, as the liberal press did not neglect to point out, that the Russians were our allies during this period and the CPUSA was a patriotic institution, it is difficult to conceive how Silvermaster and some of the other men named by Miss Bentley could have hung on as long as they did without the services of some good angel or an extraordinary lack of interest on the part of our security services. The Committee would naturally be inclined to put the more sinister interpretations of the facts, and the facts seemed to bear out some of their suspicions. As a result of the investigations set off by Miss Bentley's charges, eleven State Department employees were abruptly dismissed from their jobs."[48]

Silvermaster, in testimony before the Committee, objected to "this diabolical conspiracy" and called Elizabeth Bentley "a neurotic liar," but refused to answer questions under his Fifth Amendment rights against self-incrimination. Note well that Silvermaster called Bentley a "neurotic" liar, and not just an everyday liar. Alger Hiss also impugned the character of Whittaker Chambers, on the grounds of psychological instability.

The government, in fact, would soon have proof for Bentley's accusations against Silvermaster. Between 1947 and 1950, many cabled telegrams between Moscow and New York were decrypted by the National Security Agency's VENONA project.

The FBI placed a "mail cover"—i.e., a surveillance of incoming mail to Robert Brady, after his resignation and return to Berkeley. This surveillance lasted from "February 15, 1943, until March, 1943."

Robert Brady worked two full-time jobs (June 28, 1943-January 13, 1944), as a Professor at UC, and as a labor-management consultant with the War Manpower Commission for the Permanente Metals Corporation at the Richmond Shipyards. The second job caused the FBI to reopen the investigation of Robert Brady. This investigation continued from November through April, 1945.

After fifteen years as an Associate Professor of Economics, Brady finally secured a tenured position at UC Berkeley in October of 1944.

After two quiet years, in 1947, the FBI reviewed Brady's case and found that "no indication subject engaged in Communist Front activities since 10/44; no indication whatsoever of Communist Party activities. Informants indicate California Communists do not trust Brady to follow the Communist line in his lectures." The redacted, three final paragraphs of this report have been re-classified, on 2/9/98, under the NSA and CIA Acts. (Bufile# 100-3368-57)

On April 15, 1947, the Special Agent in Charge from San Francisco's FBI Field Office recommended that "In view of the fact that subject has never engaged in any Communist Party activities, and has not had any contact with Communist front organizations since October, 1944, it is recommended that his Security Index Card be cancelled." (Bufile #100-3368-58)

As late as 1950, FBI files show confusion over the date of Mildred Edie's marriage to Robert Brady. The FBI was unable to locate the marriage records themselves, but found that "Who's Who for 1948-1949 reflects that this marriage took place on April 14, 1936." (Bufile #100-3368-59)

On May 19, 1947, following the recommendation from the San Francisco Field Office, the FBI cancelled Robert A. Brady's Security Index Card.

However, the Brady case was reopened, not as a "Security Index" matter, but in connection with espionage, in 1953. By then, Robert Brady had made his suicide attempt, and had recovered for a brief time. It should be remembered that, by this time, the government had been investigating the charges of Miss Elizabeth Bentley, who confessed to acting as a courier for the spy ring run by Nathan Gregory Silvermaster, Brady's student at Berkeley. The government had decrypted numerous cables between Russia's embassy in New York and Moscow Centre, and these documents verified in some detail Bentley's charges. Also, this summary was crafted during the time when J. Robert Oppenheimer's security clearance was under review; it was revoked in June of 1954.

In a heavily censored FOIA release, a letter from J. Edgar Hoover to Kimbell Johnson, Chief of the US Civil Service Commission reveals that the government was taking another look at Brady's trips to Russia in 1931 and 1935. Several names are deleted, for reasons of privacy, in the FOIA release of this letter, and it is plain that the government was interested in a specific individual who accompanied Brady on one or both of these trips to Russia.

Next, a four-page internal FBI memorandum to agent William A. Branigan, dated 9/28/55, with a subject heading "MRS. ROBERT BRADY. SECURITY MATTER-C." Whatever the contents of this memorandum (Bufile# 100-3368-63), it is apparently too hot to be released, even today, to the general public. (This document is included in the appendix.) The entire contents of the memo are *classified* under "the National Security Acts of 1947 and the CIA Act of 1949" as of February 9th, 1998, and deemed not yet releasable under the Freedom of Information Act. The contents of this letter will be released perhaps in the year 2023.

As my official appeal of the censorship of this file has been denied, I can make only conjectures as to what might be in Mildred Edie Brady's FBI files, that would require this strict, ultra-secret classification today. Obviously, there is some information with an international espionage component, relating to the National Security Agency and the CIA, but beyond that I can only speculate.

The Brady files released to me conclude with Bufile # 100-3368-64, a "Correlation Summary" of some very specific information gathered by the FBI regarding Robert Brady. This report is dated January 16, 1956. The government's scrutiny is directed to Brady's mentor status with several of his Communist students back in the 1930s who were subsequently identified as agents for the Soviets. One of them was Nathan Gregory Silvermaster.

The 1956 report is of a very different character than the documents previously filed under "Brady" by the FBI. The earlier documents all pertained to the alleged "subversive" status of the Bradys, and that meant: the Bradys were publicly identified, and open Communist agitators in the 30s and 40s. This 1956 summary breaks new ground in that many of the references are to "Espionage—R[ussia]."

Here follows a list of the most notable features of the FBI's 1956 report on Robert Brady:

1. The FBI was aware that Brady made many trips to Europe (Germany, Russia, Switzerland, France) in 1930, 1931, 1934, and 1935.

2. References are made to several bright Communist students of Brady's, including William Henry Taylor, Nathan Gregory Silvermaster, and Norman Chandler Bursler. One or more of these students

accompanied Brady on his research trip to Russia under Brady's grant from the Carnegie Foundation to study bureaucracies.

3. Brady had encouraged Silvermaster (and other unnamed students) to apply for government service. Silvermaster listed Brady as a reference on his application to the Farm Security Administration.

4. Brady's association with J. Robert Oppenheimer, and his friend, Haakon Chevalier, are mentioned. On September 17, 1944, an FBI agent observed Robert and Mildred Brady driving around the Bay Area with Oppenheimer's "recruiter" Haakon Chevalier.

5. References are made to the FBI's surveillance, including "Tesur"—that is, "telephonic surveillance." The FBI wiretapped the phones of the Silvermaster residence in Washington, DC. (2/17/46).

6. The report reads: "On February 20, 1946, Bob, believed to be Robert A. Brady, contacted Helen Silvermaster relative to his arrival in town (Wash., DC) and his plans while in town. Bob further indicated that he would take his sabbatical year of 'life' in 1947 and 1948 and take his family to England."

7. Sometime between that tapped phone call, in February of 1946, and May of the same year, there was a reunion between Robert Brady and the Silvermasters in Washington, DC.

8. On August 26, 1946, a mutual friend of the Silvermasters and the Bradys called Helen Silvermaster and confirmed that Brady "still planned to go to London on his sabbatical leave."

9. "On May 7, 1947, Mildred Brady contacted Helen Silvermaster. Robert and Mildred Brady were living in Berkeley at that time, and intended to go to England possible [sic] in February, 1948."

Then, the May 26, 1947 issue of *The New Republic* was published. Mildred Edie Brady set into motion the prosecution of Wilhelm Reich by writing the article, "The Strange Case of Wilhelm Reich." She sent the clips to the FDA and demanded action. She maintained tight contact with the Silvermasters since the 1930s. She routinely flew coast to coast during the war, between New York and San Francisco, the two most active cities for Soviet espionage at the time. She worked under the cover of the Consumers Union, and she was an active Soviet courier, as best as I can tell.

The FBI knew all this at the time of the injunction, the contempt trial and during Reich's imprisonment. However, the FBI could never reveal their knowledge of Brady's espionage, since that might compromise

Secret Life of Mildred Brady

the ultra secret VENONA documents. It was official FBI policy to not acknowledge Reich's allegations of a Communist conspiracy against him.

Notes

1. *Conspiracy, An Emotional Chain Reaction, Item 389 p. 15.*
2. The Communist International, or Comintern, was an umbrella-group of national communist parties, through which Russia coordinated the communist movement as a whole.
3. Klehr, Haynes and Firsov, *The Secret World of American Communism.* Yale University Press, New Haven, 1995.
4. Reich, Wilhelm and Moise, William. *Red Thread of a Conspiracy. Wilhelm Reich Biographical Material; History of the Discovery of the Life Energy.* Documentary Supplement No. 2, A-XII-EP. Orgone Institute Press, Rangeley, 1955. Available from the Wilhelm Reich Museum Bookstore.
5. *Conspiracy, An Emotional Chain Reaction, Item 389 p. 16.*
6. Strangely, there was a Charles A. Wood on the board of the National Labor Relations Board council that arbitrated the dispute in the Consumers' Research strike of 1935.
7. Jerome Eden, Lois Wyvell, and Mary Boyd Higgins are notable exceptions as writers who have confronted the issue of conspiracy against orgonomy.
8. Consumers Union President Colston Warne's account in *Notable American Women, The Modern Period: A Biographical Dictionary,* Belknap Press/Harvard Univ., 1980.
9. *Consumers Report,* Oct. 1965.
10. The term, "Fifth Columnist," is a classic example of the CP jargon of the era, now incomprehensible to the contemporary reader. "Fifth Column" refers to the fascist encirclement of Madrid during the Spanish Civil War; four columns of troops surrounded the city to the North, South, East and West, while a "fifth column" of infiltrators and saboteurs operated within the city.
11. Michael Straight recalls knowing D.S. Gillmor while they served in the Army Air Corps.
12. Indeed, Kim Philby, already a veteran of double-agency for the Soviets, withdrew from contact with his control officers during these years because he realized that any information he passed on to them might well end up in the hands of the Nazis.

13. *La Marseillaise* (1937 - Jean Renoir), *Hiroshima, Mon Amour* (1959), *Last Year at Marienbad* (1961 - Alain Resnais), *Z* (1969 - Costa-Gavras), *State of Siege* (1972 - Costa-Gavras).

14. *Theory of War.* Knopf, NY, 1993, p. 126.

15. FBI file #100-3368-64, Robert Alexander Brady, Correlation Summary, 1/16/56, page 4.

16. Kuczynski, Jürgen, *Freunde Und Gute Bekannte.* Gespräche mit Thomas Grimm, Berlin: Schwarzkopf & Schwarzkopf, 1997.

17. See Joseph Albright and Marcia Kunstel's *Bombshell; The Secret Story of America's Unknown Atomic Spy Conspiracy.* Times, New York, 1997, p. 78.

18. Kuczynski, Jürgen, *Freunde Und Gute Bekannte.* Gespräche mit Thomas Grimm, Berlin: Schwarzkopf & Schwarzkopf, 1997, p. 214f.

19. Ibid., p. 270.

20. See *Conspiracy, and Emotional Chain reaction* (Statement by Angelica M. Haymes, 12/7/1948) for report of the Brady's attendance of this lecture. Dexter Masters, Mildred Edie Brady's lover, had told a Columbia University student, Matthew Lipman, that "Miss Brady had attended a Psychoanalytic Congress sometime in 1934, where, Miss Brady says, Dr. Reich made the soundest speeches, but the signs of psychosis were already in Dr. Reich's behavior; he was living in a tent on the hotel lawn, with a dancer, and wore a dagger." See *Reich Speaks of Freud,* p. 106 for Reich's discussion of this rumor. Masters also said that Consumers Union had obtained an orgone accumulator.

21. Reich had been expelled from the IPA a year before the Lucerne congress by a secret vote of the German Society, organized by Max Eitingon. Reich was not informed of his expulsion until he arrived at the Congress.

22. Letter from Robert Brady to Fred Schlink, Sept. 1935.

23. Letter From Robert A. Brady to Frederick J. Schlink, of Consumers' Research, Inc., September 21, 1935

24. Brady, Robert A. *The Spirit and Structure of German Fascism.* Viking Press, New York 1937, 1st ed. With a Foreword by Harold J. Laski. ["Every liberal, every progressive needs to know the things in this book." - *The Daily Worker*, 1938.] The foreword to Brady's book was written by the guiding light of the "Fabian School" of Socialism, Harold J. Laski, a professor of political science at the London School of Economics. In 1934, a seventeen year-old Michael Straight discussed his plans for the future with Professor Laski, who offered him entry into the London School of Economics to study despite the fact that Straight's open curriculum at Dartington Hall had not prepared him for the requisites of the University. Laski was a frequent contributor to *The New Republic* from the time of its endowment by Willard and Dorothy Straight, Michael's parents.

25. Brady, p. 133. In ironic concurrence with Emmanuel Josephson's populist analysis of the Rockefellers' role in the institution of the "company union."
26. Ibid. p. 156.
27. Ibid. p.400.
28. Chevalier, Haakon Maurice, *Oppenheimer: The Story of a Friendship.* Braziller, NY 1965. p. 24
29. Gregory Heifetz was an important Soviet agent. It is unknown whether he had any relation with Ottilie Reich Heifetz, the wife of Wilhelm Reich's brother, Robert.
30. "Personnel Security Board, Atomic Energy Commission: Hearings of J. Robert Oppenheimer." p 158.
31. "Conspiracy; An Emotional Chain Reaction," Item 27. Emphasis in the original.
32. "Conspiracy; An Emotional Chain Reaction," Item 45. Emphasis in the original.
33. Author's interview with Reich's daughter, Eva Reich, MD, in 1998, confirms this.
34. Item #178 in *Conspiracy, an Emotional Chain Reaction.*
35. *Reich Speaks of Freud.* Noonday Press, NY, 1968, p. 106.
36. "To All Active Supporters of Democracy and Peace—An Open Letter calling for greater unity of the anti-fascist forces and strengthening of the front against aggression through closer cooperation with the Soviet Union, released on August 14, by 400 leading Americans." -*Soviet Russia Today*, September, 1939. The Non-Aggression Pact was signed between Germany and the Soviet Union on August 31, 1939.
37. "Grade Label Plan is Denied by Brady," Food Field Reporter, July 7, 1941.
38. Brady, Joan. *The Unmaking of a Dancer,* Harper & Row: New York, 1982, p. 29. Joan Brady's novel is a good piece of writing, and I recommend it to anyone. Joan devotes important sections of her book to the home life of the Bradys.
39. Ibid., p. 24-26.
40. Brady, Joan. *Theory of War,* Knopf, NY, 1993, p. 142.
41. Brady, Joan. *The Unmaking of a Dancer,* Harper & Row, NY, 1982, p. 202.
42. Joan Brady's letter to the author, accompanying written answers to my interview questions. May, 6, 1997.
43. FBI File #100-10536, February 18-19, 1942, Los Angeles Field Office.
44. FBI File #100-3368-47, Internal Security-Hatch Act, Case of Robert A. Brady, May 4, 1942. p. 19.

45. FBI File #100-3368-47, Internal Security-Hatch Act, Case of Robert A. Brady, May 4, 1942. p. 20.

46. Statement of ROBERT ALEXANDER BRADY, made in presence of Special Agent Galen N. Willis, Federal Bureau of Investigation." April 14, 1942. FBI File 100-3368.

47. Ibid., p. 27.

48. Goodman, Walter. *The Committee.* Farrar, Straus & Giroux, NY, 1968, p. 259.

Chapter Six:
The Einstein Affair: 1941

When Wilhelm Reich first discovered the specifically biological energy he called orgone, he waited years before publishing anything about it. He knew the difficulties attached to such a thoroughgoing revision of textbook physics, and was loathe to commit himself until he had verified the phenomena under a variety of experimental protocols. One such experiment, called To-T, measured the heat in an orgone accumulator (To) compared to the ambient air temperature (T). An orgone accumulator, or ORAC, is a box constructed of alternating layers of metal and wool with a steel interior and a wooden exterior, and is generally warmer than the outside temperature. I have verified this phenomenon myself. Moreover, the temperature difference varies as a direct result of changes in the atmosphere. A change in the weather can increase or decrease the temperature differential. I found the experiment to work as an accurate predictor of changes in the weather, just as Reich reported.

Think about it: if you had an empty box, and left it in your garage, wouldn't you expect that volume of air to be the same as the room's air temperature, over time? If I could construct a box such that it remained warmer by even one degree for extended periods of time, wouldn't you want to discover the source of that heat? I've had many people scoff at me at the very possibility of this "free energy," or they have said that there must be some very simple explanation for this phenomena, but no one has offered me a adequate reason for why this is so.

Albert Einstein was one physicist who found the question intriguing enough to invite Reich to his home to demonstrate the effect. Reich had written him a cautious letter in the hopes that Einstein would recognize the experiment's results. Reich's ORAC violated the Second Law of Thermodynamics, the law of entropy, which anticipated only an equillibrium of temperature between the ORAC and its environment.

Wilhelm Reich and the Cold War

Albert Einstein had become the most famous scientist in the world after his Theory of General Relativity gained support from new astronomical observations in 1919. He had published a paper on Special Relativity in 1905. He became a full professor of physics in 1919 and received a Nobel Prize in physics in 1921.

Einstein's friend from Vienna was Friedrich Adler, the son of the founder of the Austrian Social Democratic Party, Viktor Adler. Friedrich Adler, also a physicist, shot and killed the bellicose Austrian Prime Minister Count Stürgkh in 1916. Friedrich Adler was sentenced to death for the assassination, but the sentence was commuted when Adler was ruled insane—in part, because of his academic criticisms of Einstein's theory of relativity.[1] He became a hero of the labor movement for eliminating the warmonger Count. Adler would subsequently become a supporter of the German resistance group, the Org, and is mentioned as a reference in Karl Frank's statement to the State Department.

Albert Einstein arrived in the US in 1933. The FBI suspected that Einstein's assistant, Helen Dukas, had been an asset of Soviet intelligence as early as 1929. During that time, Einstein was living in Berlin. "During this time Dr. Einstein allegedly had two secretaries in his office, both of whom were Communist sympathizers. One of these secretaries decoded the Soviet espionage messages and in turn passed them on to a courier."[2] This secretary, Helen Dukas, screened Einstein's mail in the US, since "She had a fairly free hand in deciding which letters he should read and answer, which might divert him, and which to put straight in the files—for the record—unseen by him."[3]

Throughout the Weimar years, Einstein had drifted steadily to the left, with clear sympathies with Russia and ambiguous statements in support of the purges. A recent book published in Germany, based on government files, demonstrates that Einstein was a naive dabbler in using his famous name in connection with dubious politics. "Blinded by illusions about the social conditions in the Soviet Union in 1931 Einstein showed understanding for the political trials in the Soviet Union."[4]

It is clear from the FBI files on Einstein and Dukas that they were under fairly close observation by FBI agents, especially after the secret 1939 letter signed by Einstein, and other physicists, urging President Roosevelt to develop an atomic bomb in view of the German capacity to do the same. Einstein would never play much of a role in the

The Einstein Affair: 1941

Manhattan Project, beyond serving as a figurehead for a group of emigre scientists lobbying for the construction of the bomb. In August, 1940, the Rockefeller Foundation gave the major funding support, over $1 million, to the University of California at Berkeley, for the construction of a particle accelerator.

Even if Einstein had endorsed the discovery of orgone, the collaboration may not have produced the results Reich had hoped for, which was simply financial support to continue the research. Wilhelm Reich first wrote to Albert Einstein on December 30th, 1940:

> "Several years ago, I discovered a specific, biological energy which in many ways behaves differently from anything that is known about electromagnetic energy. The matter is too complicated and sounds too improbable to be explained clearly in a brief letter. I can only indicate that I have evidence that this energy, which I have called orgone, exists not only in living organisms but also in the soil and in the atmosphere; *it is visible and can be concentrated and measured*, and I am using it with some success in research on cancer therapy."[5]

Einstein responded to Reich's letter promptly, on January, 6, 1941: "I am very willing to discuss your scientific concern with you, and propose that you come to see me some afternoon, having made an appointment in advance."[6]

Reich made an appointment with Helen Dukas, Einstein's secretary, for Monday, January 13th, 1941.

Reich knew nothing of Einstein's secret letter to Roosevelt, nor the hidden push to develop nuclear weapons, when he traveled to Princeton with several devices with which to demonstrate the orgone energy. That morning, Reich outlined his thinking about this exciting opportunity in his diary. He noted that he was "famous as a psychiatrist" but "doubted as a biologist."[7] And, "As a physicist, I am regarded as nothing. Therefore I am going to Einstein."

"Orgone constitutes the 'field' that Einstein is searching for. Electricity, magnetism, gravitation, etc., depend on its functions." Reich had both reason for optimism and a realistic attitude and was well prepared for Einstein's reaction, either way.

What other possibilities that might have run through Reich's mind as he made the drive from Long Island to New Jersey, have not been recorded. Upon his return from Princeton late that Monday night, Reich described in his diary his long, four-hour discussion at Einstein's home as a true meeting of minds. "The first genuine and fruitful scientific conversation in ten years!"[8] Einstein observed the visual phenomena of the scintillating light in the orgonoscope, but could not rule out "the subjective element."[9] The two men discussed Reich's "trigonometrical diagram of the spinning wave," and finally Reich mentioned, hesitatingly, the apparent creation of heat inside orgone accumulator. Einstein said, *"That is impossible. Should this be true, it would be a great bomb!"*[10] They discussed the implications of the experiment, and Einstein asked Reich to return to Princeton, bringing an orgone accumulator set up for the experiment. According to Reich, Einstein told him that if *"the fact were true, he would support my discovery."*

Einstein met once again with Reich on February 1, 1941, when Reich delivered the apparatus necessary for conducting the To-T experiment to verify the temperature difference. Einstein's biographers have attempted to explain the meetings by pointing to Einstein's simple goodheartedness, well-known naiveté, or lack of good judgment. That's untrue; Einstein was intrigued. "Otherwise," Reich wrote to his friend, A. S. Neill, "he would not have listened for nearly five hours."[11] It's impossible to explain his second invitation to Reich, that he bring his instruments back to Princeton, where he and Einstein could verify the temperature difference, other than the obvious conclusion that there was a serious interest in the matter on the part of Einstein. That evening, Reich wrote in his diary,

"In the event that everything goes well with Einstein: a) I will propose elaborating my theory of biological energy with his help—i.e., I will become his pupil."[12]

Einstein's immediate reaction to his first meeting with Wilhelm Reich, if he recorded it, is unknown to me. My requests to Einstein's Archive went unanswered, and in fact, this was the only archive that did not respond at all to my inquiries during the entire period of my research.

We know that Einstein took some time to confirm the positive temperature difference. After studying the orgone accumulator for a week, Einstein wrote to Reich on February 7th: "To start, I took a sufficient number of readings without any changes to your apparatus set-up. The box-thermometer regularly showed a temperature of about 0.3-0.4 degrees [centigrade] higher than the one suspended in the air."[13]

The Einstein Affair: 1941

This is important to remember: whatever the validity of Reich's conclusions, the phenomena he observed were real and corroborated by Einstein. The temperature effects of the orgone accumulator have been confirmed, and reconfirmed, over the six decades since its discovery.

Einstein himself was unable to account for the temperature difference himself, but, *deus ex machina*:

"One of my assistants now drew my attention to the fact that in rooms, of the usual sort, the temperature at the floor is always lower than the one at the ceiling. Now, if one deals with a fixed horizontal plate, the underside of this plate is in closer touch through convection with the floor and the top side with the ceiling. In fact, I found that on the underside of the table in the middle of the room the temperature averaged about 0.6 degrees lower than the top side. This factor turned out to be the decisive one, as is clear from the following:

"I first took the metal box out of the enclosure with the windows," Einstein wrote, describing how he did what most people do when confronted with the warmth, by taking the accumulator apart, "put it directly on the table, and placed the thermometer on top of it, reducing the size of the rather wide opening which the earthenware pipe had occupied with a folded page from a newspaper.

"Results (after many trials): the temperature difference disappeared or in any case was no more than ± 0.1 degrees. This seems to lead to a striking conclusion: if you leave off the windows, the effect disappears! But I was not satisfied with this, for the metal was now in better contact with the table top, which was cooled down from below.

"To eliminate this factor, I now proceeded to put the metal box (again without its windows) not directly on the table but instead atop a padding made of one of your sheets of blue paper that I had folded together in a suitable manner. Thereupon, I obtained the original effect immediately and repeatedly. The temperature of the thermometer on the box was 0.3 and 0.4 degrees higher than that on the thermometer suspended in the air.

"After these experiments, I regard the matter as being completely explained. The temperature difference has nothing to do with the windows and the metal box, but is solely due to the horizontal tabletop.

"I hope that this will awaken your sense of skepticism, so that you will not allow yourself to be deceived by an illusion that can be easily explained. Please have someone pick up your instruments, since they are of some value. They are undamaged.

"With friendly greetings,
[signed] A. Einstein"[14]

An assistant offered an explanation that the effect was due to "convection" between the air above and below the table upon which the accumulator had been placed.

Reich responded with a long letter that described the protocols under which he answered this objection. He buried the accumulator underground, removing the possibility of such convection. The temperature differential increased under these conditions. Reich also wondered why Einstein seemed so ready to give up after the first plausible explanation, without following it through logically. He concluded the letter with an agonizing and heart-rending appeal for respect and consideration. Reich did not receive it. Einstein's office discontinued the correspondence. It is unlikely that Einstein, in fact, ever saw Reich's letter regarding the assistant's objection, since his secretary, Helen Dukas, filtered his mail at the gate.

Reich's letter to Einstein, in response to the "convection" idea, is the most eloquent example of scientific dialog I have ever read. Indeed, Reich's description of the issues involved, the experimental protocols to test the objection, and the design of new experiments to shed light on the question, so inspired me when I first read *The Einstein Affair* (available as a complete document from the Wilhelm Reich Museum) that I set up the experiment at my home. I confirmed the basic fact, for myself, and have sought a rational explanation that fit into established physical science, without success. Like so many of Reich's discoveries, this has been completely ignored, but never experimentally refuted.

High-technology has offered new ways to confirm the temperature difference and in 1996, British television producer Jon East brought his orgone accumulator to a facility with an infrared thermal imaging camera.[15] Kenn Thomas, editor of *Steamshovel Press* magazine, who was coincidentally in London for a conference, accompanied Jon East to the demonstration and videotaped the proceedings. These cameras can register small temperature differences and express them onscreen

The Einstein Affair: 1941

as gradients of color corresponding to precise temperatures, so that the warmer surfaces register as red, while the cooler temperatures show up as blue, along a spectrum. Among those present was a medical doctor in general practice, an infrared camera design engineer, and two consultant physicians from Kings College Hospital in London.

The results were dramatic and immediately apparent visually on the thermal imaging monitor. Within minutes, the ORAC warmed up beyond the ambient temperature of the room. The engineer and physicians present at the demonstration were able to immediately rule out convection as the explanation because the thermal imaging camera simultaneously registers surface temperatures on the screen, in real time. "Convection currents were discussed and ruled out as being responsible for the phenomena because the other surfaces at the same height immediately next to the ORAC maintained exactly the ambient room temperature on all sides."[16] Like Einstein had done in 1941, the attendant engineer wanted to take the ORAC apart. The British observers suspected they might locate hidden batteries!

In February of 1944, when rumors circulated that Einstein had tested the orgone accumulator and found it worthless, Reich's translator and colleague, Theodore Wolfe, wrote to Einstein and informed him that to set the record straight, he would publishing the full correspondence including Einstein's statement that the temperature differential had been clearly observed. The rumors about Einstein's debunking of orgone energy were conveyed by Edith Jacobson to a physician who knew Reich. (It's a curious thing to me that this rumor came from the direction of the German underground group, the Org, as Jacobson had been a key member. Einstein's relationship with his friend Friedrich Adler was another point of contact with this group.)

Einstein responded to Wolfe's letter, and told him haughtily that he could not lend his name to Reich for "advertising purposes." Reich responded heatedly, informing Einstein that no such advertising had occurred and that, to the contrary, Reich's enemies in the psychoanalytic community were using Einstein's name to impugn Reich's work. Einstein replied that he had not "circulated unfavorable opinions about your endeavors." Reich was satisfied by Einstein's conciliatory statement, his assurance that he had not been talking loose while failing to answer Reich's replies to his objections, and above all, that Einstein had not rejected "my specific claim that he [Einstein] has confirmed the phenomenon of the temperature difference [...].[17] For the time being, Reich was satisfied with Einstein's reply, and did not immediately publish their correspondence. He later did so in 1953 when legal and political pressures mounted against him. Had he not done so, it would be very difficult to document the entire exchange without this text, and the whole interaction would probably be described by skeptics as a fantasy on Reich's part.

Einstein's "Assistant" Leopold Infeld

Einstein routinely referred to his associates, even his most talented colleagues, as "assistants." One such "assistant" was Leopold Infeld, the coauthor with Einstein of *The Evolution of Physics*. Reich eventually came to suspect that Infeld may have squelched the Reich-Einstein relationship in the bud.

Around the time that Reich was demonstrating the orgone accumulator at Princeton to Einstein, Leopold Infeld wrote the following appreciation of his colleague Einstein in 1941:

"IN AMERICA I saw for the first time in my life Negro dances and plays which were full of fire and vital force. The Savoy dance hall in Harlem changes into an African jungle with burning sun and richly growing vegetation. The air is full of vibration. Vital force emanates from the loud music and the passionate dancing until the whole atmosphere becomes unreal. In contrast the white people look half alive, ridiculous and humiliated. They help to form the background against which the primitive, unbounded vitality of the Negroes shines more brightly. One feels that any pause, any interval is unnecessary, that this intensive motion could go on forever.

I often had this picture in mind while watching Einstein work. There is a most vital mechanism which constantly turns his brain. It is the sublimated vital force. Sometimes it is even painful to watch. Einstein may speak about politics, listen kindly to requests and answer questions properly, but one feels behind this external activity the calm, watchful contemplation of scientific problems, that the mechanism of his brain works without interruption. It is a constant motion which nothing can stop. Other scientists have a switch which allows them to turn off or at least to decelerate the mechanism by a detective story, exciting parties, sex or a movie.

The Einstein Affair: 1941

There is no such switch in Einstein's brain. The mechanism is never turned off."[18]

Nearly a decade later, Infeld returned to Poland, where he had been born in Warsaw. He supported the Soviet puppet-state there. Reich noted the following news article many years after his disappointment with Einstein. In the 40s, Reich had never ascribed Einstein's withdrawal from discussions to any covert tactics. But news of Infeld's defection to the Communist camp, in the later context of the 1950s, caused Reich to speculate that Infeld had been the "assistant" mentioned by Einstein in his objections to Reich's experiment.

"From *The New York Times*, March 17, 1950:

SCIENTIST'S LEAVE STUDIES
Einstein Ex-Associate Seeks to Teach Again in Poland
    Ottawa, March 16 — Opposition leader George Drew in the House of Commons this afternoon raised the question of the propriety of permitting Dr. Leopold Infeld, a former associate of Dr. Albert Einstein and at present a teacher of mathematics at the University of Toronto, to return to Warsaw to organize certain educational programs in cooperation with the Communist Government of Poland.
    Mr. Drew said that Mr. Infeld, who had been given hospitality as a refugee in Canada and the United States, had gained considerable knowledge of the latest discoveries in the atomic field.
    He had several times stated that he would return to Europe if and when a "progressive government was established in Poland," and he returned there last summer and taught in the Universities of Warsaw and Kracow, Mr. Drew said. Now, Mr. Drew added, he had applied for a sabbatical year during which he would receive half pay from Toronto University in order to permit him to work at educational organization under the Polish Government."

Why did Infeld leave Canada and return to Poland? He had always been pro-Soviet. In his posthumously published autobiography, Infeld wrote, "Basically, the Polish emigres in Canada divided into two groups. The first was composed of progressives—communists and their sympathizers or non-party people who understood that the future of Poland was necessarily bound up with the Soviet Union. The second group, much larger, was made up of violent opponents of the Soviet Union."[19] In Poland, Infeld was given an entire research institute, and worked for the Soviet atomic munitions program, which was then administered by security chief Lavrenty Beria. He had cause to briefly regret his decision to return to Poland, but soon set aside his misgivings.

> "My eyes were opened to what was happening only when, in the last months of Stalin's life, it was reported that over a dozen Jewish doctors had been jailed in the Soviet Union. They were supposed to have been bribed by pounds sterling and English razor blades to kill Stalin and other public officials. It was a shocking accusation. Then, for the first and only time in my life, I regretted returning to Poland. But soon the Stalin era ended—Stalin died. I was afraid it might be still worse, but the atmosphere slowly began to improve, and, since then, I have never regretted my return."[20]

All of this taken into consideration, there is not enough evidence to support Reich's suspicion, in 1950, that Infeld had been the "assistant" who offered the bogus objection that derailed the prospects of Reich working with Einstein back in 1941. I could not establish that Infeld was in fact this "assistant."

Einstein's FBI case file was closed in 1955. "Additional investigation is not warranted in view of the long lapse of time since Einstein's office was allegedly used by the Soviets, the lack of corroborating information, and the fact that personnel involved are scattered in many countries and in many cases deceased. Therefor, both the Dukas case and the Einstein case are being closed."[21]

In *The Einstein Affair,* Reich was more critical of Einstein than Infeld, for his failure to follow through the scientific dialogue. Reich's final note in *The Einstein Affair* puts it in perspective:

> "Einstein succeeded in fascinating the first half of the twentieth century just because he had emptied space. Emptying space, reducing the whole universe to a static nothing, was the only theory that could satisfy the desert-like character structure of man of this age. Empty, immobile space and a desert character

structure fit well together. It was a last attempt on the part of armored men to withstand and withhold knowledge of a universe full of life energy, pulsating in many rhythms, always in a state of development and change; in one word, functional and not mechanistic, mystical or relativistic. It was the last barrier, in scientific terms, to the final break-down of the human armoring."

– Wilhelm Reich, *The Einstein Affair*, document E-36, December 22, 1952.

Notes

1. Brian, Denis. *Einstein; A Life.* John Wiley & Sons, 1996, p. 20.
2. *Einstein; A Life.* p. 419.
3. *Einstein; A Life.* p. 378.
4. Grundmann, Siegfried. *Einsteins Akte. Einsteins Jahre in Deutschland aus der Sicht der deutschen Politik*, Berlin: Springer-Verlag, 1998, pp. 330-1.
5. *American Odyssey*, p. 46.
6. *The Einstein Affair, Wilhelm Reich Biographical Material; History of the Discovery of the Life Energy, (American Period, 1939-1952) Documentary Volume A—XI—E;* Orgone Institute Press, Rangeley, 1953. Translation from the original German by Patrick Diehl.
7. *American Odyssey*, p. 54.
8. *American Odyssey*, p. 55.
9. *The Einstein Affair*, letter from Albert Einstein to Wilhelm Reich, February 7th, 1941. Translation from the original German by Patrick Diehl.
10. *American Odyssey*, p. 199.
11. *American Odyssey*, p. 198.
12. *American Odyssey*, p. 60.
13. *The Einstein Affair*, letter from Albert Einstein to Wilhelm Reich, February 7th, 1941. Translation from the original German by Patrick Diehl.
14. *The Einstein Affair*, letter from Albert Einstein to Wilhelm Reich, February 7th, 1941. Translation from the original German by Patrick Diehl.
15. East, Jon. *Infrared Thermal Imaging of the Orgone Energy Accumulator's To-T Effect.* Privately published, London, 1996.
16. Ibid.
17. *American Odyssey* p. 227.
18. Infeld, Leopold, *Quest; An Autobiography*, New York: Chelsea, 1941, p. 271.

19. Infeld, Leopold. *Why I Left Canada: Reflections on Science and Politics.* trans. Helen Infeld. McGill-Queen's University Press, Montreal 1978, p. 32.
20. Ibid, p. 84.
21. *Einstein; A Life.* p. 422.

Chapter Seven: Reich's Final Years

"Have we the nerve and the will? Can we carry through in an age when we will witness not only new breakthroughs in weapons of destruction, but also a race for mastery of the sky and the rain, the ocean and the tides, the far side of space, and the inside of men's minds?"
— John F. Kennedy, 1959 "New Frontiers" campaign speech.

Given the array of forces operating against Wilhelm Reich in America throughout the forties and fifties, it's amazing that he was able to conduct any positive work at all. In this chapter I'll describe some of the experimental work published by Reich at the time of this severe harassment. Beginning with the Oranur Experiment, much of Reich's research during the last years of his life bore serious implications for the new national security state. Having become more comfortable with the earlier discoveries of the bions, orgone accumulator, and cancer research, Reich pursued the root that unified all cosmic functioning, a course he had steered relentlessly since 1919.

Published after his death, the compendium of Reich's articles in the book *Selected Writings* (1960) offers an excellent overview of the development of orgonomy. For readers seeking more information about the scientific aspects of Reich's ideas, *Selected Writings* provides a start. Touching on disciplines ranging from psychoanalysis, political psychology, sociology, mind-body continuum, anatomy, biology, biogenesis, physics, astronomy, weather modification, and world religions, Reich's writings on orgonomy recall the works of earlier natural philosophers before the development of sharp specializations in academic fields.

Holding it all together was the *common function principle,* Reich's "red thread" that motivated all of his research. He established a systematically new way of thinking he called *orgonomic functionalism*, a research tool rather than a philosophy. Reich introduced energetic concepts of growth, movement and yearning into the field of theoretical physics. He believed that physics should follow biology, medicine and psychiatry, rather than the reverse.

Did Reich make mistakes, errors, and misjudgments during his later years? Yes he did. Had he lived, would he have revised them according to new findings? Of course he would have, just as he revised his earlier misjudgments about the social and political movements. This perhaps is the greatest tragedy of his untimely death; we'll never know how some of these intriguing concepts may have played out. Some of the most important developments of establishment science were the product of the majority taking extraordinary measures to answer minority views and critiques.[1]

We may never learn about some of Reich's discoveries. In the summer of 1948, for example, Reich demonstrated an orgone energy motor to many of his colleagues. Myron Sharaf could not remember exactly how it worked, but "I do recall that it involved the use of an orgone accumulator to a wheel; concentrated orgone energy was triggered by a small amount of electricity, an amount insufficient to rotate the wheel without the accumulator. I also recall that when the wheel was rotated entirely by electricity, it had a steady grinding motion. When powered by the combination of orgonotic and electrical energy, it ran smoothly and quietly; but its speed varied depending on the weather—more rapidly on dry, clear days, more slowly when the humidity was high."[2] Reich never published the design for the motor, but Eva Reich also witnessed it. Sharaf's biography discusses the orgone energy motor mainly in the context of his college friend, William Washington, whom he had introduced to Reich, and who subsequently would work with Reich on the mathematics associated with the motor principle. When Washington disappeared with the orgone energy motor in 1949, Reich worried that this might have been an indication of foul play, or red fascist thievery. Perhaps Reich read more into the situation than was warranted; Washington, who as it turned out had lied about his academic record, might simply have been a psychopath. But I found it galling to read Sharaf's summation that the whole episode "provides a nice example of the paranoid aspect of Reich's psychic functioning" – when in reality Sharaf bore much of the responsibility for the situation, since he brought Washington in.

Reich's Final Years

What follows is a discussion of a few of Reich's experiments that touched directly on tricky problems such as radiation poisoning, weather modification, and UFOs. Again, placing Reich's work in its true historical context, we'll see that Reich's views were either shared by others at the time, or have been vindicated since, no matter how "bizarre" they seemed then and now. I'll examine the evidence that suggests that not only were US government officials interested in Reich's activities, but there is the likelihood that his weather modification technique was absorbed into the "black projects" of the military.

Reich's last published book, *Contact With Space,* has not been widely distributed and it's worth reviewing here. Written under the pressure of the FDA's legal onslaught against Reich, his report reads like dispatches from a battlefront. He was convinced that the Earth was under an assault by UFOs, and their operators had somehow solved the anti-gravity puzzle. Reich sensed that UFOs drew their power from the Earth's rapidly depleting reserves of orgone energy and that at least some of the pollution associated with industrialization was in fact a result of the "exhaust" of UFOs. His observations of UFOs seem positively tame to some of the stories current today. Yet Reich's work has been completely avoided by ufologists.

I've seen things in the sky I couldn't identify, even things that looked like classical "flying saucers." Yet my inability to identify such objects never stirred me to report them to the authorities. I'm nearsighted.

It is difficult to imagine that the universe is empty of life everywhere save for Earth. Early in his career, noted astronomer/celebrity skeptic Carl Sagan of Cornell University once calculated that life on other planets is a statistical probability. The frightening, sheer terror associated with some of the alien contactee/abductee stories goes a long way, in my view, to explain and understand the intense level of ridicule and derision directed at serious UFO researchers. Yet, it's true and only common sense, that if the Earth has been repeatedly visited by beings of vastly superior technology, most people, myself included, wonder what anyone can do about it.

Everybody talks about the weather, but Wilhelm Reich did something about not only the weather, but did something about the UFOs when he disabled them with his cloudbuster. Perhaps we should take another look at Reich's work in this area.

The Oranur Effect: 1951

> "The USA faced a dangerous situation in the first days of December, 1950, when the disaster in Korea had struck with the evil attack of the Chinese communists; with the hands of the USA bound by a pledge not to bomb their hinterland in Manchuria; with the English allies still doing business with the red dictators; with the helplessness in the face of the red fascists who were far superior in the use of all of the most refined methods of the emotional plague, and with the terrible experience of the Chinese aggressors making propaganda through the UN right in the middle of the USA, while their forces marched in Korea. The USA was left holding the bag." —Wilhelm Reich[3]

It's true that the Korean War presented a very dangerous situation not only for the participants, but for the world, which was still adjusting itself to the possibility of an all-out nuclear war. Because of his work with orgone accumulators and cancer diseases, and because radiation sickness bore a direct relationship to cancer, Reich felt it was urgently necessary to investigate the possibility that orgone energy might be of some use not only to the US military effort, but to protect civilian populations against radiation poisoning due to the fallout from a nuclear exchange. (Nuclear war was a very real possibility, held in check by the restraint of American leadership. Of course, the USSR could not at that time launch enough missiles to prevent their own annihilation; but some American leaders wanted to push the button while they could. See Documentary Appendix for an intimate discussion between John J. McCloy and Dwight Eisenhower about Soviet-American relations during this time.)

There was little understanding of the biological effects of nuclear radiation in the 1950s. Many people remember that soldiers were sent directly into test sites shortly after the dust cleared from nuclear explosions. They wore no protection and were merely dusted off afterwards. The US Department of Energy (DOE) has recently released many documents about the true nature of America's nuclear heritage. Energy Department Secretary Hazel O'Leary offered full disclosure about the years of chronic abuse of an unwitting population of human guinea pigs by the scientific establishment and the military. There were numerous reports of scientists illegally injecting human beings with plutonium, without the subjects informed consent. Throughout the forties and fifties the military dropped radioactive dust over vast areas of the Western States. To put this into perspective, the military essentially turned each citizen into a huge cohort of experimental subjects in an on-going test of the biological effects of radiation poisoning. President Clinton distanced himself from DOE Secretary Hazel O'Leary, who authorized the release of new information, characterizing her forthcoming posture as "very emotional."

There is little public understanding today of the true nature of Reich's research. However, the military, as of 1948, was fully advised of his findings, as was the AEC.

Reich had first made contact with the Atomic Energy Commission on April 30, 1948, to discuss unusually high Geiger counter readings in connection with his research. It would still be another three years before Reich embarked on the Oranur Project, a controlled experiment dealing directly with the biological effects of radiation poisoning. In between, Reich kept the AEC completely informed of his research via meetings, letters and phone calls, as he grew closer and closer to an essential national security issue (i.e., keeping the public in the dark about the real danger associated with radioactivity), while simply trying to figure out why he was getting such unusually high readings on his Geiger counters.

In 1950, Reich asked himself whether it was possible to neutralize the effects of nuclear radiation by direct application of an orgone energy accumulator.

Simply put, Reich found that there is *no shielding possible* against the biological effects of nuclear radiation. On January 5, 1951, Reich placed a minute sample of radium (1 milligram in the form of a needle) inside a powerful orgone accumulator. Reich's shocking report, which can be found in the book *Selected Writings*, details how radiation sickness is a function of the organism's response to the invasive insult, and not a direct result of the radiation poisoning itself. Thus different people may be more susceptible to very minute doses, while others may feel no noticeable effects; each person's reaction is different according to their own emotional and biological structure.

In re-reading his original documentation, I was impressed with Reich's ability to distinguish between observed facts, corroborated by others, new theories drawn from and supported by these facts, and finally speculation based upon insufficient evidence. Of utmost importance in his research method was an awareness of the attitude of the observer, basic trust in one's own perceptions and observations. Although he alerted them when his Geiger counters told him something was amiss, he did not trust the response offered by the AEC: don't worry, everything is fine. In this as in other matters he seems to have been virtually alone.

Reich's intent for the Oranur Experiment was to investigate the treatment of radiation sickness with the orgone accumulator. He had some success in treating terminal "lost-cause" cancer patients with the medical device (although he never claimed having found a cure.) Since it was well known that radiation sickness could lead to leukemia, Reich planned to investigate the matter at his laboratory in Rangeley, Maine.

The animal testing began on January 5, 1951: the plan was to "irradiate some of our mice with radium." Originally Reich planned to produce radiation sickness in the mice by injecting radioactive liquid. These mice would then be treated with the orgone accumulator. Instead, Reich attempted to produce the radiation sickness by exposing the mice directly to radium needles.

He decided to test the effect of orgone-charged radium on lab mice in comparison to untreated radium. In preparation for this, a sample of the radium was placed in an extremely powerful 20-layer accumulator for five hours, and after that for one hour per day for the following week. Soon Reich subjectively noticed a change in the atmosphere, which he described as heavy and oppressive. On February 11, forty of the mice were found dead. A week later, Eva Reich suffered *anorgonotic shock*, and nearly died. Finally, on April 12, Reich finally realized that the source of the oppressive atmosphere was the radium needles that had been kept in a steel safe in another building. This in itself was a kind of orgone accumulator and when Reich removed the radium needles from the safe, he found that the Geiger counter readings returned to normal.

From the very start of the actual experiment the research shifted to a new focus: the exposure of one of the two radium needles to highly concentrated orgone. Reich expected that orgone energy would calm down radiation, but he found that the very opposite occurred: ionizing radiation drove orgone energy into a frenzy. Oranur was planned as a biomedical experiment, but circumstances shifted it to the realm of physics.

Reich would subsequently state that "it is the organismic OR [orgone] energy within living bodies which continues to react to the NR [nuclear radiation] material for months and even years."[4] In checking background counts around the safe in which the orgone-charged radium had been disposed, he discovered a more disturbing phenomena: that the steel and concrete enclosure itself comprised an orgone accumulator and that the Oranur effect was still evident.

It's difficult to discount Reich's documentation of all this, given the complete records and corroboration of his coworkers. In addition to the subjective impressions of a persistent, overcharged atmosphere that continued long after the experiment had been concluded, there were objective indications such as the aberrant fluctuations on the Geiger counters. *The New York Times* (2/3/51) reported that there were unusually high background radiation levels recorded from Rochester, New York to Canada during the last week of January.

There was a bright side. Both the experimental mice that had survived and the workers involved with the project exhibited full recovery and more: they were far better able to withstand the effects of radiation in subsequent work. Reich had found what he was looking for: a method of *immunizing against the effects of radiation.*

We know from Reich's documentary evidence volume, *Conspiracy: an Emotional Chain Reaction,* that Mrs. Eleanor Roosevelt wrote to physicist J. Robert Oppenheimer to ask about the Oranur Experiment. (This was two years before Oppenheimer lost his security clearance and he was still closely associated with the Atomic Energy Commission.) Oppenheimer replied to Mrs. Roosevelt on January 15, 1951: "I am afraid that the evidence of this paper makes me suspect that the undertaking is a hoax; I have been unable to find out anything about it that is reassuring." Later, when Reich had been apprised of Oppenheimer's statement, and when Oppenheimer was called in to discuss his contacts with the CPUSA (including Mildred and Robert Brady) and stripped of his clearances, Reich understandably suspected that Oppenheimer had obstructed interest in orgonomy while at the AEC.[5] I do not know how Mrs. Roosevelt had learned about the Oranur Experiment so quickly, but apparently one of Reich's circle of friends in New York had forwarded the previously published plan of the Oranur experiment to her, which she forwarded to Oppenheimer. Oppenheimer could not evaluate the report on the Oranur Experiment in matter of hours, and he could not believe the evidence of unusually high background radiation produced under the circumstances. So he called it a

Wilhelm Reich and the Cold War

"hoax."

The persistent, oppressive conditions lasted for months in the wake of the experiment, and the region was plagued with black clouds of deadly orgone (DOR). Reich was desperate to clear the atmosphere at Orgonon, willing to try anything. It was another logical accident that provided Reich with a new tool called the "cloudbuster. From the necessity of the moment it evolved into "cosmic orgone engineering" (CORE), and Reich would soon publish results of his weather modification experiments with the cloudbuster. These results showed as much promise as anything else the government had tried, and certainly the government was in a mood to try anything. This power to create atmospheric conditions at will is of utmost strategic military significance and economic importance. (In 1999, a weather futures market opened at the Chicago Mercantile, where investors and corporations can hedge on mean seasonal temperatures and rainfall.)

In 1952, Reich had constructed the first full-size version of the cloudbuster, essentially a rack of long metal tubes, up to twenty feet in length, grounded into running water by means of flexible metal conduits. It was mounted on a pivot-and-gear platform to allow the tubes to be pointed in different directions and angles. The cloudbuster draws off stagnant orgone energy and can be used by the operator to selectively increase or decrease the energetic potential in the atmosphere.

Reich first noticed the effect that hollow metal tubes had on the waves on the surface of Mooselookmeguntic Lake in Maine in 1940. He had been using a pipe to observe the movement of atmospheric orgone over the water, and that "casual pointing of some pipes at the surface of the lake seemed to affect the movement of the waves."[6]

Tom Ross, the long-time caretaker at Reich's estate, "Orgonon," in Rangeley Maine, recalled designing the first working cloudbuster. "Reich told me what he wanted, and I was to figure out the best way to do it. I said, 'you know, that reminds me of anti-aircraft artillery from World War I.' And Reich said, 'that's right, on the Italian front we noticed that when we ran out of ammunition, and left the guns pointing toward the sky, the clouds would disperse and the sun would shine. '"[7]

Reich took the prototype cloudbuster from Ross and tossed the cables into a well and pointed the half-dozen or so pipes at the dark clouds over Orgonon, and they dissipated, according to everyone who was there.

Reich's Final Years

Dorothy Kilgallen on Cloudbusting

Dorothy Kilgallen was the well-known celebrity journalist who appeared for many years on the TV show "To Tell the Truth." (She died on November 9, 1965, of an apparent overdose of barbiturates, after writing investigative articles about the Kennedy assassination. That same evening, the power grid went down over large portions of the eastern seaboard, in conjunction with UFO sightings around New York. She was the last person to interview Jack Ruby, Lee Oswald's killer.)

In the early fifties, Kilgallen and her husband, Richard "Dick" Kollmar, had a morning radio talk show in New York City. On November 9, 1953, Reich cloudbusting was the subject of derisive banter on the "Breakfast with Dorothy and Dick" program on station WOR.

Josephine T. Brenner, a patient of a medical orgonomist, Dr. Duvall, caught the show and later wrote the gist of the conversation down and sent it to him:

> "Dick: Energy is drawn from the vast reaches of outer space slowly dissolving the bands of unbearable muscular tension, then the energy departs into outer nothingness, leaving wonderfully soothing relaxation and freedom from nervous tension!
> Dorothy: What *is* that Yogi?
> Dick: No, *this* is *Orgonomy*!
> Dorothy: Orgonomy! -Where did you get this?
> Dick: From a bass player at a party! (much laughter)
> Dorothy: No! Tell me more!
> Dick: Yes, Orgonomy as expounded by Dr. Wulfgang von Reich.. Dr. Reich it seems is the only living student (or exponent) of Dr. Freud (more laughter) - But ah! Dr. Reich has gone further than Dr. Freud - in fact it seems Dr. Reich has gone *way* beyond *himself*, Dr. Reich can use his energy to bring forth rain from the heavens."[8]

According to Kilgallen's biographer, Lee Israel, Kilgallen and her husband (who had done the wildly popular radio show since 1945) were both heavy drug and liquor abusers, and they often had been up all night careening around New York's "cafe society" before their "Breakfast" show, broadcast from their living room at 8:15 in the morning.[9] Dick kept a tank of oxygen at home to refresh himself just before the show, after a long night of smoking and drinking left him breathless. Their exchange about Reich, quoted above, could serve as another example of the way his work made sick people uncomfortable; yet the interesting thing is that this exchange occurred just a few months after Reich invented the cloudbuster, so word must have traveled quickly for the bass players of New York to know about it. It was all fun and games, but years afterward, in 1965 prior to her death, Kilgallen openly asked then-President Johnson about the Pentagon experimenting with cloudbusters as a form of mind control in one of her nationally syndicated columns:

> "MEMO TO PRESIDENT JOHNSON: Please check with our State Department... the leaders of our Armed Forces or our chief scientists, to discover what, if anything, we are doing to explore the ramifications of 'cloud-busting' which in its refined stages means thought-control... and could change the history of the world. We could catch up if someone in command gave the word before it was too late." (December 22, 1964).[10]

What was Kilgallen talking about? In what way could cloudbusting be considered "thought-reform" and how could it help the US "catch up" to the Russians' perceived advantage in the field of mind control?

Perhaps we should consider the comments Reich made about how the "emotional desert" mirrored the formation and character of actual deserts. "With the spreading of the global desert, civilizations go under, life perishes completely in the affected realm, man wither tries to escape or he adjusts, too, to the life in the desert on rare spots of green, called 'oases.' The continuous presence of death (DOR atmosphere) and the ever-present dull awareness of the inevitable end is characteristic of both life in the desert and life in armored man."[11] Like plants in the desert—spiny, prickly, and leathery—humans too adapt to the conditions, Reich believed. He used a smaller version of the cloudbuster, called a *medical DOR-buster*, on human organisms. And if the desert making process could be reversed in the atmosphere, allowing softer forms of life to thrive, perhaps atmospheric cloudbusting can be seen as a kind of "thought-reform"—not in the sense of the CIA's MKULTRA mind control program, nor the Chinese treatment of prisoners of war in Korea, but as a way to improve life-affirming conditions over whole regions. Most of the intractable armed conflicts around the world, particularly in the Middle East, are at least in part kindled by water and its availability. We know that summertime heatwaves in urban areas spark increased violence. Could the cloudbuster be useful in decreasing social tensions?

For more information about the formation of deserts, and desertified humanity, read James DeMeo's *Saharasia; the 4000 BCE Origins of Child Abuse, Warfare and Social Violence in the Deserts of the Old World*. DeMeo's work has extended and expanded Reich's original ideas about desert formation and "the emotional desert."

Before leaving this topic, I would be remiss if I failed to mention D. Ewen Cameron, the president of the American Psychiatric Association (APA), who told a patient of Reich's in 1953 that orgone therapy was a "pure fake" and that the APA would be bringing "charges of fraud against Dr. Reich."[12] Cameron is best remembered today for his sadistic work with patients at the Allen Memorial Institute at McGill University (Montreal, Canada) in the course of mind-control experiments at the behest of the CIA's MKULTRA program.

The *New York Times* reported: "In the early 1950s, Dr. Cameron developed a theory that people with neurotic thoughts and behavior patterns could have their ideas and behavior changed by listening to repeated taped messages up to 16 hours a day for days or weeks at a time. He called his technique 'psychic driving' and published an account of it in the *American Journal of Psychiatry*."[13]

During the years 1957 until 1961, subjects at Allen Memorial were given massive doses of LSD and other psychotropic drugs, they were electroshocked at 75 times the usual intensity, embalmed in barbiturates that caused them to sleep for days, forced to wear football-helmet style headsets with speakers that piped in tape-recording of the subjects own voice, repeating over and over again statements made in the "therapeutic setting." This type of treatment represented the "cutting edge" of orthodox psychiatry at the time, and Cameron was not at the fringe of the profession but its leader; he would become the president of the World Psychiatric Association in addition to his duties as head of the APA.

That a man like Cameron stood in judgment of Reich staggers the imagination. History has not judged Cameron kindly, even though he died in 1967 before he was ever called to account. In October, 1988, the

US Justice Department agreed to pay the victims of Cameron's treatment a total of $750,000 in a settlement of a civil lawsuit against the CIA.

Directly below Cameron's office at Allen Memorial, a biologist named Bernard Grad worked, literally under Cameron. Grad had a keen interest in Reich's work, and while Cameron tortured his patients, Grad was reproducing Reich's experiments with treating cancer-mice with orgone accumulators. I spoke with Grad in 1998, when he gave a lecture at the Orgone Biophysical Labs near Ashland, Oregon. Grad did not work on the MKULTRA side of Cameron's operation, and told me that he felt that Cameron had his patient's best interests at heart. Cameron knew of Grad's interest in Reich but didn't know that Grad was confirming Reich's cancer work directly under his nose.

Apologies for the digression, but I wanted to insert a commentary about the reigning attitudes toward Reich around 1953, before continuing with a discussion of the history of weather control, and the events surrounding Reich's expedition to Tucson.

Operation Weather Control

Since Plutarch, who observed that rain often followed big battles, the connection between war and weather has been traced closely. During the American Civil War, it had been conjectured that heavy artillery fire caused rain clouds. A highly popular book in the US, Paul Edwards' *War and Weather,* first published in 1871 and republished in 1890, aroused popular interest in rainmaking with cannon fire. In 1891 the US Congress approved $9,000 to study rainmaking in Texas using explosives.[14]

The results of blowing up the sky to make rain were mixed. The image of a rainmaker as a ballistic con man is thoroughly ingrained in the heartland psyche of the American people.

Reich conceived of the weather as a function of the pulsation of the underlying orgone energy envelope of the planet. Like an organism, the atmosphere was sensitive to insults. When pollution in the form of smog, pesticides or ionizing radiation reached critical levels, the atmosphere reacted accordingly: the natural process of contraction-and-expansion was inhibited. Reich first observed this during the

Oranur experiment, when he discovered that the effect of orgone energy became deadly when it was triggered by ionizing radiation. Deadly orgone, or DOR, is visible in the stagnant haze that shrouds much of the world today, no longer isolated to urban areas and atomic testing grounds as it was in the fifties.

Perhaps DOR is related to a weather feature meteorologists call a *fohn*: a "very dry warm wind coming down on the leeward side of a mountain range after having lost most of its moisture on the windward side. Brings fair weather, but is associated with harmful effects on the nervous system - causing severe migraine, nervous tension, deterioration in the state of many sick people, increased suicide and accident rates etc."[15]

On July 6th, 1953, the first public cloudbusting operation took place in Ellsworth, Maine. Two days later, Reich reported, Air Force planes appeared over Orgonon, "trailing what appeared to be research equipment behind them." Nearby was an important US Air Force base. Eisenhower was filmed standing on the tarmac there during a refueling stop for his flight.

Rainmaker

"Shortly after the white man adapted explosives to evil purposes, it was remarked that 'it always rained after a battle.' This led to serious proposals after the Napoleonic Wars and again after our Civil War to carry out rainmaking experiments by shooting explosives into clouds or detonating them at the ground. Some inconclusive experiments were tried. The idea of explosives for rain production has cropped up recurrently up to the present day. A project in New Zealand in 1906-1907 attracted widespread attention; in 1932 Professor Kleinschmidt of Stuttgart felt compelled to write a painstaking article to point out the nonsense in the vociferous proposals for hail "shooting" by some of his countrymen. Today the shooting is bigger and the urging voices seem louder, but it must be said in behalf of the proponents that at least in the H-Bomb we are approaching meteorological scales of energy."

–Horace R. Byers (University of Chicago) at the Conference on the Scientific Basis of Weather Modification, Tucson, Arizona, 1956.

Wilhelm Reich and the Cold War

In 1953, in a closed meeting of scientists at the University of Arizona in Tucson, dozens of meteorologists, physicists, and statisticians discussed with military and government officials how to evaluate and assess weather modification techniques. (There was also mention of pursuing solar energy as a "peaceful" harnessing of the atmosphere.) "Cloud-seeding" had been tested with encouraging results, and privately operated rainmaking outfits had been offering their services to farmers for years. Lew Douglas had first become interested in cloud-seeding when he hired Dr. Irving Krick, a private cloudseeder and the proprietor of Water Resources Development Corporation, who earned over $1 million annually on his operations,[16] to bring rain to his land outside Tucson. Israel was using the technique with such fervor that it was impossible to set a baseline for controlled experimentation there. Air Force officers were included because of the obvious military applications of weather modification. Fort Huachuca, an Air Force base that developed electronic warfare capacities, sent personnel to the meeting.

President Eisenhower signed the Weather Control Act into law in August, 1953, forming the President's Advisory Committee on Weather Control. He immediately appointed Lewis W. Douglas, owner and president of the Southern Arizona Bank & Trust Company, to the blue-ribbon weather modification committee.

Wilhelm Reich decided to contact Lew Douglas, who had been empowered with official sanction to investigate the effectiveness of numerous untested weather strategies, including cloudseeding. Sixty years after the US Congress first funded weather modification, the power to control the atmosphere remained an alluring goal to powerful political interests. Lew Douglas was at this time an important advisor to President Eisenhower on foreign policy, while Lew's brother-in-law, John J. McCloy, served as the President's informal political advisor. All of Reich's communications with the White House must be viewed in light of the cordial reception Reich received from Lew Douglas.

Who Was Lew Douglas?

Lew Douglas was a powerfully interesting man. Born near Bisbee, Arizona, in the rough-and-tumble days of the 1890s, to an educated family of mining metallurgists, Douglas was the sole heir of the Phelps-

Dodge copper fortune. His father, whose name is borne by the Arizona town of Douglas, sent young Lew back east to attend prep school at Montclair Academy and then Amherst College. After serving in World War I (where he was injured and gassed at Argonne and awarded the Belgian Croix de Guerre), he returned to Arizona where his father expected him to work his way up in the copper mines, starting in the pits.

Apparently, a taste of the wider world ill-prepared Lew for the copper mines, and he informed his father that he wanted to try his hand in politics. He had just married Peggy Zinsser (a Smith girl from Hastings-on-Hudson) in 1921, and his close Amherst buddy John J. McCloy became his brother-in-law when Lew introduced him to Peggy's sister, Ellen. (The Zinsser sister's father, Hans Zinsser, was a Harvard professor of Biology and closely associated with the Rockefeller Foundation. His book, *Rats, Lice and History* is considered a classic work.)

Douglas was elected to the Arizona House of Representatives in 1923, and went onto the US Congress in 1927. In 1933, he served as FDR's Director of Budget, but soon left government service in protest over the New Deal big-spenders to become Vice-President of American Cyanamid Company.

The New Deal would produce an economic recovery "as empty as a blown-out eggshell," he was reported to have said at the time. His book, *The Liberal Tradition*, attacked the expansion of government powers. He was convinced that the Roosevelt administration had been infiltrated by Communists from the Harvard Law School. "Jews too were part of the conspiracy to destroy the capitalist system. More than once, Douglas spoke to friends about 'Hebraic influence' and blamed the New Deal's faults on the Jewish race: 'Most of the bad things which it [the administration] has done can be traced to it. As a race they seem to lack the quality of facing an issue squarely.'"[17] When the Republicans nominated Alf Landon for President in 1935, his first choice for Vice-President was Lewis Douglas, even though Douglas was a Democrat. Douglas became an informal advisor to Landon's campaign after party regulars vetoed him in favor of a Republican.

During World War Two Douglas was Deputy Administrator of the War Shipping Department, putting him at the nexus of the emerging military-industrial complex. W. Averill Harriman brought him into the Lend-Lease program. After the war, he worked with Lucius D. Clay and the German Control Council, and served another stint with Averill Harriman in London as well, setting up a nascent European federation of states. Meanwhile his brother-in-law McCloy separated the good Nazis from the bad at Nüremburg (quietly importing the good ones into the United States.)

Lew Douglas led a charmed life, judging from his posthumous entry in *Who Was Who* (1971-1980). The listing for his accomplishments, government positions, and appointments to corporate directorships runs easily three times the average notable's.

His directorships at one time or another included: United Bank International, Mutual Life Insurance (president), Southern Arizona Bank & Trust (president), American Cyanamid, Western Bancorporation, Christiana Oil, Southwestern Research, General Motors, *Newsweek*, Continental Oil, and Transamerica (which merged with Douglas' Southern Arizona Bank & Trust) and many others. He also held a major interest in the Happy Jack Uranium Mine in Utah.

Douglas was also a director of the Council on Foreign Relations between 1940 and 1964. Those twenty four years rolled out the Pax Americana. (He gave a commencement address at MIT in 1953 calling for foreign investment, free trade, and other Anglo-American Establishment goals.) Along with OSS chief Allen Dulles, he served on the Budget Committee. He would have been the first head of the World Bank in 1946, but declined so that his brother-in-law, John J. McCloy could assume the post. By this time, Walter Lippmann called Douglas "the youngest of the elder statesman."

Lew Douglas' other "public interest" and charitable foundation activities included directorships on the Rockefeller Foundation and General Education Board (1935-1960), Resources for the Future, Inc. (to manage natural resources), the Sloan-Kettering Institute for Cancer Research, the Channel Tunnel Study Group, the US Chamber of Commerce, and he was an original member of the Alfred P. Sloan Foundation (among many other such trusteeships).

In 1947, Douglas was appointed to represent the United States as Ambassador to Britain, the most important post-war diplomatic position. Professor Carroll Quigley, whom we have met in the discussion of Michael Straight, remarked that Lew Douglas was a prime example of what he termed "the Anglo-American Establishment." Quigley wrote, "It might be mentioned that the existence of this Wall Street, Anglo-American axis is quite obvious once it is pointed out. It is reflected in the fact that such Wall Street luminaries as [...] Lewis Douglas [...] were appointed to be American ambassadors in London."[18]

In England, Douglas made quite an impression during his four year stay, especially by the way he carried on his work after hooking himself in the eye while fly-fishing on the River Trent. After he appeared in public wearing an eye-patch, the dashing eye-patch look became popular in American advertising, too. Hathaway Shirt Company featured a model with a black eye-patch, inspired by Lew Douglas.

His eye injury, along with nagging problems stemming from his exposure to chemical warfare agents in 1915, eventually forced him to return to the US. Save for brief periods of work in Europe with US Secretary of State John Foster Dulles, he spent the rest of his days in his sunny home state of Arizona.

Weather Control Act

Douglas' name turned up in the newspaper when he spoke on the subject of "Our Weather" at the Arizona Cattle Grower's Association on January 24th, 1954, at the Tucson Masonic Temple, and when he applied for a water permit for a real estate development he was putting in the barren environs outside Tucson.

That Lew spoke on "Our Weather" was indicative that he had not relinquished his drive to take on new responsibilities, but that the new arena for mastery had shifted; from international relations and global confederations to cosmic environmental modification. The Weather Control Act of 1953 (Public Law 256 - August 13, 1953) had mandated that he and the President's Advisory Committee "shall make a complete study and evaluation of public and private experiments in weather control for the purpose of determining the extent to which the United States should experiment with, engage in, or regulate activities designed to control weather conditions."

Tuesday morning, January 26th, 1954, the *Arizona Daily Star* ran an editorial entitled "Prospecting the Atmosphere — Why Not?" Three days before, the University of Arizona had announced it had received a $150,000 grant (the equivalent of more than $1.5 million today) from the Alfred P. Sloan Foundation. Dean Rusk, at the time directing the Rockefeller Foundation, assured Douglas that the Foundation would also fund the venture. Douglas was a trustee of the Rockefeller Foundation.

The funds were earmarked to establish the Institute of Atmospheric Physics. "It means that here in Arizona a lot of brains are going to be organized to investigate not merely the weather, but the physics of the atmosphere, with the hope that mankind can make better use of possibly unused power to be found there." The paper thanked Lew Douglas, "who has probably spent more time organizing it on a scholarly basis, and getting that initial gift of $150,000 from the Sloan Foundation, than he has on his many other interests."

The Institute of Atmospheric Physics would play host to annual gatherings of meteorologists, cloud physicists, and statisticians from around the world in an effort to study the basic properties of the atmosphere, with an eye toward rainmaking in arid lands. Time-lapse movie cameras were to be placed around the valley, and "radiocameras" (radar readings of cloud formations) would record cloudseeding experiments. Air Force planes and crews were assigned to the University of Arizona to assist in the study.

An accompanying article mentioned that Lew Douglas had hired a cloudseeder in 1952. Perhaps he had some doubts as to whether it really worked, and wanted a full-scale study.

In March of 1954, Douglas was called back to Europe for a few weeks of high-level talks with British officials as an economic advisor to John Dulles. Before he left, he warned of the need of a new water code. "There is no natural resource in the state of Arizona more vital to its people and their well-being than its water. The rate at which the underground water supply is being drawn upon makes the problem of conserving it one of the most serious [...]"

July 24th, the IAP took its first "cloud census" using the radar and time-lapse cameras. On August 9th, scientists sponsored by the President's Commission on Weather Control met for the first time at the University of Arizona. They warned the public against expecting "mail order" weather anytime soon, while holding out the promise of increased rainfall, defusing dangerous storms and "basic natural research." They knew their financial angel, Lewis Douglas, was a man who was used to results.

Reich's Final Years

Seeds of Rain: Early developments in weather modification

Kurt Vonnegut, one of the most popular American authors of postwar fiction, mentions his brother, Bernard, in *Welcome to the Monkey House*; "My only brother, eight years older than I, is a successful scientist. His special field is physics as it relates to clouds. His name is Bernard, and he is funnier than I am."[19]

Bernard Vonnegut studied electrical effects on rainfall at SUNY in Albany, New York. He was a participant at the 1956 IAP conference. Louis Battan, who co-directed the IAP with James McDonald, wrote that Vonnegut's claim, "electrification precedes the formation of precipitation" was not accepted by all atmospheric scientists; "these are revolutionary notions."[20] Vonnegut pursued an "energetic" approach to the atmosphere, which did not fit with particle theory. Yet Vonnegut was the discoverer of a cloudseeding technique that uses silver iodide crystals to coax precipitation from clouds, and the technique is still widely used today, particularly over ski-resort areas in the Sierra Nevada mountains.

One widely accepted theory of rain formation at that time suggested that dust particles in the air attract moisture under certain atmospheric conditions and that precipitation is caused by these particles. Reich referred to this notion in *Contact* rather dismissively and asked whatever came of the IAP's studies of particles in the air. Indeed, as best as I could determine, there is still a lot of debate in the scientific community about how exactly raindrops form. In a standard textbook, *Physical Meteorology*, Henry G. Houghton wrote in 1985, "At the time of this writing there are many clues to the physical processes involved in the heterogeneous nucleation of ice from the liquid and from the vapor. However, there is as yet no comprehensive theory that relates all or most of the known properties of ice nuclei to their activity."[21]

Vincent Schaefer and Irving Langmuir, working for General Electric Labs at Mount Washington, NH, discovered that dry ice, under supercooled conditions, creates cloud formations. (Langmuir, Nobel Prize winner for chemistry in 1932, had pioneered electric amplification, sonar, and smoke screens during WWII.) Their first dry ice test occurred on November 13, 1946. They cut the corporate trademark of the GE lightbulb into a cloudbank soon thereafter (see photo section). Bernard Vonnegut, working with Schaefer and Langmuir as an icing nuclei expert, sought a substance with a suitable structure for seeding.

Wilhelm Reich and the Cold War

He discovered silver iodide as a possible "seed" for ice nucleation when he paged "by scanning a tabulation of crystalline substances for low solubility and crystal lattice dimensions close to those of ice."[22] That is to say, Vonnegut chose silver iodide after noticing that its crystalline structure resembled the morphology of a snowflake. (I find this quite fascinating but leave it to scientists to explain how shape, size and structure influence precipitation. It sounds much like "morphological resonance" to me. Indeed, Irving Langmuir argued that "To assume that a hurricane could not be successfully modified by even a single pellet of dry ice is like assuming that a very large fort could not be set on fire by a single match."[23])

In February, 1947, Langmuir signed contract with Army Signal Corps, who were later joined by the Office of Naval Research, the Air Force, and the US Weather Bureau, to provide funding for the cloud-seeding studies called "Project Cirrus." Early results were promising enough and soon commercial cloudseeders began offering their services to farmers and ranchers. Langmuir openly believed that Project Cirrus had caused floods in the Kansas River in July, 1951. The lack of hard scientific evidence that the cloudseeding effect was actually causing any additional rainfall caused Lew Douglas, along with many others, to call for the formation of an institute that would take a critical look at the promising method, and thereby the IAP was born.

Whether we point cloudbusters at the sky or disperse chemicals in the air, the atmosphere is an extremely complex system and it is very difficult to determine causal effects. Where is the baseline? How do you know it wasn't going to rain anyway?

When the IAP announced in the mid-fifties that they determined that cloudseeding caused up to eighteen per cent increased precipitation, they were met with a chorus of statisticians from the University of Chicago who pointed out the obvious defect of such a claim, without at least a 100-year baseline of observed precipitation trends. In other words, the cloudseeders should shut down operations for such time as that baseline could be determined—perhaps shut down for decades.

Without question, the IAP was under pressure to produce quick results as well as stimulate research into natural science in the abstract. In the Douglas archives, I found a moving exchange of letters between Lew Douglas and atmospheric physicist James E. McDonald, in the aftermath of a meeting of the IAP when the statistical objections were raised, where both men agreed that the subjective observation that cloudseeding really worked was equally valuable to the statistical studies—a point made emphatically by Reich in his books.

At the end of the day, while cloudseeding is today regarded as proven effective under certain conditions, it is perfectly useless if there are no clouds at all, and such are the conditions for the most part in the desert southwest under drought. Furthermore, all of the IAP's early studies were skewed by the fact that Reich was in the same area cloudbusting, and I know from personal experience that the cloudbuster has an effect on the atmosphere.

I also know that today the IAP possesses hundreds of time-lapse films, and detailed weather observations from the period when they were actively studying cloudseeding. Some of these films were taken during the time when Reich was cloudbusting in Tucson, but these particular films were missing from the files when I was allowed to inspect them. The staff at the IAP, cordial and helpful to my research, told me that professors used the films in their classes to this day. I was permitted to remove a few of the old films temporarily and convert them onto videotape; they are quite lovely to watch the clouds' accelerated stream across the field of view over the desert sky. The department staff told me they had recently considered throwing the whole collection away, and I urged that they be preserved. Perhaps someone with openness to Reich's ideas and with knowledge of meteorology will compare the IAP's records with the detailed protocols that Reich presented in *Contact With Space*. Perhaps some of the staff at the IAP in 1955 noticed a correlation at the time.

Battan also noted that the Russians quickly picked up the technique of cloudseeding (By the mid-seventies, the Russians were using two and a half tons of silver annually on cloudseeding.[24]) Battan wrote, "In 1958, they [the Soviets] seeded clouds with tracer bullets impregnated with silver iodide. There is no record that this technique was given a special name. Perhaps it could be called the 'cloud blaster' to go along with the Air Force's Cloudbuster."

UFOs in the Atmosphere

James E. McDonald, one of the first top-rank scientists to argue for the reality of the UFO phenomena, was the director of research at the Institute of Atmospheric Physics.

Born May 7, 1920, in Duluth, Minnesota, he died June 13, 1971 in

Tucson Arizona, of a suicide. McDonald served in the US Navy from 1942 until 1945, working in intelligence and aerology.

McDonald was the associate director of the Institute of Atmospheric Physics from 1954 until 1971. The IAP was founded in 1953 as a result of the President's Advisory Committee on Weather Control (1953-1957), pursuant to the Weather Control Act of 1953. Since the project had military implications, each participant underwent security checks by the FBI. In 1966, McDonald was appointed as a member of the Advisory Panel for ESSA-Navy Project STORMFURY.

Eva Reich told me that when the TV crew came down to Tucson and filmed Reich's cloudbusting operations, McDonald was on hand, and both he and Reich were interviewed. Eva was there. When they went home to watch the film on the news, "the screen just went black."

Jerome Eden republished Dr. McDonald's 1968 testimony before Congress, with his comments added in the margins.

By some strange coincidence, the original copy of the IAP's report on weather modification handed to me by Bill Sellers was James E. McDonald's personal file copy. It has a handwritten inscription on the cover "J.E.M. - permanent file copy."

McDonald and Lew Douglas had a close working relationship. Wilhelm Reich was not the only rainmaking UFO watcher in Tucson back in 1954.

Reich's Real Estate Broker

Lew Douglas had written to Wilhelm Reich and requested a personal meeting, according to *Contact With Space*. Although this meeting would never take place, Reich wrote "his banking institution helped along during the entire expedition in a most friendly and cooperative fashion."[25] On my research trips to Tucson, I was able to verify this cooperation and I have every reason to believe that Douglas wanted to meet Reich, but could not.

Reich and his daughter Eva, then thirty years old, arrived at the Spanish Trail Motel in Tucson on October 29th, 1954. "With the help of the Southern Arizona Bank, a house with 50 acres land around, suitable for our purposes, was found and on October 31 we began establishing ourselves."

Reich's Final Years

Planetary Valley Forge

On October 10th, 1954, just before he left Maine for Arizona, Wilhelm Reich disabled two UFOs with his Space Gun, a cloudbuster used in conjunction with denatured radium produced in the Oranur Experiment. When Reich wrote to President Eisenhower regarding his alarm at the intense UFO-activity around the sites of his cloudbusting operations, the White House directed him to the Air Force and the Central Intelligence Agency. Reich wrote to Roscoe Hillenkoetter, the Agency's first director. Hillenkoetter, after leaving the CIA, became an outspoken proponent of the reality of UFOs, and once wrote:

"The unknown objects are operating under intelligent control. It is imperative that we learn where the UFOs come from and what their purpose is."[26]

Eva Reich recalls when (on August 12, 1954) the Air Force established AFR 200-2, its regulation on reporting UFO sightings to Air Technical Intelligence Command (ATIC). "It needs to be pointed out that Reich published the Air Force questionnaire for the first time in *Contact With Space*. The interest of the Air Force in UFOs was being totally suppressed at that time. When you reported a UFO, they came to you with this questionnaire, and told you it was secret, and had you fill it out. Then they took it away; what happened with it, nobody knows. When Reich published the questionnaire in *Contact With Space*, he revealed a big secret."

Reich sent Bill Moïse, Eva's husband, on ahead to report to the Air Technical Intelligence Command headquarters in Dayton, Ohio. Reich had been in contact with the Air Force since January, when he was asked to fill out a long questionnaire about his early UFO sightings. In March, Reich sent a ten page "Survey on Ea" to the Air Force in Portland, Maine, including gravitational equations worked out ten years before in experimentation with pendulum swings. He believed that the propulsion systems of the UFOs could be understood only in terms of orgonomic science.

Moïse called ATIC at Dayton and he was given an appointment on October 14th to meet with General Watson, ATIC's commanding officer on the base. When Moïse arrived in Dayton, on his way out to Tucson, he was informed that Watson was unavailable, having been called away to a meeting with CIA officers. Watson called Moïse at his hotel, apologizing, and asked if Moïse could meet with his deputy. Moïse agreed and met the next day with Deputy Commander Col. Wertenbaker, Capt. D. M. Hill, USAF and "Mr. Harry Haberer, civilian, working with the Air Force in regard to the history of UFOs." In all likelihood, Haberer was one of the CIA officers who had just flown in from Washington.

Moïse described Reich's interactions with the UFOs, while Wertenbaker listened with serious intent. After the meeting, Wertenbaker promised to accept Reich's orgonometric equations, by which Reich had described possible methods of anti-gravity propulsion, for safekeeping.

Moïse traveled on to Tucson, where Reich met him on October 29th. The University of Arizona's weather modification study was well underway, and Air Force jets crisscrossed the sky, leaving contrails at various levels of the sky in conjunction with the time-lapse photography. Reich reported Air Force jets appearing during his cloudbusting operations on the way out west, and had been using their contrails in his own atmospheric studies in Tucson. Reich wondered, "Whether the Air Force had actually such problems in mind, I cannot tell."

It's important to remember that in the mid-fifties, jet-propelled airplanes were rare birds, and by far the majority of commercial and military planes had propellers and they did not emit contrails. As the 16-mm films of the IAP's experiments show, these ideas were definitely in the Air Force's mind. Color films of wafting checkerboard jet contrails sit today in a half-forgotten file in the Physics and Atmospheric Science Building at the University of Arizona. They show that Air Force jets were used in the government's weather modification studies just exactly as Reich had described. Against the pale blue open sky of the Arizona desert, a criss-cross pattern drifts across the screen's field of vision; what a perfect way to observe the movement of the atmosphere.

Bob McCullough

Robert C. McCullough, who worked for Reich as a biologist and chemist after the Oranur Experiment, died on November 26th, 1995, at the age of 74. I had a brief correspondence with McCullough, who passed on before I could ask him about his experiences in any detail. McCullough's article, "The Rocky Road to Functionalism," published in an issue of the *Orgone Energy Bulletin*, was an honest and forthright appraisal of the difficulties in understanding Reich's scientific methods. Reich's daughter, Eva, described McCullough as a "large, bumbling, patient man with a heart of gold." She said Bob was a very knowledgeable person who was "a source of information for my father. Bob read a wide number of scientific journals and was always finding morsels of new information which had some bearing on orgonomy." He had a stroke after suffering the effects of getting too close to an operating cloudbuster, which permanently damaged him.

Bob McCullough was one of the few people who stood up for Reich at his trial. According to Trevor James Constable, McCullough was so crushed by the judge's decision that he never fully recovered. McCullough went to work for the US Army in the Chemical & Biological Warfare Division, with a security clearance, at Dugway Proving Grounds in Utah, the biowarfare experimental station most famous for its release of some biological agent that killed thousands of sheep in the area. Constable accounted for Bob's decision to accept such life-negative work to some measure of Bob's loss of faith in people after Reich's death in prison. I asked Trevor if the Army knew of Bob's history with Reich. "Oh sure," he replied, "they had no problem with it." What where his duties? "Well, he was under an oath of secrecy and couldn't discuss what he was doing, but he inferred that some of it was pretty horrible."

McCullough later quit the Army and moved to Costa Rica in a failed attempt to start a coffee hacienda in the early eighties. "The effort nearly killed him," Trevor said.

McCullough attended a lecture by Trevor and introduced himself, back in the fifties, after Trevor had published his first book on UFOs, called *They Live in the Sky*. McCullough told Constable that he had seen forms, similar to what Constable called "critters," in Tucson on the Arizona expedition. From that meeting, a collegial relationship grew over the years, and Bob and Trevor exchanged audiotape letters instead of written correspondence, and out of this came Constable's interest in weather engineering. Constable, in turn, introduced McCullough to Rudolf Steiner's work, which he said was a big part of Bob's outlook in the last 20 years of his life. Bob guided Trevor through the process of building a cloudbuster.

On a personal basis, Trevor recalled Bob as a man with "truly phenomenal intuition. He was what Dr. Reich called a 'functional thinker.' In other words, he could follow living processes and relate everything to the living. It was a delight to have any kind of conversational intercourse with him, because you didn't feel as though you were swimming in fuel oil - which is the way it is with most of the conventionally trained people. He had a good conventional training. He was a biologist in the best sense of the word. He wanted to know everything: he wanted to know why things moved and what the source of the movement was, which is what they never study in conventional biology. They stop it before they study it. 'Kill it'."

McCullough was an early member of the Borderland Sciences Round-Robin letter, back in the fifties. "He became a kind of reference source for people inquiring about Reich," Constable said. "He was also much more 'catholic' in his approach to things. In other words, he was not a doctrinaire Reichian. He was extremely sensitive in a bodily sense. He could pick up the fields and emanations in the atmosphere and the weather engineering gear. He could detect the periphery of a field with his tongue. And at this stage, where we really don't have meteorology that we need, the human organism has to serve in that capacity until meteorology does come to hand. Bob could detect life-fields. He was one of the two or three most sensitive people I have known."

The Spacegun

> "We speak now in strange new terms of harnessing the cosmic energy; of making the winds and the tides work for us; of purifying sea water for our drink; of creating new and unheard of synthetic materials to supplement or even replace our old standard basics; of mining ocean floors for new fields of wealth and food; of disease preventives to expand life into the hundreds of years; of controlling weather for a more equitable adjustment of heat and cold, of rain and shine; of space ships to the moon; of the prime target in war no longer the armed forces of an enemy, but instead his civil population; of ultimate conflict between a united human race and the sinister force of some other planetary galaxy."
>
> – Douglas MacArthur, Manila, Independence Day, July 4, 1961.[27]

By November, 1954, Reich was settling in at Tucson. Joining Reich in Tucson were his son, Peter Reich, his daughter Eva, her husband, Bill Moïse, and Robert C. McCullough, Reich's research associate in Oranur Chemistry.

On December 7, Michael Silvert arrived in Tucson on a airplane flown by charter pilot Harry Hudgins. The plane carried the denatured radium, or *orur,* tethered behind in a football-sized container, and its arrival was filmed by KEAN-TV from Phoenix. The orur was to be used to potentiate the cloudbuster and augment its effect. Three weeks later, on December 30, a US Marshall appeared at Reich's residence, without a warrant, asking to inspect "everything." Reich told him to leave, and he did.

Silvert returned to the East Coast after delivering the Orur. In January, 1955, he shipped a number of orgone accumulators from Maine to New York, a direct violation of the terms of the injunction that had been decreed back on March 19, 1954. Reich had complied with the basic terms of the injunction insofar as he was able, and had not shipped orgone accumulators across state lines. He was completely taken by surprise by Silvert's letter informing him of the violation.

There was nothing Reich could do about it, and there were pressing matters at hand. On January 17, Reich used the cloudbuster to clear fog from the Tucson Municipal Airport, in which Lew Douglas had a heavy investment. Toward the end of January, Reich met with farmers and other local community members to discuss the University of Arizona's Institute of Atmospheric Physics cloudseeding projects. In February, a local TV station broadcast featured Reich's work with the cloudbuster.

Reich called his cloudbuster a "spacegun" after he began to use the radium needle from the Oranur Experiment with it. He found that the orur greatly increased the effect of the cloudbuster. Mounted on the back of a blue Ford pickup truck, the spacegun was driven to different locations. Reich called the process of using orur with the cloudbuster *orurization.*

Although he clearly stated that he was speculating, Reich guessed that the spacegun might not only heal the atmospheric effects of atomic testing, but might also prevent nuclear bombs from being detonated. "It appeared not impossible that in due time the atmosphere would be immunized against atomic explosions as one immunizes living systems against infection. Only those ignorant of the functional identity between organismic and atmospheric OR energy would be shocked by such an idea. A highly orurized atmosphere would siphon off any kind of DOR energy from atomic blasts or Ea [UFOs]. We had the impression that there were difficulties in setting off the A-bomb."[28] I found an article in the *Arizona Daily Star* that described the events to which may have Reich referred, and indeed at least one A-bomb test had been postponed due to "the weather."[29] In April, 1955, a test was repeatedly postponed at Yucca Flat, Nevada. This test was planned to determine the effect of the blast and fallout of a bomb, and whether a city might be able to withstand one. The newspaper reported that "Civil defense people are also testing how to survive in an area quite close to 'ground zero'[...] The Army is placing men even closer, some only 2,600 yards from the blast."

The newspaper report continued, "Aircraft, too, have been exposed to the after-effects of the burst with a group of air force scientists flying through the boiling, reddish-brown cloud produced. The airborne observers are measuring the amount of radiation which penetrates the human body by attaching sensitive film badges outside and then swallowing a similar badge on a string." Given the nature of these experiments, I cannot imagine that the test was called off due to the wind blowing the wrong way. Thousands of people had gathered in anticipation of watching the above-ground blast, and had to be turned away in disappointment. If Reich's speculation was correct, the military implications of the spacegun were staggering.

It cannot be stated emphatically enough that the US government was aware of Reich's cloudbusting. Two members of the IAP who were military weather researchers at the Cambridge AFB and had attended many of the symposia in Tucson later developed their own device to be used for fog removal at airports. "In 1966, William W. Vickers and James F. Church, working for the US Air Force, published an article dealing with this subject. They had set out to determine optimal cloud seeding techniques. In the course of their work over a three week period, they developed a Dry Ice seeding machine called the Cloudbuster."[30] It was a comically different device from the one used by Reich. (See photo section.) Then there was Harry Hudgins, the charter pilot who flew Silvert and the Orur into Tucson. He also flew military officials from the IAP around Tucson in the course of their experiments.

Cigar-chomping Air Force General Curtis LeMay, who is widely believed to have been the inspiration for the character who commandeers an Air Force Base in the film *Dr. Strangelove,* was commander

Reich's Final Years

of the Strategic Air Command during the mid-fifties. LeMay was directly involved in the operations of the IAP, providing bomber jets and other aircraft for the study. After his retirement from the military, LeMay provided the use of his home to Trevor James Constable for a series of smog removal experiments in 1990 in Southern California.[31] (Constable's weather modification methods, while inspired by Reich's cloudbuster, use completely different devices.) I point this out not to imply that LeMay was personally interested in Reich in the fifties, even though it might be true, but that top military brass can be open to these unorthodox approaches.

My guess is that cloudbusting was viewed by the military with the same mixture of interest and skepticism with which they approached cloudseeding, which was similarly unproven and controversial. Finally, there is recent evidence that the military has since that time employed cloudbusters as Reich designed them.

In their book, *Left at East Gate*, Peter Robbins and eyewitness Larry Warren explore one of the strangest UFO incidents in modern memory, the Bentwaters Air Force Base case in December 1980. Bentwaters, located near Suffolk, England, is a US base that, ironically, was one of several bases in England that Lew Douglas first established when he was Ambassador to London.[32] Larry Warren was a security officer at Bentwaters who claims that UFOs invaded the base. The base commander issued a statement that tends to support Warren's story.

Peter Robbins, who has long been interested both in orgonomy and ufology, and who now edits the on-line site Ufocity.com, showed Larry Warren a picture of Reich and a cloudbuster. "As I carefully studied the picture, I tried to figure out why it looked so damned familiar. Christ, we had those things at Bentwaters! I'd seen the same kind of device on the base back in 1980, but had thought it some kind of communication device. Peter seemed shocked when I told him and started bombarding me with questions. All I could tell him was that the things I'd seen were made up of long pipes, painted black, and mounted on concrete bases. They had a multitude of large cables running from them into concrete framed wells dispersed around the base. There was no question in my mind that they looked like the one in the picture. Peter seemed intrigued and disturbed by my observation."[33]

We could discount this story as anecdotal, but Peter Robbins found two or three other people who saw such devices at Bentwaters. It's significant, because in October 1987, in the forest near that same base, a freak storm occurred where winds estimated to be over 100 miles per hour completely leveled the trees in the area. Robbins talked to local Suffolk residents who said this storm just wasn't "natural."

In *Contact With Space*, Reich writes of his concurrent observation of UFOs with cloudbusting operations. In the Bentwaters case, Larry Warren and many other military personnel state that the air force base came under an invasion of UFOs in 1980. According to one top official, these UFOs "compromised the ordnance." In other words, the UFOs either enabled, or disabled the nuclear weapons stored at Bentwaters. The evidence compiled in *Left at East Gate* is detailed and compelling. It would suggest that the Air Force might have finally heeded Reich's warning that this planet is under attack.

This "war of the worlds" scenario was given added weight in 1997 when the former US Army liaison with the National Security Agency, Col. Philip J. Corso, published his amazing memoir, *The Day After Roswell*. Like every other eyewitness testimony, Corso's has been debunked as unbelievable, but I'm not so sure he wasn't telling the truth. In any case, we now have a top-level intelligence officer from the US government whose detailed observations of alien technology closely mirror Reich's in *Contact With Space*. Corso sates flatly that the entire Cold War buildup, from nuclear arms on up to the Strategic Defense Initiative, was not so much a competition with the Soviet Union, but an anticipation of defending the planet from space aliens. He corroborated the existence of a small, highly classified group of government officials, including the President, called "Majestic-12" that was formed in 1947 when a crashed UFO was recovered. Corso wrote:

> "They would form nothing less than a government within a government, sustaining itself from presidential administration to presidential administration regardless of whatever political party took power, and ruthlessly guarding their secrets while evaluating every new bit of information on flying saucers they received. But at the same time, they would allow disclosure of some of the most far-fetched information, whether true or not, because it would help create a climate of public attitude that would be able to accept the existence of extraterrestrial life without a general sense of panic.
>
> "'It will be' General [Nathan] Twining said, 'a case where the cover-up is the disclosure and the disclosure is the cover-up.

Deny everything, but let the public sentiment take its course. Let skepticism do our work for us until the truth becomes common acceptance.'" [34]

Wilhelm Reich and MJ-12?

The only hard document that I found that conclusively proves the communication between Lew Douglas and Reich, a single half-page memo, deserves a bit of scrutiny (see appendix for a facsimile of this memo). It is a note typed by Douglas' secretary reflecting a telephone call received from Bill Moïse requesting a meeting with Douglas to discuss weather modification with Reich's cloudbuster.

Here is the full text of the "Douglas-Moïse Memo":

> WILLIAM MOISE (MAN FROM MAINE INTERESTED IN WEATHER CONTROL) WILL BE IN N.Y. EARLY WEDNESDAY AND HOPED HE COULD SEE OR SPEAK TO YOU. I EXPLAINED YOUR SCHEDULE WAS VERY FULL AND IT WAS NOT LIKELY THAT YOU COULD SEE HIM. WILL YOU TALK TO HIM ON THE 'PHONE. YES____ NO [X]
>
> MR. MOISE CALLED TWICE FROM MAINE AND HOPED URGENTLY THAT HE TALK WITH YOU ABOUT A NEW APPROACH FOR COMBATING THE DESERT DROUGHT.

[Ed. note: Stapled below, a subsequent note reflects further efforts on Moïse's part to speak with Douglas:]

> MR. WILLIAM MOISE OF THE ORGONE INSTITUTE, ~~RANGELEY~~ HANCOCK, MAINE CALLED TODAY TO SEE IF HE COULD SPEAK WITH MR. DOUGLAS. HE HAD JUST COME FROM WASHINGTON AND HAD SPOKEN TO PEOPLE IN THE DEPT. OF AGRICULTURE, WEATHER BUREAU AND IN MR. GARDNER'S OFFICE ABOUT WEATHER CONTROL. THEY SUGGESTED THAT MR. D. MIGHT BE INTERESTED IN INFORMATION HE HAD. HE MENTIONED THAT HE HAD COME TO N.Y. FOR PURPOSE OF MAKING RAIN BUT HE WOULD BE LEAVING FOR MAINE AGAIN POSSIBLY TOMORROW. HE SAID HE

MIGHT WRITE MR. D. IN TUCSON OR CONTACT HIM AT SOME LATER DATE.

[On this stapled addendum to the memo, a handwritten notation follows:]

7/27/54 telegraphed Moïse - *suggested that he write LWD a letter*

On the surface, the memo looked like a polite kiss-off to Moïse from "Mr. D" - Lew Douglas, but upon further inspection, several points became apparent to me. First, it does establish the basic fact that Reich's people made repeated attempts to establish contact with Lew Douglas. Moïse had been directed to Douglas by Charles Gardner, Jr., who was the Weather Bureau's liaison with the Institute of Atmospheric Physics as well as the Executive Secretary of the Advisory Committee on Weather Control. (Reich reprinted a letter from Gardner to Moïse dated March 21, 1955, in *Contact With Space,* when Gardner thanked Moïse and told him "we appreciate being informed of your activities."[35]) Finally, Douglas sent a telegram to Moïse on July 27, 1955 asking him to write him a letter.

Certainly Moïse or Reich would have accepted the invitation, and written that letter to Douglas, and in *Contact with Space* Reich wrote that he and Douglas had a cordial exchange of letters, but never met. Through research at the Douglas archives in Tucson, I was able to determine that Douglas was in and out of the hospital at the time of Reich's stay in Arizona, having developed sciatic nerve problems, as well as having undergone surgery for intestinal cancer that was probably related to his exposure to chemical warfare agents in WWI. He was unable to meet with Reich. Douglas' appointment calendar is completely blank for weeks at a stretch during the whole time Reich was in Tucson. For the same reason, Douglas was unable to attend important sessions of the IAP during that time.

The discovery of the memo from Moïse to Douglas conclusively establishes that Reich did in fact, as he claimed in *Contact*, establish friendly communications with high government officials. Beyond that, we can speculate about the ultimate importance of this relationship. So we will.

My friend and fellow researcher, Kenn Thomas, and I have carried on a heated debate over the significance of the proximity of the dates on

Reich's Final Years

this "Douglas-Moïse Memo," as I will call it, and the infamous "Cutler-Twining Memo," the only official archival reference to an "MJ-12" group in the National Archives. Thomas, a professional archivist, has held the actual "Cutler-Twining Memo" in his hands at the new archive facility in Arlington, Virginia. Thomas points out that Robert Cutler, who wrote the memo dated July 14, 1954, was formerly an expert in Psychological Operations with the CIA, and that Cutler was deeply involved with the crafting of Eisenhower's "Atoms for Peace" initiative. Thomas noted that the date of the "Douglas-Moïse Memo" is ten days before the "Cutler-Twining Memo." In the latter memo, Cutler advised General Nathan Twining, (alleged to be an MJ-12 member) that a previously scheduled meeting of the "Special Studies Project" would be changed to occur *"during* the already scheduled White House meeting of July 16, rather than following it as previously intended." Something important was up and the President would be there.

Thomas suggests that the timing of this sequence of events might indicate that Douglas, as a member of Eisenhower's "kitchen cabinet," may have been privy to or associated with the MJ-12 group. I agree that Douglas, one of the most powerful men in American politics at the time, would have known about MJ-12 if it existed. Thomas argues that Douglas, having been briefed about the meeting of MJ-12 members on July 16 at the White House, developed a more serious interest in Reich's planned operations in Tucson on the basis of Reich's observations of UFOs. This would explain the sudden change in attitude on July 27, 1955, when Douglas sent a telegram to Moïse inviting further correspondence. After all, Douglas had hired a "UFO nut," James E. McDonald, to head the IAP in 1954.

But wait a minute, my boggled mind rebels against Thomas' eloquent Dagwood sandwich of conspiracy theories. Are the MJ-12 documents themselves for real?

Stanton Friedman, a physicist and well-established UFO writer, has covered the MJ-12 documents in his book, *Top Secret Majic*. One of the objections raised by skeptics was that no other documentary reference could be found to corroborate the existence of any "MJ-12" group, but the "Cutler-Twining Memo" (which has been authenticated by the National Archives) debunks the debunkers on that score. And the content of the memo is consistent with the idea of a highly-classified gathering of wise men in the Oval Office. Both the recipient and the author of the memo were the kind of officials who would be privy to such a meeting.

To say the least, however, the MJ-12 documents have a squirrelly provenance. Essentially, they came over the transom to UFO researcher Jaime Shandera. And couldn't the Cutler-Twining Memo have been smuggled into the Archives as a hoax? "When I did the tour of the National Archives," Thomas told me, "they went to great lengths at the research room to explain how it was impossible to sneak anything out. The tour guide was dumbfounded by my question about sneaking things in."

Perhaps future research will provide answers to the open question as to Douglas' relationship with Reich. It's worth speculating about just to stimulate the discussion.

Wilhelm Reich in Roswell?

Did Wilhelm Reich visit Roswell, New Mexico seven years after the alleged crash of a "flying saucer" in 1947?

Pitfalls await researchers interviewing witnesses. They are often elderly and they forget; they are flattered to be interviewed and inflate their own importance in events; they might be crazy liars, for all you know. Sometimes, you'll hear a story that sounds unbelievable, and then through the process of verifying the claim, you learn a serendipitous truth. And so it was, that I rediscovered that Wilhelm Reich passed through Roswell and then spent the night in Ruidoso Downs, New Mexico in 1954 on his way to Tucson, Arizona.

I would have never noticed this simple fact, had I not heard a unlikely story from a man I met in Tucson, Arizona, in 1995. His name is Joe Blankenship, a colorful character who has lived in Tucson since the early forties. He is widely known as an oddball, but he is well liked in Tucson, and something a local celebrity. His interests range from free energy and alternative medicine to mining, local history, UFOs, architecture and orgone energy.

Joe told me a great many things that sounded so outlandish, incredible, and unbelievable, that I thought he must be crazy or a liar, but then he offered me proof, or I found some independently. I was never able to catch him in a direct lie. He is an old man with a sharp mind, without formal education. He gets words wrong, makes mistakes. Even when he's telling the truth, it sounds like he's pulling your leg. He's told many researchers that he was in Tucson in 1954, and knew Reich when he was

Reich's Final Years

there. I am not sure, but I tend to believe that he did meet Reich, even if all the details might be sketchy.

Joe probably doesn't sleep more than three or four hours a night, but I could hardly keep up with his frenetic energy as he drove me all over Tucson showing me local landmarks and sharing his knowledge of local history. I was extremely skeptical when Joe Blankenship told me that he and Reich had gone to Roswell to investigate the crash of a "flying saucer."[36]

Joe obviously loved a good story and regaled me with one after another. One in particular sounded patently absurd to me. Joe said he had been invited to President Richard Nixon's funeral. Given the national importance of this event, attended by President Bill Clinton and three former US Presidents, and given that Joe had an eighth-grade education, looked every bit the rumpled "desert rat" who had just crawled out of a hole in the ground—let's just say I had my doubts about this story.

But Joe produced a front-page section from the *New York Times*, with a full-color photo of the man himself, Joe Blankenship hunched under a plastic tarp in the rain, seated in the front row of Richard Nixon's funeral.

Joe explained that he had been a friend of Nixon's mother, who hired Joe to physically move her house in Whittier to Mount Lemmon in Arizona. Mrs. Nixon had another son who had died of tuberculosis, and Joe said that they talked about Wilhelm Reich and how the orgone accumulator might have helped this lost son. This all seemed preposterous to me. Yet there was the picture of old Joe, in the front row of Nixon's funeral.

To check Joe's story about Reich in Roswell, I talked to Eva Reich. She did not recall anyone named Joe Blankenship, and never remembered her father talking about him, nor any interest on Reich's part in the alleged crash at Roswell. She believed she would have remembered his name, since every morning she attended a meeting of the crew to discuss "order of the day."

I pointed out that Joe made no claim of being actively involved, and when I asked her if Reich went out driving he might have spoken with guys like Joe, she said indeed he did. He had a new Chevy Impala and drove around for several hours each morning. I doubt if Joe would have come up in any "order of the day," being a kid of 22 years.

Eva also verified that Reich had an interest in dowsing, something

Wilhelm Reich and the Cold War

Joe had asserted to me and a fact that is not widely known. She remembered that Reich had an aged dowser come over to Orgonon in Maine and dowse there. She remembered no dowsers in Arizona, though.

Furthermore, she remembers almost no local Tucson resident from that time period, save for one maid. It was Eva's sad duty to fire her because Reich suspected her of stealing the silverware.

She also said that Reich did not tell her everything he was working on. She said that it is entirely possible that Reich may have met Joe Blankenship (who definitely knows and knew everything going on in Tucson, which was a very small city, with a small population, at that time.)

Eva has always said that she still did not fully understand everything Reich was doing in Tucson, and that she was there as a "soldier." Eva is no social butterfly and never was. But Reich absolutely loved to discuss things with everyday people in the community. He always, throughout his life, had true respect for decent working people (e.g. Tom Ross, who revered Reich) and by the end of his life preferred their company to that of his highly educated colleagues. Eva says that Reich would often drive around all day, meet local people and discuss everything under the sun. (Joe, on our travels, often spoke to total strangers with easy familiarity. He's an outgoing guy.)

So there is a possibility that Reich took time to drive several hours into the next state to visit Roswell. Since the government had issued at least three different stories about what had happened at Roswell (first reports said specifically a "flying saucer" and later reports said it was a "weather balloon," and recently it purported that the incident could be explained by a crashed test-flight of an experimental aircraft filled with mannequins), Reich may have believed that what happened at Roswell was being treated by the USAF as a war-time military secret. This would be completely consistent with Reich's own attitude toward the UFO phenomena, and it would have been natural for him to be silent on what, if anything, he learned of the crash at Roswell.

I would have never printed Blankenship's comments had I not first discovered, by studying *Contact With Space*, that Reich was indeed *in* Roswell over the course of his journey to Tucson. In fact, he says clearly that he spent the night in Ruidoso, NM. After passing through Roswell, it's along the only road to Tucson from the East.

Reich reported:

"Although it was very hot as we neared Roswell, New Mexico, no OR flow was visible on the road, which should have been 'shimmering with "heatwaves".' Instead, DOR was well marked to the west against purplish, black, barren mountains, in the sky as a blinding grayness, and over the horizon as a grayish layer. The caking of formerly good soil was progressively characteristic and eventually the caked soil prevailed over the vegetation, which now consisted only of scattered low bushes, while grass disappeared.

"After the desert valley it was a relief to spend a night in Ruidoso, New Mexico, in the Sierra Blanca Mountains (near 7,000 feet). Here a strong, reactive secondary vegetation had sprung up, again more marked on the western slopes."[37]

Therefore nobody can say that Reich was never in Roswell. Moreover, Ruidoso is much nearer to the alleged crash site at a ranch near Corona, NM. Since he spent the night there, and he always liked to mingle with people, he may have picked up the story. Because the government, we now know, put out a "cover story" about what happened, perhaps he assumed that the government was treating the event as a national security issue, which exactly how Reich believed the UFO crisis should be handled——as a wartime emergency.

Did the local Roswell-Ruidoso populace have any knowledge of what happened? Stanton Friedman has published many interviews with local residents, so it's clear that many people there did know something about whatever went on.[38]

So, taking all the facts I have into account, I think it is an intriguing possibility, but an unlikely one. In some ways, it makes perfect sense. Yet I have no direct evidence, other than Joe's testimony, Reich's physical presence in Roswell, the wide publicity the event received, to prove the story.

One thing is certain: if Joe had not told me that story, I would have never looked into it and discovered that Reich passed right through Roswell. A delicious irony, if nothing else.

Joe Blankenship helped me understand the basic background of Tucson. He definitely knows that town's history, its famous citizens (personally) and the military-industrial interests operating there during the Cold War. On these matters, I credit Joe with an extremely high degree of reliability, since I was able to crosscheck his information with other sources.

Wilhelm Reich and the Cold War

D.S. Gillmor and "The Condon Report"

"The concreteness of my father's discoveries was in contradiction to most researcher's vagueness. Back then if you said you saw a UFO, you were crazy. But he was guided by his observations, and he was right."
 – Eva Reich

Published in 1968, "The Condon Report"[39] was heralded as the *Final Report of the Scientific Study of Unidentified Flying Objects*, was intended to debunk the earlier work done by Edward J. Ruppelt, former head of the US Air Force's Project Investigation Flying Saucers - better known as "Project Bluebook."[40] Major Ruppelt had written a study that was honest and forthright enough to raise serious questions about the phenomena. It appeared to Reich that Ruppelt was talking about Reich's own UFO investigations in certain parts of the book, according to his court records.[41]

In a "historical correlation" presented to the court, Reich pointed out that in chapter fifteen of Project Bluebook, called "The Radiation Story," Ruppelt described scientific efforts to set up Geiger counters to pick up any radioactive traces of "green fireballs" that had been sighted around the test site – UFOs in need of ID. Reich was incorrect here, since the record now shows that Ruppelt in fact was talking about a group of scientists at Los Alamos who, like Reich, had been intrigued by the possible relationship between UFO sightings and radioactive fallout. While Reich can be forgiven for reaching the wrong conclusion about this experiment - since Ruppelt states that he changed the names, dates and circumstances of those involved - it is intriguing that physicists working at the Los Alamos facility set up experiments identical to those Reich described in *Contact With Space*. "In the fall of 1949, at some unspecified place in the United States, a group of scientists had set up equipment to measure background radiation, the small amount of harmless radiation that is always present in our atmosphere," Ruppelt described. "This natural radiation varies to a certain degree, but will never increase by any appreciable amount unless there is a good reason."[42] Ruppelt tells of the scientists noticing a marked leap in background counts, when coincidentally a colleague walked in and reports seeing "three silvery objects moving in a V formation." Mind you, this is at *Los Alamos*, not Orgonon. "After the two scientists briefed their excited colleagues on the unusual radiation they detected, the three men asked each other the $64 question: Was there any connection between the two incidents? Had the UFO's caused the excessive radiation?" The following year, the scientists, adopting the cover of a "Mineral Club" for rock hounds, thus explaining their presence in the mountains without revealing to the public their actual interests, set up Geiger counters in an east, west, north, south configuration. Sure enough, they recorded as much as 100 times normal background counts at the same time UFOs were independently sighted. The Los Alamos scientists recorded four such events in a two-month span in 1950.

The Condon Report on UFOs (conducted by the University of Colorado under contract to the United States Air Force) was funded by the CIA, according to James E. McDonald, in Congressional testimony in 1968.[43] Condon's debunking of Ruppelt's report is *not* supported by the actual contents and data presented in the half-million dollar government study. The Condon Report was edited by Mildred Edie Brady's former employer, D. S. Gillmor, who was a wealthy progressive like his wartime acquaintance, Michael Straight.

Edward U. Condon was a strange choice to head the CIA's whitewash report on UFOs. Condon had his own problems with the redhunters in Congress, where he was called "one of the weakest links in our atomic security." Appointed to head the National Bureau of Standards by Secretary of Commerce Henry Wallace, Condon, a physicist, made rare and valuable contributions to the Manhattan Project since its inception in 1939. (Both Robert A. Brady and Fred Schlink of Consumers' Research had, at one time, worked for the Bureau of Standards; it should also be remembered that this Bureau had managed parts of the Manhattan Project). Yet, in public testimony, Elizabeth Bentley named Gregory Nathan Silvermaster,[44] a government economist, as Condon's Soviet contact. Confirmatory evidence of the charges against Condon were insufficient, and the Truman administration came to his defense. The only further information investigators could dig up was Condon's impolitic associations with leftists while at the University of California during his tenure there between September 1943 and February 1945 – Condon had attended parties with the Bradys, the Oppenheimers, and Haakon Chevalier. Even so, by 1951, Condon was forced out of his job at the Bureau of Standards. In 1954, the US Navy granted Condon a security clearance, but suspended it almost immediately. Condon thus was forced to resign as director of research at Corning Glass. Like Condon, D.S. Gillmor made no secret of his iconoclastic leftism, but in 1966 the CIA had confidence in them to perform the duties involved in the *Final Report of the Scientific Study of Unidentified Flying Objects*.

Wilhelm Reich v. USA (1955-1957)

What can be said about the contempt trial and verdict of *Wilhelm Reich versus the USA*? It has become, in the last twenty years or so, a celebrated case in the annals of freedom of expression in America. English Professor Jerome Greenfield, in his book *Wilhelm Reich versus the USA* published in 1974, has ably covered the particulars of Reich's legal case, and this book was recently been revised and translated into German. (The German edition is even more exhaustive because I proofread the English edition for Greenfield prior to the German publication.) In Greenfield's view, the injunction was unconstitutional. Like Reich's biographer, Myron Sharaf, Greenfield obtained many of the government files pertaining to Reich's case through the Freedom of Information Act, but both authors neglected to obtain Mildred Edie Brady's FBI files, nor did they research the backgrounds of the conspirators against Reich. As political liberals, they sounded almost embarrassed by Reich's claims of a Communist conspiracy. I can understand their skepticism, but the end result of their omission is that Reich's entire defense has been viewed as a symptom of his delusion.

Greenfield's book covered the legal case to the point of exhaustion and I won't go into it much further here. I would like to emphasize that the FDA and the Justice Department spent over two million dollars to obtain a conviction against Reich, and that US federal agencies that set that kind of priority against a target rarely come up empty-handed. As for Reich's legal strategy, the current trustee of the Wilhelm Reich Infant Trust Fund, Mary Boyd Higgins, has offered her thoughts on the case and I believe they are worth quoting here. The following remarks were recorded during a symposium at Orgonon in 1988.

> "There were two legal actions involving Reich and the FDA. The first one occurred when the FDA brought a complaint into the federal court asking the court to enjoin or prevent Reich from doing certain things. Reich was sent a copy of this complaint and

given a certain amount of time in which to respond. This as standard legal procedure."

Reich considered hiring a lawyer and having the case thrown out as unconstitutional; but Michael Silvert, who had broken the terms of the original injunction by transporting the accumulators across state lines, argued that to appear in court would be a capitulation to falsehood. Reich considered all sides of the issue, Higgins explained:

"He made the decision not to go into court. He wrote a statement explaining to the judge why he felt he couldn't go into court and he sent this as his response to the complaint. In so doing, he said he was fully aware that his response might not be accepted and he might be totally enjoined in all his activities.

"There was a possibility that the court might accept that statement as a legal response. It did not, and the complaint was turned into an injunction and issued 'in default.' That is, the court acted as if Reich had completely ignored the complaint and entered no response.

"After the injunction was issued, Reich continued with his work. He refused to give FDA agents access to anything. He wouldn't talk with them; he would have nothing to do with them.

"Then, as you know, a student of Reich, Dr. Silvert, came here to Orgonon in January, 1955 when Reich was in Arizona and he transported books and accumulators panels to New York City. The only person who was here at that time was the caretaker, Tom Ross. I've asked Mr. Ross what happened and whether he called Reich to check with him before Silvert moved the materials. He didn't. And he didn't because he thought Reich had left Silvert in charge. So he felt Silvert had a right to do this, even though he questioned it. He asked Silvert if he had discussed this with a lawyer and Silvert replied: 'Well, lawyers are like doctors. You can take their advice or not.' We'll never know if that meant he had talked with a lawyer or he hadn't. And there are a lot of unanswered questions we'll never know the answers to. But we do know that no effort was made to contact Reich, and Silvert took the material to New York.

"I'm sure you realize that much of the injunction per se is unconstitutional. You can't burn books in this country. You can't force people to do certain things. But the part of that injunction that can be enforced is that part dealing with the shipment of accumulators in interstate commerce, because the FDA is authorized to control medical devices in interstate commerce. That's what makes Dr. Silvert's action so significant. He flaunted the one aspect of that injunction that could be easily enforced.

"After this happened, the FDA moved in very quickly, and Reich was charged with contempt of court. This was the second action taken by the FDA against Reich. The first action was a civil action; Reich was told he wasn't supposed to do something. In the second action, he was accused of having disobeyed the injunction that had been issued by the court. He was thus in contempt of court. This is a criminal charge and Reich could no longer decide whether or not he would go into court. He was forced to do so."

Higgins concluded:

"Reich's experimental work remains largely untouched, its validity to be decided upon in 'prolonged, faithful bona fide observations in friendly exchange of opinion.' But his historic decision not to allow it to be judged by litigation has been validated conclusively: His books are available throughout the world, his archives are secured for future generations, and his Maine observatory stands proudly as The Wilhelm Reich Museum. Orgone energy accumulators, the 'devices' designed to concentrate cosmic Orgone Energy, and thus to make it available to further research in medicine, biology and physics, are being built today in many lands, and there is no statute which prevents anyone from investigating or utilizing the natural law on which they function. The injunction is untouched, but as Reich discerned, "Such an injunction would mean practically exactly nothing at all."[45]

I can only add to this one important confirmation. Eva Reich was present in Tucson in 1955 when Wilhelm Reich received a letter from Michael Silvert informing him that he had broken the main terms of the injunction by shipping orgone accumulators across state lines. Eva told me that her father was "completely surprised by Silvert's actions, and understood that it was a very serious matter." Reich had nothing to do

Reich's Final Years

with the violation of the court order.

Unbelievably, the 59 year-old scientist, researcher and teacher was sentenced to two years in prison and, as with many people his age, this proved to be a death sentence.

Prison and Death: 1957

"They let my father die in prison, when they knew the whole time there really were such things as UFOs. I stopped paying allegiance to the United States, I mean, I live here, but I think they are pigs."
-Eva Reich

Although they are sporadic and brief, documents revealed in a recent FOIA release of Reich's prison files[46] indicate many details about the tragic final months before Reich died in Lewisburg, Pennsylvania Federal Penitentiary on Sunday, November 3rd, 1957.

Prison officials derided Reich's marital status (his wife at the time, Aurora Karrer, was "common law"): "Since the defendant does not believe in marriage as an institution, he has never developed any strong home ties. Where he lives is mostly a matter of convenience to the defendant and if he has any sentimental attachments they are not connected with the home as an institution."

Following sentencing recommendations, Reich was ordered to undergo psychiatric evaluation. Among these psychiatrists, there was a general agreement that Reich suffered from "paranoid schizophrenia." They felt constrained to say that this assumption was "not based on personal observation."[47] (See the 1957 entry of the chronology toward the end of this book for a detailed and interesting sequence of events surrounding Reich's psychiatric evaluation.) "Reich gave no concrete evidence of being mentally incompetent." Yet they warned: "it is felt that Reich could easily have a frank break with reality, and become psychotic, particularly if the stresses and environmental pressures become overwhelming." "In his discussions he unraveled a rather intricate and somewhat logical system of persecutory trends, particularly regarding the Rockefeller Foundation[48] 'which made me a tool of its socioeconomic interpersonal relations'."

A copy of an article from *True* magazine called "The Marvelous Sex Box" was filed dutifully in Reich's files. Reich's parole officer commented that he had an "unusually keen interest in getting an early release." (I must ask what the officer meant by "unusually keen?") The parole officer, Frank Walker, Jr., reported "Reich went on to state that he was the victim of an international conspiracy and that is why he was sent here... that his situation was unique and different because the fate of the planet was involved." Reich's appeals for early release were denied. The Supreme Court dropped his final writ without comment.

Reich was suspicious when two volumes of specially-prepared defense briefs, called "Vol. IV and V of the Suppressed and Secret Evidence" were sent to him in prison with "different colored binding and handwriting that is familiar to me but which I cannot identify" on one of the covers. Reich believed that someone had tampered with private legal briefs.

Reich openly suspected that a conspiracy of "Hoodlums In Government" had been behind the decade-long, multi-million dollar prosecution. Indeed, his own personal lawyer, Peter Mills, turned up on the prosecution team, and much of the evidence[49] provided by the FDA failed to take into account Reich's experimental protocol. But the real hoodlums were among the nonprofit sector and directed the FDA and other federal agencies in a very subtle manner toward Reich.

Jail must have been pure hell for Reich, an unwilling martyr who had never crossed paths with the law before and who was already well isolated before his sentence. In an effort to find someone inside who could hear him out, he began speaking with the prison chaplains. He introduced himself as the author of a book called *The Murder of Christ* and began with saying, "Due to my discovery of the cosmic orgone energy and its social implications, I am in deep trouble, emotionally and socially." Reich was reaching out to the only officials in the prison who dealt with emotional life.

To Reverend Silber, Protestant minister, Reich wrote, in a handwritten "Inmate Request to Staff Member":

> "It was clear from the very beginning that prayer, and now lyrics, were subverted by such use of *stupidities and evasions* on our part, especially by the staid reluctance to talk bluntly & take the bull by the horns. The bull is really no more than a few *slimy tapeworms* eating away at our emotional guts. It is high time to start giving social power to the established functions of Love,

Reich's Final Years

Work & Learning as bastions against the tapeworms.
– *(signed) Wilhelm Reich, Sept. 14, 1957."*

Lewisburg prison was chock-full of interesting people in the mid-1950s. Alger Hiss served his time there after his conviction of perjury. He described his meetings with Frank Costello, the mobster from New York, with whom Hiss was friendly in Lewisburg. Costello had sought out Hiss for legal advice. Hiss told him that, according to his limited knowledge of trial law, Costello would be entitled to a new trial.

Harvey Matusow, a Communist who gave false testimony in the McCarthy hearings, and later recanted, was in Lewisburg. He now admits he had initially gone to the McCarthy people and became a paid informer in a scheme to discredit the investigation. For what it's worth I interviewed Matusow for this book, but obviously his testimony must be viewed with skepticism. He said that Reich stood out in the prison yard and looked up into the sky, saying "They're coming for me." Matusow inferred that Reich was talking about UFOs, but at the time, Reich had not been associated with UFOs publicly. He hadn't publicly distributed *Contact with Space*, only sent them to trusted associates and to the courts and the government agencies concerned. Reich's observations on UFOs were unknown outside the limited circulation of Reich's research journals, I remain skeptical that Matusow could have known about this aspect of Reich's work. Even so, the story has gained wide currency since Matusow published his piece "The Day Reich Died," in the *East Village Other*. Matusow claims that he had been handcuffed at one time to Frank Costello, who funded his art magazine after his release. We could not find the photo in the archive, and Matusow suggested we might not have the right edition of the paper.

"I was good friends with Alger too," Matusow told me. "After he served his time, he worked at Davidson-Bluth, selling printing. I didn't know him at Lewisburg, but when I saw him after my release, I went up to him and said, 'I bring greetings from J-1 Block,' which made him go pale. He knew it referred to the section of Lewisburg where we both had been. He wasn't getting many printing jobs at Davidson-Bluth. So I gave him as much as I could. I gave him about two years worth of work. I kept him alive." Hiss got by with the help of his friends.

One of the most intriguing characters who was at Lewisburg at this time was the rebel publisher, Sam Roth, a publisher who had been convicted of obscenity in 1957, fined $5,000 and sentenced to five years

Wilhelm Reich and the Cold War

in prison.[50] While in Lewisburg, Roth appealed his conviction all the way to the Supreme Court, and in April, 1957, the Court upheld Roth's conviction, while for the first time defining standards for obscenity. Previous to *Roth v. USA,* a single illegal word could result in the banning of a book or film. In *Roth,* the Court modified obscenity laws by setting forth the following test:

"...whether to the average person, applying contemporary community standards, the dominant theme of the material taken as a whole appeals to prurient interest." (Justice William J. Brennan, for the majority.)

The *Roth* decision led to an irreversible revision of sexually oriented material. Gay Talese, who devoted a chapter to Roth in his book, *Thy Neighbor's Wife,* wrote of the situation facing Sam Roth in Lewisburg: "Gradually, as one obscenity conviction after another was reversed by the Supreme Court, as banned novels and erotic art films were suddenly redeemed by Roth, the name became more easily recognized as an italicized legalism than as a reminder of the illegal man now residing in prison at Lewisburg; and ironically, Roth could have received through the mail into his cell most of the books that had contributed to his being there."[51]

Born in the Austro-Hungarian region of Galicia amidst the Carpathian mountains in 1895, Roth emigrated to the US in 1904. A poor student but an avid reader, Roth became a follower of the anarchists Emma Goldman and Alexander Berkmann in his youth.

In the 1920s Roth owned a bookstore in New York City, where he began to dabble in publishing. Without the permission of the author, Roth serialized James Joyce's *Ulysses,* which had been banned in the US, in his own periodical *Two Worlds Monthly.* He began a mail-order operation and illegally sold "obscene material" and was jailed on many occasions. Yet Roth was no ordinary pornographer, and cast his business activities as a principled battle for freedom of expression.

Along with *The Perfumed Garden* (a fourteenth-century ritualized Arabian love guide) and the *Kama Sutra* (the Hindu sex manual), Roth purveyed banned literary classics such as D.H. Laurence's *Lady Chatterley's Lover.* There was an insulting biography of Herbert Hoover, a theological tome *The Woman Who Was Pope,* and a hoaxed book allegedly written by Friedrich Nietzsche, wherein he confesses to an incestuous relationship with his sister, published in 1951 as *My Sister and I.* (In 1965, George David Plotkin confessed to having

Reich's Final Years

written the "Nietzsche" book on assignment from Sam Roth.[52])

Wilhelm Reich had read *My Sister and I* as soon as it appeared and was apparently taken in by the ruse because he refers to it in his book, *The Murder of Christ*.[53] Elsewhere in an essay called "The Silent Observer," Reich wrote that Nietzsche surpassed himself. "Never did he tell the truth as clearly, as penetratingly, as in *My Sister and I*."[54] (The book is a fairly convincing hoax and he wasn't the only one fooled; professors of philosophy were convinced that Nietzsche wrote at least parts of the book.)

Harvey Matusow, who was a "very close friend" of Roth's in Lewisburg, says that Reich and Roth sometimes ate lunch together in the prison cafeteria.[55] One can only imagine what the two men spoke about; perhaps they worked toward a definition of obscenity (Reich hated pornography for the way it strips sexuality of its emotional content), or discussed the issues of freedom of speech pertaining to their legal cases, or maybe they talked about their youth in Galicia.

Habeas Corpus

Wilhelm Reich was found dead in his cell at the 7 a.m. head count in Lewisburg Federal Penitentiary on November 3, 1957, only seven days short of his parole eligibility. Fully clothed except for his shoes, his body was found lying on the prison cot. Dr. Lacovara came to the cell at 7:15 a.m. and took the corpse to the morgue. At 9:20 a.m., Eva Reich was notified of her father's death. The Warden told her that an autopsy should be performed. Eva opposed this at first, but told the warden she'd call back when she had a chance to speak with her family. In the interim, the Warden called the Director of the Bureau of Prisons in Washington, DC, and Dr. Janney, who discussed the matter and agreed that Warden Taylor should order an autopsy. In shock and sadness, Reich's immediate family decided not to have an autopsy performed. Eva phoned the Warden, telling him that Reich "probably died of a broken heart." At 2:15 p.m., Dr. Lavocara observed Dr. Willard Christman, a pathologist, as he conducted the autopsy. In Lacovara's opinion, Reich died of "coronary insufficiency with calcified aortic stenosis and generalized arteriosclerosis." The official cause of death was an heart attack, and the autopsy showed enough formaldehyde in Reich's system to interfere with testing for other compounds. All "specimans [sic] sent to the National Institute of Health at request of Dr. Janney." On November 4, J.C. Taylor, Warden, filed Reich's autopsy report with the Director of the Bureau of Prisons. The next day, Reich's corpse arrived in Rangeley, Maine. On November 19, L.L. Ashburn, MD (chief of HEW's Section on Pathological Anatomy) advised Lewisburg Prison's chief medical officer that he had no facilities to check the stomach contents from the autopsy. Microscopic studies "reveal marked myocardial scarring and a severe bronchiopneumonia."

After Reich's death, Mrs. Aurora Reich requested information about a packet of unsent letters that Reich wanted published after his death. "You may one day read that (not sent) letter [...blank line: erased...] large publishing house my "Silent Observer" or "Creation": The 3rd Volume of the History of Orgonomy. This volume is nearly completely conceived and constructed." This reference, found in Reich's letter to Mrs. Reich dated September 16th, 1957, apparently refers to the never-published book Reich mentioned having written while in jail. This manuscript has never turned up. Reich sent it to the Department of Education, to be forwarded to Mrs. Aurora Karrer Reich, where she worked.

Mrs. Reich's lawyer wrote to the Warden at Lewisburg in an effort to obtain the missing manuscript, as well as to report that on her last visit, Reich had mentioned "that he had asked for aspirin and had been given two pink pills instead. ...I wondered whether you would permit us access to the prison hospital and dispensary record concerning the decedent." In the margins of the letter in the records, the Warden pencilled in: "NO."

In the final analysis, the real tragedy in Reich's life and death was the silence and obscurity that greeted his important discoveries. As replication after replication of Reich's experimental findings pile in today, there is still a conspiracy of silence around the facts.[56]

Reich's Final Years

Notes

1. For an example of the way dissident views can spur the majority on to new developments, see James Strick's article on Adrianus Pijper in "Swimming Against the Tide: Adrianus Pijper and the Debate over Bacterial Flagella, 1946-1956." *Isis,* 87:2 (Chicago: The History of Science Society, June, 1996).

2. *Fury on Earth*, p. 354n.

3. *Selected Writings*, p. 358.

4. *Selected Writings,* p. 410.

5. KGB General Sudoplatov wrote: "Oppenheimer reminded me very much of our classic scientists who tried to maintain their own identity, their own world, and their total internal independence. It was a peculiar independence and an illusion, because both Kurtachov [USSR's scientific leader on a-bombs] and Oppenheimer were destined not only to be scientists but also directors of huge government-sponsored projects. The conflict was inevitable; we cannot judge them, because the bomb marked the opening of a new era in science, when for the first time in history scientists were required to act as statesmen." *Special Tasks*, p. 209.

6. *Selected Writings*, p. 438.

7. Interview with Tom Ross videotaped by Marilyn Bacon (December 28, 1990), in the author's possession.

8. *Conspiracy: An Emotional Chain Reaction,* 1954, No. 417.

9. Israel, Lee. *Kilgallen.* New York: Delacorte, 1979, p. 177.

10. Lincoln Lawrence, *Were we Controlled*, published as *Mind Control, Oswald and JFK*, edited by Kenn Thomas, Kempton, IL: Adventures Unlimited, 1997, p. 162.

11. *Selected Writings,* p. 462.

12. *Conspiracy: An Emotional Chain Reaction,* item 439B.

13. *New York Times*, Nov. 7, 1988.

14. Humphreys, W. J. *Rain Making and Other Weather Vagaries.* Baltimore: The Williams & Wilkins Company, 1926.

15. Breuer, George. *Weather Modification; prospects and problems.* Cambridge, 1980. p. 170.

16. Halcy, D. S. *The Weather Changers.* Harper & Row, New York, 1968, p. 95.

17. Bird, Kai. *The Chairman,* p. 101.

18. Quigley, Carroll. *Tragedy & Hope*. p. 953.

19. Vonnegut, Kurt. *Welcome to the Monkey House.* New York: Dell, 1950, p. ix.

20. Battan, Louis J. *Harvesting the Clouds; Advances in Weather Modification.* New York, Doubleday, 1969, p. 47.

Wilhelm Reich and the Cold War

21. Houghton, Henry G. *Physical Meteorology,* Boston: MIT, 1985, p. 220.
22. Houghton, p. 220.
23. Breuer, George. *Weather Modification; prospects and problems.* Cambridge: Cambridge University Press, 1976. 1980, p. 12.
24. Breuer, p. 130.
25. *Contact With Space,* p. 132.
26. For an excellent discussion of this subject, see Peter Robbins' article, "Wilhelm Reich and UFOs," *Journal of Orgonomy,* 24:2 and 25:1.
27. *A Soldier Speaks, Public Papers and Speeches of General of the Army Douglas MacArthur,* edited by Major Vorin E. Whan, Jr., Frederick Praeger Publishers, 1965, p. 349.
28. *Contact With Space,* p. 238.
29. *Arizona Daily Star,* April 29, 1955, Sec. A, p. 2.
30. Battan, Louis J., *Harvesting the Clouds; Advances in Weather Modification.* Doubleday, New York: 1969. Photos, including USAF "cloudbuster" dry ice dispenser for fog removal.
31. Constable, Trevor James. *Loom of the Future.* Bayside, Borderland, 1994, p. 119.
32. "Churchill demanded that he be given information about UFOs, but his own people denied him. They gave him the whole MJ-12 slogan that was going to be the 'truth' for the rest of the world. Who's in charge? Who is running the world? If Churchill didn't find out the truth, who's behind it? Somebody is running the show, and it's not who we think it is. " - Eva Reich, author interview.
33. Warren, Larry and Peter Robbins. *Left at East Gate.* New York: Marlowe & Co., 1997, p. 227.
34. Corso, Philip. *The Day After Roswell.* With a foreword by Senator Strom Thurmond. New York: Pocket Books, 1997, p. 75.
35. *Contact With Space,* p. 253.
36. "Operation Weather Control, Part Two," from *Flatland Magazine* #13, page 30. Interview with Joe Blankenship, 8/31/95.
37. *Contact With Space,* pp. 119-120.
38. Friedman, Stanton, *Crash at Corona.* New York: Paragon House, 1992.
39. Condon, Edward U. and Daniel S. Gillmor (ed.) *Final Report of the Scientific Study of Unidentified Flying Objects* (Conducted by the University of Colorado Under Contract to the United States Air Force) New York: Bantam Books, 1968. 965pp.
40. Ruppelt, Edward. *The Report on Unidentified Flying Objects.* NY: Doubleday, 1956.
41. see Ruppelt's comments about the "Mineral Club."

42. Ibid., p. 264.

43. "Symposium on Unidentified Flying Objects." House Science and Astronautics Committee (90th Congress, 2nd Session). Committee Print No. 7. July 29, 1968.
http://nicap.org/books/1968Sym/1968_UFO_Symposium.pdf

44. Silvermaster was a Russian-born economist who served at the Farm Security Administration.

45. *Newsletter for Friends of the Wilhelm Reich Museum*, #24 (Fall, 1988).

46. My thanks to Kenn Thomas of *Steamshovel Press* for obtaining these documents.

47. Documents dating from early FBI assessments of Reich as a national security risk contained Otto Fenichel's privately spoken slander. It was repeated in the prison's subsequent "diagnosis."

48. Nelson Rockefeller paid for a "feasibility study" for *The New Republic*. From the Rockefeller Archives.

49. Today, the FDA has responds to FOIA requests for this evidence by declaring it no longer resides in their files.

50. Talese, Gay, *Thy Neighbor's Wife,* New York: Doubleday, 1980, p. 74-89.

51. Ibid., p. 89.

52. For debunking of *My Sister and I,* which was reprinted by "Amok Books" a few years ago as the "real article," see *Saturday Review,* (May 24, 1952); *Partisan Review* (May/June 1952); *Philosophical Review* (January, 1955).

53. *Murder of Christ,* p. 3.

54. *Orgonomic Functionalism,* Vol. 1, Rangeley: Wilhelm Reich Infant Trust Fund, 1990, p. 98.

55. Author's interview with Harvey "Job" Matusow 10/31/96.

56. To cite a few:

- Braid, B. and R.A. Dew: "Reich's Bioelectric Experiments: A Review with Recent Data" *Annals of the Institute for Orgonomic Science* 5, September 1988, pp. 1-18.

- Harman, R.A.: "Current Research with SAPA Bions" *Journal of Orgonomy,* 21(1), May 1987, pp. 42-52.

- Moss. T.: *The Body Electric. A Personal Journey into the Mysteries of Parapsychological Research, Bioenergy, and Kirlian Photography*, J.P. Tarcher, Los Angeles, 1979.

- Kolokolstev, S. : "An Accumulator Of Subtle Energy". Aura-Z, Moscow, 1993. 2:85-87.

- Baker, C.F., R.A. Dew, M. Ganz, and L. Lance: "Wound Healing in Mice: Part 1" *Annals of the Institute for Orgonomic Science* Vol. 1, September 1984, pp. 12-23; Part 2, ibid. 2, September 1985, pp. 7-24.

- Müschenich, S. & Gebauer, R. "Die (Psycho-) Physiologischen Wirkungen des Reich'schen Orgonakkumulators auf den menschlichen Organismus" [The (Psycho) Physiological Effects of the Reich Orgone Accumulator on the Human Organism] University of Marburg, Germany, Department of Psychology, Dissertation 1986.

- Hebenstreit, G. "Der Orgonakkumulator nach Wilhelm Reich. Eine Experimentelle Untersuchung zur Spannungs-Ladungs-Formel" [The Orgone Accumulator After Wilhelm Reich. An Experimental Investigation About the Tension Charge Formula] University of Vienna, Austria, Dissertation 1995.

- Lappert, P.W. "Primary Bions through Superimposition at Elevated Temperature and Pressure" *Journal of Orgonomy*, 19(1), May 1985, pp. 80-91.

- Harman, R.A. "The Pendulum Experiment Reconsidered" *Journal of Orgonomy*, 18(1), May 1984, pp. 29-41.

- Eden, J. *Orgone Energy - The Answer to Atomic Suicide,* Exposition Press, New York, 1972, p. 156.

- Deutsch, B.G. and P.S. DeCamp "Rediscovering Melanor, 1979-1980" *Journal of Orgonomy*, 14(2), November 1980, pp. 163-170.

- DeMeo, James. *Preliminary Analysis of Changes in Kansas Weather Coincidental to Experimental Operations with a Reich Cloudbuster,* University of Kansas, Geography-Meteorology Department, 1979. Reprinted Natural Energy Works, 2012.

- DeMeo, James: "OROP Israel 1991-1992: A Cloudbusting Experiment to Restore Wintertime Rains to Israel and the Eastern Mediterranean During an Extended Period of Drought," *Pulse of the Planet*, 4:92-98, 1994.

- Eden, J. *Planet in Trouble - The UFO Assault on Earth*, Exposition Press, New York, 1973, p. 214.

Documentary Appendix:

1. Document 20 (1936)*

Cadres Department memorandum on "Trotskyists and other hostile elements in the émigré community of the German CP."

4 cop[ies]. mkh.
4. IX. 36.   Top secret

1936
To c. Sergeyev[i]
Carry out and report
on the <u>results</u> of the verification of
the German émigrés.
9. 36. GD[ii]

                To: <u>Com. Dimitrov.</u>
                        <u>Com. Manuilsky</u>
                        <u>Com. Moskvin.</u>

The Cadres Department is sending to you a memorandum "On Trotskyists and other hostile elements in the émigré community of the German CP."

We here mention only the most typical cases. Actually, the work on exposure is in progress, and a much larger number of these kinds of elements has already been revealed.

                                      Chernomordik
                                      /Chernomordik/[iii]

(Continued next page)

---

* English Translation of Russian Original, from Yale University Archives: http://www.yale.edu/annals/Chase/Documents/doc20chapter4.htm

5 cop[ies]. mkh.
2. IX. 36. **Top secret.**

MEMORANDUM
ON TROTSKYISTS AND OTHER HOSTILE ELEMENTS
IN THE ÉMIGRÉ COMMUNITY OF THE GERMAN C.P.

Among the German émigrés in the USSR there are people who were known in the CPG as active Trotskyists and factionalists before their arrival in the USSR:
[...]
10)   OTTO KNOBEL (OTTO BRANT)[xxv] – a former worker of the YCL of Germany who worked in the International Publishing House for Youth in Germany and who, in 1933, emigrated to Paris without the party's consent. [He] was not granted political émigré status. [Knobel] returned to Germany and then emigrated again. In Paris, [he] was involved with the Trotskyists. [He] moved to Copenhagen to work, where he was not connected to the party and did not establish connections with the émigré community. In Copenhagen, [he] worked in the publishing house of Wilhelm Reich[xxvi] who had been expelled from the CPG for Trotskyism. According to him, he broke with Reich over personal differences. However, according to some party comrades, he went to Berlin with Reich's consent and, a month later, in late 1935, came to the USSR via Inturist without connections and without the party's permission. He was so close to Reich that he read [Reich's] letters to Trotsky and even mailed them himself.
    In April 1936, the Cadres Department reported Knobel's past to the responsible organs. His last address [was]: Hotel "Novomoskovskaia."
[...]

[xxv] Otto Knobel (Brandt). Born in 1908 in Schwerin, Germany, he was a member of the CPG from 1929. In June 1935, he moved to the USSR and taught at the Karl Liebknecht German school in Moscow. He was arrested on 1 October 1936 and, on 22 June 1937, sentenced to five years in a corrective labor camp.

[xxvi] Wilhelm Reich (1897-1957). A psychoanalyst who specialized in the problems of sex and sexual education, he joined the CPG in 1930 and was expelled in 1934.

Documentary Appendix

2. Silvermaster VENONA Decrypts (1944)

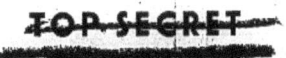 

- 2 - S/NBF/T1 77

1. ALBERT asks for word to be passed to you that EL'[b] and he [D% profoundly]
   [2C groups unrecoverable]

2. ROBERT is sincerely overjoyed and profoundly satisfied with the reward [given him][c] [Q% in accordance with your instructions]. As he says his work for us is the one good thing he has done in his life. He emphasised that he did not take this only as a [C% personal] honour, but also as an honour to his group. He wants to see the reward and the book.

3. DORA[viii] is very uneasy about the fate of her relations [and][c] again asks news of their whereabouts.

No. 914
21st November
MAY[MAJ][ix]

T.N.: [a.] Given in Latin letters in the original.
      [b] I.e. the Russian name for the letter 'L'.
      [c] Inserted by translator.

Comments: [i] Lt.-Gen. P.M. FITIN.
          [ii] Nathan Gregory SILVERMASTER.
          [iii] Brigadier-General Patrick Jay HURLEY, appointed U.S. Ambassador to CHINA in December 1944.
          [iv] U.S. Department of State.
          [v] Office of Strategic Services [O.S.S.].
          [vi] LONDON.
          [vii] The British.
          [viii] Helen (WITTE) SILVERMASTER, wife of ROBERT.
          [ix] Probably Stepan Zakharovich APRESYaN, Soviet Vice-Consul in NEW YORK.

This two-page item is an example of the NSA's newly released VENONA documents, which decoded Soviet spy communications during the war and contributed to the discovery of hundreds of moles in the US government. Here, a cable describes Nathan Gregory Silvermaster receiving an award for his service to the Soviet Union. Silvermaster and his wife, Helen, were close friends of Mildred and Robert Brady.

TOP SECRET DINAR VENONA

Reissue (T()

From: NEW YORK
To: MOSCOW
No: 1619-1620

20 November 1944

[Part I]

To VIKTOR[i].

Your telegram No.4582[a]. Felix SOMARY, an Austrian of Jewish origin, became a Catholic, married to an Austrian Catholic; was a financial adviser to the German government during the last war, broke with the Kaiser on the question of submarine warfare.

SOMARY is a partner in the firm BLANKART and Company (Geneva and Zurich)[ii] together with Jacques BLANK[iii] and Arthur EITLER[iii]. The Vice-Consul of the COUNTRY[iv] in Zurich is an official of NABOB's[v] Department. OSTROW[vi] in his time consulted SOMARY about financial questions. S. is reported to be a friend of Archduke OTTO and he assisted a commission headed by General BUTTERWORTH[vii].

[20 groups unrecoverable]

S. was a member of the German Finance Control Commission in Belgium where he was an associate of SCHACHT[viii]. He was subsequently associated with a German bank in Tokyo.

S. left Germany before 1939 and settled in Switzerland, from there in 1939 he travelled to the COUNTRY[iv] on a Swiss passport. After that he returned to Switzerland and again went to the COUNTRY in 1941 via the COUNTRYSIDE [DEREVNYa][ix] on a Swiss quota visa. According to information dated February this year S. is advising TOWNSMEN [GOROZhAN][x], firms and corporations in respect of claims against Germany [5 groups unrecovered] their property in Axis territory.

[Part II]

It is reported that from time to time S. submits a report on his activities to NABOB's Department[v]. He also frequently consults, they say, with Leon FRASER, (former head of the Bank for International Settlements) who spent a considerable part of the war in Switzerland.

S. is reported also to have been associated with CALIPH [KALIF][xi]. He is represented by a lawyer of the firm WILLKIE, OWEN and GALLAGHER[xii]. There was information that S. is an unofficial representative of SCHACHT and of German industrial-financial circles in the COUNTRY.

Documentary Appendix

TOP SECRET DINAR VENONA

```
        ROBERT[xiii] received this information from the
        FELLOW-COUNTRYMAN [ZEMLYAK][xiv] Norman BURSLER[xv].

                        [16 groups unrecoverable]

        ---MAN[b] irregularly since the other fears for his own safety
        and [C% is reluctant to come to our aid.]

                        [8 groups unrecovered]
        Later [C% NORMAN] will try to find out in detail about S. and
        his activities.

        No.908                                        MAY[MAJ][xvi].
```

Notes: [a] Not available.

[b] ---MAN: This may be NORMAN, i.e. Norman BURSLER.

Comments:
  [i] VIKTOR: Lt. Gen. P. M. FITIN.

  [ii] i.e. BLANKART & Cie.

  [iii] Jacques BLANKART and Arthur REITLER are the correct renderings of these names.

  [iv] the COUNTRY: U.S.A.

  [v] NABOB: Henry MORGENTHAU Jr., U.S. Secretary of the Treasury.

  [vi] Mr Walter W. OSTROW, U.S. Vice-Consul in Zurich.

  [vii] Gen. J.B. BUTTERWORTH, Chief of the Conservation Branch, Services of Supply, United States Army.

  [viii] Dr. Hjalmar SCHACHT, head of the German Reichsbank.

  [ix] the COUNTRYSIDE: Mexico.

  [x] TOWNSMEN: United States citizens.

  [xi] CALIPH: William Christian BULLITT, U.S. Ambassador to France 1936-1941.

  [xii] This is actually the firm of WILLKIE, OWEN, OTIS, FARR and GALLAGHER.

  [xiii] ROBERT: Nathan Gregory SILVERMASTER.

  [xiv] FELLOWCOUNTRYMAN: Member of the Communist party.

  [xv] Norman Chandler BURSLER, expert in the Anti-Trust Division of the Department of Justice.

  [xvi] MAY: Stepan Zakharovich APRESYaN, Soviet Vice-Consul in New York.

Another member of the Silvermaster spy ring, Norman Chandler Bursler, is identified in this VENONA decrypt. Bursler, like Silvermaster, was a student of Robert Brady's at the University of California in the 1930s. Brady thanked Bursler for his research assistance in his book on German Fascism.

3. Radioactivity and Plant Growth (1947)
by Henry Wallace
from *The New Republic,* October 13, 1947

Because of the accident of my talking with Einstein about my experiments with radioactivity and plant growth, in the presence of a newspaperman, a lot of publicity broke out in the American press late in September. I told Einstein that the yield of tomatoes seemed to be increased by exceedingly minute doses of radioactive material. A few days later came the story by the Japanese scientist, Takeo Furuno (see the *NR,* October 6), about the increased plant growth at Nagasaki as a result of the atomic-bomb explosion.

My own personal experiments with by-product material from a radium-processing plant began several years ago in a casual way in my sister's garden in Washington. I started the work as a result of conversations with Boris Pregel of the Canadian Radium and Uranium Corporation, who had first become interested in possibilities along this line many years ago. When I moved to New York, we set up several cement tanks where we could absolutely control both the water and the supply of radioactive fertilizer.

The differences in plant growth were not striking to look at, but the radioactivated tomatoes did give more fruit and the taste seemed to be slightly different. Carrots, watermelons, muskmelons and squash seemed to be more resistant to disease when radioactivated. Sweet corn was earlier and more sturdy.

The only comprehensive work along this line seems to have been done by a Russian by the name of Professor V. Vernadsky, whose experiments indicate that the best results were obtained with from one to 10 pounds of uranium per acre. But much smaller quantities of radium gave the same results and it proved that the results varied directly as the radioactivity of the material.

The principle by which radioactive material benefits plants is as different from the principle by which fertilizers do so, as vitamins are different from starches, fats and proteins in human and animal nutrition. Apparently the excessively minute quantities of radioactive material which enter a plant give off energy that is important to the plant cells. Pregel tells me he believes that the energy from radioactive material affects the plant cells in the same way as cosmic rays, that the effect in

Documentary Appendix

either case is a series of nuclear reactions within the living cell.

Mrs. Wallace has been worried about eating the radioactivated tomatoes. On checking into this, I find that the quantities that a plant can absorb are so small as to have no harmful effect whatever. Further experiments will be necessary before we shall be able to tell whether radioactivated crops will have a beneficial effect. Some day, after I have developed a number of inbred strains of chickens, I hope to do a little experimentation with radioactive material in either the food or the water of the chickens.

The more I think about it the more certain I am that radioactivity and cosmic rays hold many of the secrets of life. I am glad the Russians are doing constructive work on the peacetime uses of radioactivity, and I trust we too shall soon show some imagination in this field. If the preliminary indications are correct, we shall soon be producing at Oak Ridge by-product material sufficient to fertilize all the crops of the U.S. If we don't want to use it here, we should send it to Europe. The cost of shipping radioactive fertilizer is very low.

Wilhelm Reich and the Cold War

4. The Reich-Frank Letters (1948)

ILSE OLLENDORFF REICH
99-06 66TH AVENUE
FOREST HILLS, NEW YORK

Orgonon
P.O.Box G
Rangeley, Maine

den 10. Juni 1948

Lieber Karl Frank:

Ich wäre Dir sehr dankbar, wenn Du mir sobald wie möglich schreiben könntest, was Du über das "Spanish Refugee Committee", Chairman Dr. Barsky, weisst. Ist das eine kommunistische Frontorganisation oder wirklich eine bona fide antifaschistische Organisation. Wir erhielten vor kurzem eine Aufforderung, Geldspenden zu geben, und die Liste der "National Sponsors" ist eine merkwürdige Mischung, und wir möchten gern wissen, mit wem man es da zu tun hat.

Herzlichen Gruss,

Ilse Reich

June 10, 1948

Dear (Lieber) Karl Frank:
I would thank you (Dir) very much for writing me as soon as possible what you (Du) know about the "Spanish Refugee Committee", Chairman Dr. Barsky. Is this a Communist front organisation or a real bona fide anti-fascist organization. Recently we received a request to give donations and the list of the "National Sponsors" is an odd blend, and we would like to know with whom one has dealings with.
Kind Regards,
Ilse Reich

Documentary Appendix

```
            WILHELM REICH, M. D.
            99-06 SIXTY-NINTH AVENUE
            FOREST HILLS, NEW YORK
               BOULEVARD 8-5997
```

den 9. April

Herrn Karl B. Frank
210 East 15th Street
New York City

Sehr geehrter Herr Frank:-

    Frau Reich hat gestern noch am Tage vor Ihrer Abfahrt versucht sie telefonisch zu erreichen, doch ohne Erfolg. Frau Reich wäre Ihnen dankbar, wenn Sie das Buch im Laufe des Aprils zurückschicken würden.

    Mit freundlichem Gruss,

Ingeborg Schmidt
(Sekretärin)

April 9, (????)

Dear (sehr geehrter) Mr. Frank:-
Mrs. Reich yesterday tried to get you on the phone on the very day before your departure, but without success. Mrs. Reich would thank you for sending the book back during April.
Sincerely,
Ingeborg Schmidt
(secretary)

Wilhelm Reich and the Cold War

5. Moise-Douglas Memorandum (1954)

WILLIAM MOISE (MAN FROM MAINE INTERESTED IN WEATHER CONTROL) WILL BE IN N. Y. EARLY WEDNESDAY AND HOPED HE COULD SEE OR SPEAK TO YOU. I EXPLAINED YOUR SCHEDULE WAS VERY FULL AND IT WAS NOT LIKELY YOU COULD SEE HIM. WILL YOU TALK TO HIM ON THE 'PHONE. YES_____ NO __✓__

MR. MOISE CALLED TWICE FROM MAINE AND HOPED URGENTLY THAT HE TALK WITH YOU ABOUT A NEW APPROACH FOR COMBATING THE DESERT DROUGHT.

MR. WILLIAM MOISE OF THE ORGONE INSTITUTE, ~~RANGELEY~~ HANCOCK, MAINE CALLED TODAY TO SEE IF HE COULD SPEAK WITH MR. DOUGLAS. HE HAD JUST COME FROM WASHINGTON AND HAD SPOKEN TO PEOPLE IN THE DEPT. OF AGRICULTURE, WEATHER BUREAU AND IN MR. GARDNER'S OFFICE ABOUT WEATHER CONTROL. THEY SUGGESTED THAT MR. D. MIGHT BE INTERESTED IN INFORMATION HE HAD. HE MENTIONED THAT HE HAD COME TO N. Y. FOR PURPOSE OF MAKING RAIN BUT THAT HE WOULD BE LEAVING FOR MAINE AGAIN POSSIBLY TOMORROW. HE SAID HE MIGHT WRITE MR. D. IN TUCSON OR CONTACT HIM AT SOME LATER DATE.

7/27/54 telegraphed Moise - suggested that he write LWD a letter

This document records the initial contact of Lewis W. Douglas by Reich's representative, William Moïse. Subsequently Douglas and Reich corresponded and Douglas' bank assisted Reich in locating a base of operations on Arizona.

Documentary Appendix

6. Talk With Eisenhower (1955)

From a document in the Lewis W. Douglas Archive, this is a memo written by John J. McCloy, who kept Douglas and the Council on Foreign Relations' "Study Group" on Russia informed of a conversation he had with President Eisenhower on Thursday, March 10, 1955 in the Oval Office. This was a few months before Moïse contacted Douglas .

I include portions of this memo because it spoke to the kind of pressures these men were under; a soldier, a banker and a diplomat-rancher; they were humans after all and yet they held the future in their hands. War had not yet become a video game. Nuclear bombing of Russia was actively discussed at the highest levels of government, although cooler heads prevailed. This description offers a glimpse into the mind-set of Eisenhower, Douglas, and McCloy, and they felt their way into the future, back in 1955.

It was a long discussion about what to do about the Russians. As a prelude, Eisenhower told McCloy of a horrifying meeting he had with General MacArthur at the invitation of John Foster Dulles, after Eisenhower's election. MacArthur suggested that the new President

"...should go see Stalin immediately on his inauguration, for at no later time would his prestige or influence be so great; that he [MacArthur] would say to him (Stalin): these are the terms on which we must deal, this amount of disarmament must take place; *and* if Stalin did not agree we would drop the bomb. This is so radically different from the approach I [McCloy] had been led to believe MacArthur had advanced that I would like to check with Dulles. The President has a memory of an elephant and he is incapable of dissembling, but I wonder whether they could have misunderstood each other.[...]

"He talked of the difficulty we were in—the uncertainties and the dilemma which faced us. [...]First, of course, was strength; but strength was not all; we had to negotiate from strength but we had to keep seeking the basis on which to go from there—what definite agreements could be made which could be lived up to and whose implementation could be checked; agreement alone meant nothing; Roosevelt had plenty of agreements from Stalin and though Roosevelt, according to him (Eisenhower) reached almost to the point of idiocy in his concessions to the Russians and his fatuous conviction that he knew how to deal with them, yet Roosevelt had gotten some agreements in respect to the Far East, Poland, etc. that they had completely ignored. It was no good executing a few more agreements which we fulfilled and they did not. [...]

"Why is it, said he, that our allies are not critical of the Russians there, or why are they not critical of the Russians' violation of the Korean armistice or treaty? All the Neutral Commission does, according to the President, is to enforce the terms against us, as they have no means of operating against the Russians. Why do the British express concern over Matsu and Quemoy *and completely ignore Hong Kong and Cyprus?* We have difficulty speaking the same language as the British but we must or we are both in real trouble, for our alliance with Britain remains one of the world's great hopes. [...]

"The trouble was that there was such a record of violations of agreements on the part of the Russians that we could not move into the negotiations with any real faith, and this almost doomed them from the start. Yet something, he was convinced, should be tried. [...] What it all required was *faith and almost superhuman patience.* People came to him and said we must get more and more scientists and engineers; what he said we needed was more philosophers—some great new moral philosophers; we didn't have enough of them and they were needed badly. [...]

"He talked about his paintings and an ancient, life-size ibis which had been taken from a tomb in Egypt and had been given to him by Naguib; the bird of wisdom which constantly evaded mankind. It stands alongside his desk.

"These are only the high spots of a very refreshing and fascinating talk ranging over even some other things than are here recorded, and I came away again convinced that this was a deeply spiritual, knowledgeable man, well-equipped to be the President of the United States—indeed, better equipped than any man I know.[...]

JJmcC." [John J. McCloy]

Documentary Appendix

7. The Brady FBI File (1955)

The smoking gun: the following four pages of Mildred Edie Brady's FBI files comprise one of several sections that have been newly classified as "state secrets" under the National Security Act of 1947, and the CIA act of 1949. My appeal to this decision was denied. What could be so important about this information that would require such secrecy in 1998, and beyond? More than likely, the material came from the NSA's VENONA project, and the subsequent heading "Security Matter C" would indicate that Mildred Edie Brady's name appeared in Soviet cable traffic during the war years. Only a fraction of the VENONA decryptions have been released to the public.

329

Wilhelm Reich and the Cold War

Second page of the section of Brady's FBI file now declassified by the CIA and the NSA. Note that Brady is under active investigation by the FBI as late as September 28, 1955, during the time of Reich's trial.

Documentary Appendix

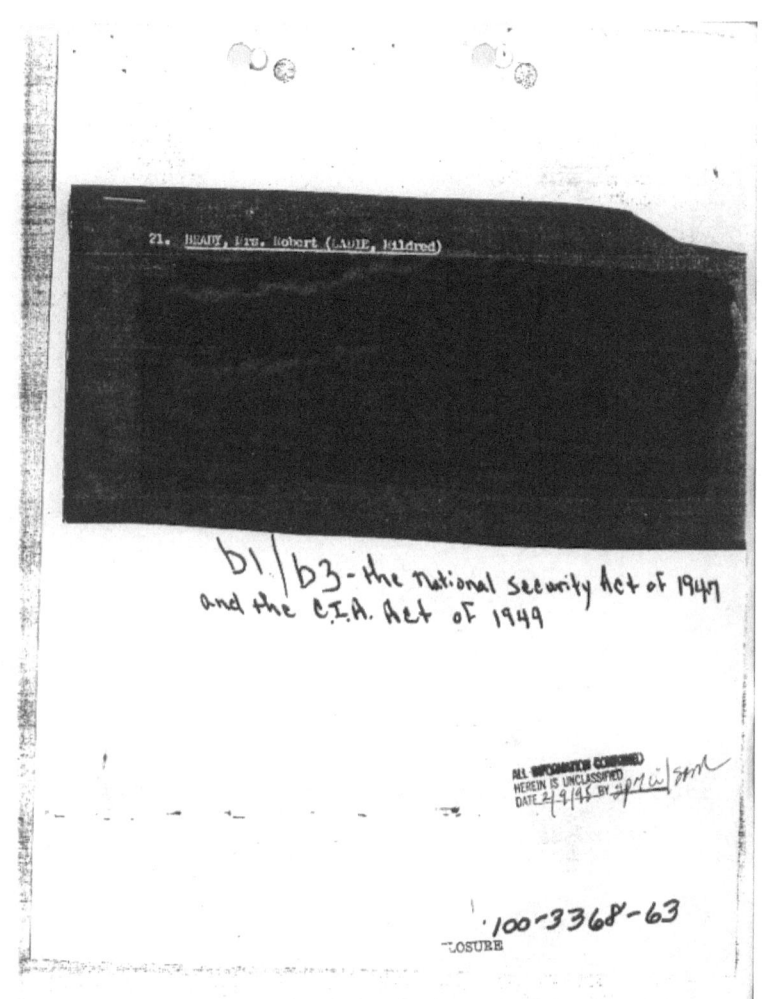

Third page of Brady's secret FBI files.

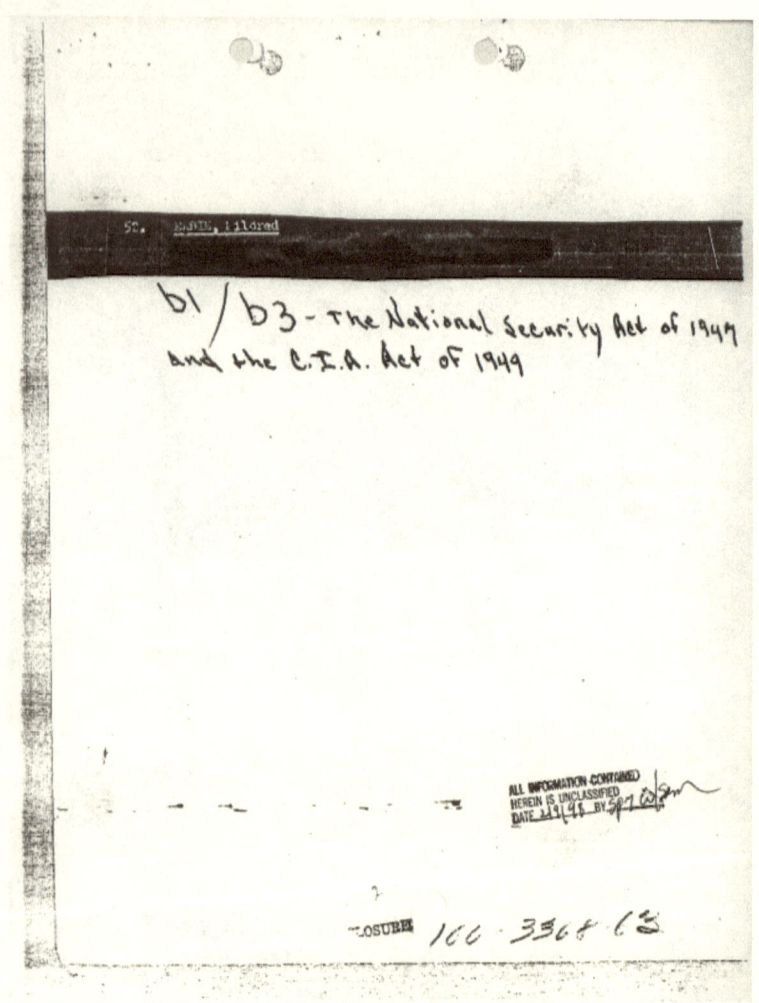

Last in a series of Brady's secret FBI files.

Documentary Appendix

8. *Lack of Due Process* under the law by the FBI in the case of Wilhelm Reich (1955-56)

In September, 1955, Michael Silvert, MD, wrote to the FBI regarding a series of outrageous intrusions by FDA agents, or people purporting to be FBI agents and FDA agents. The FBI held an official policy not to respond to any communications from Reich or his colleagues. The date of Silvert's complaint, September 23, 1955, coincides with another FBI document found in Mildred Brady's FBI file. That document, previously presented, shows that Brady was still under active investigation at the same time as Silvert's protest was written. The FBI took no action even though they had reason to believe that Brady was linked to Soviet espionage.

....

ORGONE INSTITUTE RESEARCH LABORATORIES, INC.

September 23, 1955

Federal Bureau of Investigation
Central Office
Washington, D.C.

Gentlemen:

I wish to inform you and to place on record additional instances of typical Red Fascist (Communist) activity by alleged agents of the U.S. Food and Drug Administration (FDA). They are part of a conspiracy to kill the discovery of the cosmic life energy and its Discoverer, Wilhelm Reich, M.D., Orgonon, Rangeley, Maine. Please refer to previous communication about this matter to the FBI offices in Washington, D.C., New York City and Tampa, Florida.

1. This morning, September 23, 1955, Mrs. Betty Corrode Dune of 75 Street, Brooklyn 1, New York, phoned to report two visits by alleged inspectors of the FDA. On both occasions, September 19 and September 20, she refused to open the door of her apartment and spoke to the man, whose name she did not obtain, through the closed door. He asked

to be admitted to "talk about the accumulator." He said she did not "understand the facts" and that he would enlighten her. She asked him to leave and when he persisted, she threatened to call the police. He then left. Mrs. was frightened and I reassured her.

2. In June of this year, an alleged inspector of the FDA visited Miss Lillian Okst who owns a liquor store in Port Jefferson, New York. Miss Okst, who had seen me briefly for medical care, was led to believe that the visitor was an "FBI Agent." He questioned her about her use of an orgone energy accumulator and threatened her with loss of her liquor license if she refused to cooperate. Miss Okst was greatly frightened by this visit and required considerable reassurance from me.

On September 22, 1955, at about noon, she was visited again by another alleged FDA inspector who identified himself as Mr. Feldman, a chemist. This time Miss Okst was not as frightened, but quite angry and questioned the visitor as to the reason for his visit. He said he was only interested in finding out about the "financial arrangement" which Miss Okst had made in connection with the orgone energy accumulator. He said they were in urgent need to prepare for the trial scheduled in Portland, Maine, on October 11, 1955. (The case of the FDA against Wilhelm Reich, M.D. and the Wilhelm Reich Foundation, and Michael Silvert, M.D.) When Miss Okst told him about the threat by the previous visitor to take away her liquor license, Mr. Feldman laughed, said it was "silly" and said, "He must have been in an angry mood that day." Miss Okst refused to give any information and sent Mr. Feldman away. Then Miss Okst phoned to report this incident to me, she again required reassurance. Please note that this is not the first instance of the behavior of alleged FDA agents in which they gave people the impression that they were FBI agents.

3. The following is a copy of an affidavit and an additional statement by Mr. Thomas Mangravite, 25 West Third Street, New York City.

The affidavit speaks for itself. Mr. Mangravite is associated with me in basic research as an engineer in cosmic orgone energy functions:

....

Documentary Appendix

COPY - AFFIDAVIT July 20, 1955

REPORT ON FDA VISIT TO BELL LABORATORIES

"This morning Mr. W. W. Werring, my Sub-Department Head at Bell Telephone Laboratories, accompanied me to the office of Mr. A. C. Keller, Head of the Department of Apparatus Development. Mr. J. N. McTighe of the Lab's legal Department was present. The meeting was called because Mr. J. Conway, Jr. of the FDA had visited Bell Laboratories on July 18, 1955. He identified himself as an agent of the U.S. Government, and asked for my employment record, age, address, background, government security clearance, etc.

"Mr. Conway informed Mr. McTighe that I was connected with an organization that had an injunction placed against it by the federal government, and that work I was doing concerning a "Viennese doctor" with a medical cure-all cabinet, was in violation of this injunction.

"I said that the facts had been misrepresented; however, the point for her was that his did not concern my work at Bell Laboratories.

"It was agreed that the Bell Laboratories' concern in this matter was only in being assured that I was not using Bell Laboratories' name in my outside work, and that I was not involved in something that would interfere with my Lab work or put me in jail.

"Mr. McTighe stated that while in general the company would not have released this information without my permission, it was their policy to turn it over to a federal investigator. Mr. McTighe would not give me an affidavit stating the facts of his interview with Mr. Conway.

Notarization:/s/ Thomas Mangravite
July 21, 1955
Frances Novick "

COPY - ADDITIONAL STATEMENT July 20, 1955

"I feel that the FDA visit to the Bell Laboratories to investigate my background was in no way a legitimate action.

"My age, schooling, marital status, employment, war record, security clearance, etc. is none of their business.

"These matters are so far divorced from even the doubtful authority of the injunction under which they claim to act, that this incident forms a clear example of the same blatant misuse of FDA power that has characterized the history of the conspiracy against orgonomy.

/s/ Thomas Mangravite "

We shall continue to advise you of any further development in this Red Fascist conspiracy.

Sincerely,

[signed]
Michael Silvert, M.D.
Treasurer
Orgone Institute Research Laboratories, Inc.

cc: FBI offices - New York City
Tampa, Florida

Documentary Appendix

Date: October, 6, 1956
Commissioner
Food and Drug Administration
Room 3460
Department of Health, Education and Welfare
Fourth Street and Independence Avenue, Southwest
Washington, 25, D.C.

From: John Edgar Hoover, Director, Federal Bureau of Investigation

Subject: Wilhelm Reich Foundation
  Miscellaneous - Information Concerning

Reference is made to my memorandum to you dated September 24, 1954, captioned: Orgone Institute Press, Publishing House of the Wilhelm Reich Foundation, Post Office Box 153, Rangeley, Maine," and my memorandum to you dated July 13, 1955, captioned "Wilhelm Reich Foundation, Miscellaneous - Information Concerning"

Enclosed are two copies each of letters dated September 16 and September 23, 1955, from Dr. Michael Silvert, 50 Grove Street, New York 14, New York. These communications are not being acknowledged by this Bureau, and we contemplate no further action in this matter at this time. For your information, copies of these communication have been made available to the Criminal Division of the Department of Justice and to United States Attorney Peter Mills.

Enclosures (4)

Note: See Jones to Nichols memo dated October 5, 1955, captioned "Wilhelm Reich Foundation, Miscellaneous - Information Concerning."

....

SAC, Boston (100-22450)
October 6, 1956
Director, (105 - 11461)

Wilhelm Reich Foundation
Miscellaneous - Information Concerning
(Espionage)

ReBulet of June 24 and July 13, 1955, captioned as above.

Attached are two copies each of letters dated September 16 and 23, 1955, from Dr. Michael Silvert, 50 Grove Street, New York 14, New York. These letters have not been acknowledged by the Bureau. No investigation is currently desired in connection with the allegations of impersonation violations described by correspondent.

You are instructed to make one copy of each letter available to United States Attorney Peter Mills for his information, and he should be advised that two copies of these communications are currently being referred to the Criminal Division of the Justice Department and to the Commissioner, Food and Drug Administration, Department of Health, Education and Welfare, Washington, D.C.

Enclosures (4)
cc - New York, with copy of each enclosure.
2 cc's - Miami, with 2 copies of each enclosures.

Wilhelm Reich: A Chronology (1897-1997)

1897 **March 24:** WR is born at home in Dobrzanica, district of Przemyslany/Galicia, part of the Austro-Hungarian Empire. His parents are Leon ("Leo") Reich and Cäcilie (Roniger) Reich. Leon Reich becomes manager of Cäcilie's uncle's estate in Jujinetz, in the Ukrainian part of northern Bukovina. Over the next years, Leon Reich will assume proprietorship of this beef cattle ranch. Otto Fenichel, Edith Jacobson are born. Alfred Jarry's *Pere Ubu.* H. G. Wells' *War of the Worlds.* Freud rejects the theory that neuroses are caused by real childhood traumas, in favor of the "seduction theory."

1898 Spanish-American War. Marie and Pierre Curie discover radium and plutonium. Bismarck dies. John Weatherill discovers carnotite (uranium ore) in southern Utah. Nathan Gregory Silvermaster is born in Russia. Ernest Hemingway, Bertolt Brecht are born. Birth of WR's sister, who dies soon.

1899 Karl Kraus founds the influential Viennese critical magazine, *Die Fackel.* Kate Chopin's story "The Storm." Dreser introduces synthetic medication with aspirin. Elster and Geitel demonstrate radioactivity in the decay of atoms, founding nuclear physics, introduce the term "atomic energy." **December 1:** KGB General Naum Eitingon is born in Sklov, district of Mogilov, Byelorussia (Belarus.)

1900 Planck founds quantum physics. First uranium processing mill in the US at Paradox Valley, Colorado. Robert ("Robi") Reich, Wilhelm's brother, is born. Thereafter, Cäcilie Reich falls ill and recuperates in spas off and on for two years. American author Thomas Wolfe is born.

1901 US President McKinley is shot and killed by an anarchist. Nihilistic terror in Russia. Roentgen receives Nobel for X-rays. Iskhak Abdulovich Akhmerov (aka "Michael Green," who would become Michael Straight's KGB Intelligence Officer) is born. **Dec 24:** MKULTRA "mind control" psychiatrist, D. Ewen Cameron is born. Enrico Fermi is born. WR discovers sex amongst workers at the ranch.

1902 A conspirator's classic: Lenin's *What is to be Done?* Wilhelm Fliess, Sigmund Freud's friend, publishes "On the Causal Connection between the Nose and the Sexual Organ". John Maynard Keynes, Lytton Strachey, Leonard Woolf and Clive Bell become undergraduate members of Cambridge's secret society, "The Apostles." Robert A. Brady and Annie Pink are born. WR recalls this time: five years of happy family life despite parents' arguments.

1903 Eleanor Roosevelt joins the Consumers' League and investigates working conditions in garment factories and department stores. Severe pogroms in Russia. Becquerel and the Curies get Nobel prize for their radiation research. Wilhelm Einthoven, using a string galvanometer, founds electrocardiography. First motorized air plane flight of the Wright brothers. WR's mother assumes education of her son, six years old, sheltering him from his extremely strict father.

1904 Howard A. Kelley of Johns Hopkins University obtains radium from the Curies and begins cancer treatment experimentation. **May 27:** Graf Karl von Motesiczky is born, Vienna. **August 21:** Arnold Deutsch is born. Ludwig Wittgenstein and Adolf Hitler attend the K.u.K Realschule in Linz, Austria. Freud's *Psychopathology of Every-day Life*. A governess prepares WR for public school.

1905 While being homeschooled WR begins a "breeding laboratory" at his family farm, at the age of eight. **January 22:** "Bloody Sunday" in St. Petersburg. **May 27:** Japan's victory over Russia. **June 27:** Mutiny on the Battleship *Potempkin*. Yacob Reich (Thomas Rubinstein) participates in failed Russian Revolution. **August 19:** The Czar is forced to introduce a constitution. Albert Einstein's three papers about Brownian movement, the photoelectric effect (photons), and special relativity. In Germany, *Geschlecht und Gesellschaft* (sex and society), a sex-reform journal appears.

1906 US Food & Drug Act passed, after proposal by Harvey Wiley. C. Dutton suggests radioactivity is the cause of volcanic activity.

Wilhelm Reich Chronology

Stanislav Shatsky creates the "Settlement" in Moscow, a network of self-managed children's democracies. In France Alfred Dreyfus is acquitted. Einstein formulates the law of the equality of mass and energy (e=mc2). Elsa Lindenberg is born. For one year, grandmother Roninger and her mother live at the WR's family farm. After an accident with a coach WR's father has to spend eight weeks cure at a spa. In the same accident WR suffers an hernia followed by an orchitis.

1907 I. F. Stone, journalist, is born in Philadelphia. *Silent Spring* author Rachel Carson is born, Springdale, Pennsylvania. KGB General Pavel Sudoplatov is born in Melitopol, Ukraine. Freudian psychoanalysis is introduced in Russia. Lenin flees from Russia. Stalin liberates an armored car in Tiflis, Georgia. Rasputin joins Russian royal court in Saint Petersburg. First investigations of the effects of radioactive and x-rays on biological organisms. United Press founded. **Spring:** WR's family moves to a new estate. WR installs a collection of plants and insects. WR attempts intercourse with a housemaid, reads a marriage counseling book of his father's. **December:** Lenin settles in Switzerland.

1908 Sigmund Freud founds the International Psychoanalytic Association. Arrhenius describes "greenhouse effect," expanding on an earlier paper, published in 1896, "On the Influence of Carbonic Acid in the Air upon the Temperature on the Ground." Austro-Hungary annexes Bosnia and Hercegovina. **June 30:** huge explosion, equivalent to a 40 megaton nuclear blast, in Tunguska, Russia, levels forest in circumference of sixteen miles. At the epicenter, no crater is found, and the trees are still standing. Tel Aviv is founded by Zionists. WR's father allows him, for the first time, to play with other children his own age. WR becomes sexually active. At the end of the year he overhears the adultery of his mother with his new private tutor, Sachter.

1909 Wilson uses cloud chamber to detect atomic "particles." Willard Straight meets his future wife, Dorothy Whitney, in Peking, China, where he is serving in the US Foreign office. Sachter leaves the house and after the summer holidays WR gets a new tutor.

1910 Early this year Leon Reich surprises his wife at an innocent incident with the new teacher, whereupon he calls his two young sons to account. Wilhelm reports to his father about the affair of his mother with Sachter. Meanwhile, in another room, their mother drinks toxic liquid. She recovers and endures several months of horrible abuse by Leon Reich. On **September 26** she takes poison again, and dies in early morning of the **28th**. WR and Robert are sent to the Gymnasium in Czernowitz, where they live at the home of an unmarried couple. In school WR remains a lone wolf with a single friend, Sabinski.

1911   Joseph Conrad publishes *Under Western Eyes*. Heike Kamerlingh Onnes discovers supra conductivity. Rutherford's model of the atom, nucleus circled by electrons. After the death of his mother, WR experiences his first (platonic) love with his cousin (lasting for the following seven years). He satisfies himself with lower-class girls, and by compulsive masturbation.

1912   WR visits whore-houses, where he experiences for the first time that the sexual act is not always followed by a feeling of emptiness. Unable to "rescue" a certain prostitute, he falls back on compulsive masturbation (with occasional coitus) and daydreaming, a depressive period full of rumination and book reading. First signs of an inferiority feeling. WR's father wants to marry a 20 year-old woman. WR develops fantasies to seduce his future stepmother. Wernher von Braun is born. Lenin becomes chief editor of the *Pravda*. Alfred Adler's *On the Nervous Character*. C.G. Jung's *Changes and Symbols of the Libido*. Rudolf Steiner founds the Anthroposophic Society.

1913   WR's father, Leon Reich, contracts pneumonia and WR returns from formal school to help run the farm and tend his sick father. **January:** WR contracts psoriasis, medical treatment, which includes X-rays, intensifies the condition. The 16 year-old WR is preoccupied with the Christ story. He starts to pray for a while. **October:** German youth movement, with Siegfried Bernfeld in a leading role, convene the Meissner Mountain meeting to unify the various tendencies of the "Wandervögel" or "roaming youth." Karl Frank enters University of Vienna to study medicine and psychology. Sidney and Beatrice Webb found the British socialist weekly *The New Statesman*. Stalin's *Marxism and the National Question*. Woodrow Wilson becomes president. Freud's *Totem and Tabu*. Niels Bohr's "planetary" atom model (the beginning of quantum mechanics). Geiger

Wilhelm Reich Chronology

counter invented. Irving Langmuir invents gas-filled electrical incandescent lamp. **December:** Willy Brandt is born. **December 23:** US Federal Reserve Act passed. After a depressive period, an affair with a young cook lets WR open himself to the world again. He takes part in sport activities, outings, etc. Then he falls in love with the sister of a classmate, but something holds him back at the first embrace; he runs away, never to see her again. Thereafter he is occupied with his school work and spends the nights in whore houses.

1914 WR's father becomes bankrupt, falls ill, and WR must borrow money from a relative to send his father to Tyrol for a six-week treatment. On his way home, WR learns that his father has died of tuberculosis. **May:** WR assumes management of his family farm. Outbreak of World War I. **July 31:** Austrian mobilization order. Gertrude Gaasland is born. J. P. Morgan partner Willard Straight and heiress wife Dorothy Whitney Straight found "progressive" magazine *The New Republic*. Price of a gram of radium: $150,000. Howard A. Kelley and Phelps Dodge official James S. Douglas (Lew Douglas' father) found the National Radium Institute, to promote markets for Arizona uranium ore. Nathan Gregory Silvermaster emigrates to the US from Russia at 16 years old. Russians invade Bukovina. WR is taken prisoner, part of a group of hostages bound for Siberia, but is able to bribe a Russian officer for his freedom. None of the others ever return. **Winter:** Ukrainian nationalist makes an unsuccessful assassination attempt on WR, the "Austrian." Austria retakes the region, and the assassin is lynched. (See photo in *People in Trouble*.)

1915 WR has a love affair with a Ukrainian teacher, and flight with her after a new Russian invasion. WR registers with the Army. Basic military training. **May:** Italy declares War on Austro-Hungary. First transcontinental telephone conversation, New York to San Francisco, by Alexander Graham Bell and Thomas A. Watson. Einstein's Theory of General Relativity. Robert H. Goddard demonstrates rocket propulsion in a vacuum at Clark University, Worcester, Mass. John J. McCloy meets Willard Straight at Plattsburg, New York, at a militia "preparedness" training for young businessmen, Ivy League students. In Plattsburg, McCloy turns down Straight's offer to work for J. P. Morgan interests in the Philippines. A. S. Neill publishes first book, *A Dominie's Log*. **October 5:** WR gets emergency graduation diploma (*Abitur*), then infantry training in Hungary.

1916   Radio-controlled pilotless monoplane, the "Aerial Target," flies in Britain. Friedrich Adler, physicist, friend of Einstein, son of the founder of the Austrian Social Democratic Party, shoots the Austrian Prime Minister, Count Sturgkh. **February 5 - April 14:** WR at the officers' school in Jaegerndorf in the Austrian part of Silesia. WR's dear friend Sabinski dies in his very first battle. **April 6:** The United States declare war on the Central Powers. **June 4:** end of WR's military training, goes as a corporal to the front north of Trieste. Reich falls ill with *psoriasis vulgaris* when he has to lie for three days in a wet trench, and on **Nov. 24** is he is sent back to the hinterland. WR makes sergeant. **September 1:** Michael Whitney Straight is born, Washington, DC.

1917   Brooklyn police arrest Margaret Sanger for opening a birth control center. Allen Dulles visits Vienna. **February 23-March 8:** revolution in Russia. **March 15:** Czar Nicholas II of Russia abdicates. **April 4:** WR is promoted to cadet. **April 16:** With the help of the German secret service Lenin travels from Switzerland to Russia. **May 5:** Trotsky comes from the US. **May 29:** John F. Kennedy is born. During a leave WR meets his brother's girlfriend, Ottilie Heifetz. **November 7:** Bolshevik "second" revolution. **December 1:** WR is promoted to second lieutenant. **December:** establishment of Cheka (Soviet state security police). Julius Hammer, father of Armand Hammer, sets up Allied Drug and Chemical Company with Ludwig Martens, a shell-company that is a forerunner of AMTORG, to circumvent Western embargo on Soviet Russia.

1918   Influenza epidemic kills about 20 million people worldwide, about 550,000 in the US. Civil war in Russia. Trotsky founds the Red Army. In Germany, Communist Karl Liebknecht declares the "council republic." Social Democrat Philipp Scheidemann declares Weimar Republic and violently suppresses the council republic. **Spring:** Wiener-Neustadt munitions workers' strike in Austria. **June 5:** WR is sent from Italian front to Vienna hospital due to skin disease. **July:** murder of Russian Tsar and his family. Presidential aide Edwin Mandell House calls in Willard Straight to help US with the Paris Peace Conference. **August:** WR is discharged from hospital and gets three-month educational leave

Wilhelm Reich Chronology

from the Austrian Army, briefly enters law school at the University of Vienna (one term). WR then enters medical school. Joins a secular student association, "Wiener Medizinerverein." Lives with his brother, Robert, and another unidentified man who later becomes a psychoanalyst. **November:** End of World War I. Austro-Hungarian Empire is broken up. **November 12:** Austrian Workers' Council revolt. Willard Straight dies **Dec. 1** with the flu.

1919 Hungarian Revolution: a short-lived council republic under Bela Kun. Rosa Luxemburg and Karl Liebknecht are murdered by right-wing military officers. The Comintern (the Communist International) is formed, lead by Zinoviev. *Deutsche Arbeiterpartei* (German Worker's Party, and later, the "Nazi" National Socialist Democratic Worker's Party) is founded, Adolf Hitler is its seventh member. Social Democratic Party of Austria is elected to form government. William Crookes, discoverer of the radiometer, and of "Crooke's tube," dies. Transmutation, a first: Rutherford, by radioactivity, turns one element (nitrogen) into another element (oxygen). Heisenberg's uncertainty principle. Radio Corporation of America (RCA). Prohibition in US. WR studies anatomy with Professor Julius Tandler, and biology with Paul Kammerer. **February 25:** WR attends convention of Jewish university graduates. WR begins student-run seminar on sexology with Otto Fenichel, meets Sigmund Freud. **March:** WR begins dream analysis, studies the work of Isidor Sadger and briefly enters a training analysis with him. **April 17:** New York furrier Waldemar Eitingon dies, and his brother, psychoanalyst Max Eitingon, becomes more wealthy. **May:** Julius Tandler becomes deputy undersecretary of the office for social welfare. **June 15:** Communist-led, unsuccessful "putsch" in Vienna, centered mostly around the University. **June 28:** Signing of Treaty of Versailles disarms Germany. **September**: WR begins treating psychoanalytic patients with Freud's blessing. His first patient, Lore Kahn, is the brokenhearted former lover of Karl Frank, a communist politician. **November 6:** Eddington makes astronomical observations that provide evidence for Einstein's general relativity theory, making Einstein the most famous scientist in the world. **December:** Karl Frank becomes a member of the Central Committee of the Austrian Communist Party

(KPÖ). Walter Krivitsky is chief of the Soviet "secret office" in Vienna.

1920 Paul Kammerer publishes *Allgemeine Biologie*, introducing the term "self-regulation" to describe an organism's ability to adapt to its environment. Freud publishes *Beyond the Pleasure Principle*. **February 12:** WR first invited to attend the "Wednesday Meetings" of the Viennese Psychoanalytic Association. **May 7:** James E. McDonald is born, Duluth, Minnesota. **June:** WR meets future wife, Annie Pink, at Otto Fenichel's home. Freud begins sending WR patients. WR delivers first psychoanalytic paper, "Breach of the Incest Taboo in Puberty." Karl Frank leaves Vienna for Berlin to work for the new Comintern office, under "Comrade Thomas" (Yakov S. Reich, later known as Thomas Rubinstein). Karl Frank reads Freud's writings aloud to group of Communist students at a youth center in Berlin run by Walter Löwenheim, the future founder of the Leninist Org. Socialist psychoanalysts in Berlin found a Polyclinic for the poor. WR falls in love with his former patient, Lore Kahn, a kindergarten teacher and friend of Annie Pink. Arnold Deutsch and Willi S. Schlamm, members of the Austrian communist youth organization, are befriended by WR. Schlamm visits Moscow at age 16. **August 18:** US women win right to vote. **October 13:** WR delivers paper on Ibsen's *Peer Gynt*, and is accepted as a member of the Viennese Psychoanalytic Society. **October:** Lore Kahn dies. Annie Pink seeks analysis with WR.

1921 Bjerknes's "On the Dynamics of the Circular Vortex with Applications to the Atmosphere and to Atmospheric Vortex and Wave Motion." **March 1-18:** Kronstadt Uprising suppressed by Trotsky's Red Army. "International Worker's Aid" formed in Berlin to help starving Russian masses. Lenin's New Economic Policy (NEP). Albert Einstein receives Nobel. Psychiatric Rorschach test developed. A. S. Neill founds his own school for children's "creative self-expression" in Dresden, Germany. WR breaks away from remaining older relatives. Still pursuing his medical degree, WR delivers several papers this year to the Vienna Psychoanalytic Association, including "On the childhood dreams of a later compulsion neurosis."

1922 Julius Tandler establishes the "Social Welfare Office of Vienna" - a marriage guidance bureau, advocating "marriage hygiene," and a eugenic sex program. **March 17:** WR marries Annie Pink. **April 3:** Stalin becomes Secretary General. **May 22:** establishment of Viennese Psychoanalytic Outpatients Clinic for the Impoverished, headed by Eduard Hitschmann and assisted by WR. **June 24:** Jewish Secretary of State in Germany, Walter Rathenau, assassinated by a nationalist. **July 10:** WR receives Medical Degree from University of Vienna, enters postgraduate study in neuropsychiatry at the University of Vienna Clinic, headed by Julius Wagner-Juaregg (who later won the Nobel Prize for treatment of progressive paralysis). WR studies firsthand the treatment of psychotics. He pursues his internship under Paul Schilder, who has an "energetic" orientation to psychiatry. Willi Schlamm becomes editor of *Die Rote Fahne*, central organ of the KPÖ and member of the Central Committee as well. Soviet Cheka reorganized as GPU. Otto Fenichel moves to Berlin. **October 30:** Mussolini's "March on Rome."

1923   Height of German inflation: $1.00=4.2 trillion marks. WR outlines concept of genitality in first paper. Arnold Deutsch enters the University of Vienna. **January 5:** Cloud seeding over McCook Field, Dayton, Ohio, by Prof. W. D. Bancroft of Cornell University. *The Rocket Into Interplanetary Space* by Hermann Oberth is published in Germany, beginning considerable discussion of rocket propulsion. Otto and Vera Schmidt, Russian pedologists, visit Freud in Vienna. **November 8-9:** Hitler's Munich putsch.

1924   With Freud's approval, WR organizes the Technical Seminar, where young psychoanalysts discuss treatment of cases. Otto Warburg relates oxygen deficiency with cancer. De Broglie's matter waves. Karl Frank arrested in Bavaria. Woodrow Wilson dies. **January 21:** Lenin's death. **February:** Julius Tandler demands that Vienna police ban Hugo Bettauer's magazine, which deals with practical matters of life and sexuality, *Er und Sie*. Tandler (WR's old professor) begins a public campaign against WR's friend, Bettauer. Bettauer had brought many patients to Reich's polyclinic, a prototype of the later Sexpol. **April 27:** Eva Reich is born. **Summer:** Psychoanalytic Congress in Salzburg: WR introduces the concept of "orgastic potency." **October 28:** Cloud formations are broken up over Bolling Field, Washington, D.C. by "blasting" with electrified silica in a fog-

Wilhelm Reich and the Cold War

dispersal demonstration by US Army aircraft.

1925 WR publishes first book, *The Impulsive Character*. Jerome Eden is born, New York City. Hollywood film producer Samuel Goldwyn offers Sigmund Freud $100,000 if he will collaborate on a love film about Anthony and Cleopatra; Freud declines. Willard Straight's widow, Dorothy, marries Leonard Elmhirst, and the couple moves to Dartington Hall to found an experimental school in Devon, England. A. S. Neill founds Summerhill School in Suffolk. Heisenberg's quantum mechanics. Hitler publishes *Mein Kampf*. Arnold Deutsch becomes active in Austrian Communist party youth organizations. "Germ cells" of communists within European Social-Democratic Party organizations begin with encouragement from Moscow. Siegfried Bernfeld moves to Berlin. **March:** "Bettauer Case" - Hugo Bettauer, publisher of the sex-reform magazine *Er und Sie*, murdered by Nazi anti-Semite, Rothstock, who is allowed to go free - possibly the first victim of the Holocaust.

1926 **April:** Robert Reich dies of tuberculosis. His widow, Ottilie Heifetz, and her daughter, live with WR and family in Vienna. Annie Reich receives medical degree. Trotsky, Zinoviev, Kamenev ousted from Politburo. Schrödinger's quantum wave mechanics. Austrian Social Democrats declare "Linzer Programm," foreswearing revolution. Charles A. Lindbergh's cross-Atlantic flight. WR coins the concept of "phallic-narcissist character." Karl Frank visits Moscow for six weeks. Friendship treaty between Social-Democratic Germany and Communist Soviet Union. At the end of 1926, there is a general political amnesty and Frank is able to return to Germany, from his exile in Paris, where he edited a Communist paper. **December:** WR rebuked by Freud on the issue of negative transference. WR argues that the analyst must first deal with the "character resistance" of the patient, while Freud insists on analyzing the symbolic material as it appears. Conflict with Freud begins.

1927 **January:** at the end of 1926, WR contracts tuberculosis, from which his brother had died. Recovers after a three-and-a-half month stay in a sanatorium in Davos, Switzerland. WR's colleague and training patient, Annie Angel, joins him in Davos. WR appointed to the board of the Viennese Psychoanalytic Association, over the strong objections of Paul Federn. WR outlines the technique of character analysis. WR joins the Social Democratic Party. **January 30:** Monarchists shoot into a crowd of Social Democrats going to a party meeting in Schattendorf, Austria, killing a working-class war veteran and a young child. **May:** WR publishes *Die Funktion des Orgasmus* (later published in English as *Genitality*). **June 5:** Society for Space Travel (*Verein fuer Raumschiffahrt*), known as "VfR," formed in Breslau, Germany. **July 14:** Austrian courts free Schattendorf's monarchist terrorist shooters, outraging the left wing. **July 15:** responding to the courts decision on the monarchist Schattendorf killers, Viennese workers riot and burn the Justizpalast (Municipal Courthouse) in Vienna. WR witnesses these events, and becomes disillusioned with the Social Democratic Party, which failed to protect the workers even though it had been training paramilitaries for years for just such an event. During the riots, Social Democrat police officers shoot into the crowd, with many casualties. **Late July:** WR discovers in conversation with Freud, that Freud viewed the riots much like a natural disaster such as a "tidal wave." **Summer:** WR reads Marx' *Das Kapital*. WR attempts to find common ground between Marxism and Psychoanalysis: beginning of "political psychology." **August:** Austrian Social Democrats begin a policy of appeasement with leads to their defeat in national elections. **October 12:** Wright Air Field, Dayton, Ohio, is formally dedicated. **November 1:** Austrian Social Democratic Party declares it has averted a civil war. WR's affair with recently-divorced, old friend Lia Laszky-Swarowsky, now a teacher at Eva Reich's Montessori kindergarten.

1928 Comintern rejects cooperation with all other left groups, especially the Social Democrats ("social fascists.") **March 13:** Lore Reich is born. Edith Jacobson admitted to German Psychoanalytic Society. Annie Reich's analyst is Anna Freud. WR joins Austrian Communist Party, an event widely noted. **July 28:** Orson Bean is born. **October 7:** fascist Heimwehr paramilitaries stage provocative demonstrations in Wiener-Neustadt, an industrial city just south of Vienna, and the Social Democrats and Communists organize counter-demonstrations; WR participates as a member of the medical support team for a band of 250 communist revolutionaries, who end the day in complete failure and disillusion. **October 10:** WR speaks before Viennese Psychoanalytic Society, on "Where does nudist education lead to?" Draws a straight line from nudist education to the overthrow of bourgeois society. Free sexuality is not a way to a better life, but an end in itself. **November 22:** Paul Federn makes first attempt to limit WR's influence in the Viennese Psychoanalytic Society, but is rebuffed by Freud. At this time, WR remains leader of the technical seminar, the Polyclinic, the sex-counseling centers, and has a thriving private practice. WR begins sexual-political work, including sex education for workers. Early Sexpol clinics are founded. With Marie Frischauf-Pappenheim, and six other physicians, WR founds Socialist Society for Sex-Counseling and Sex-Research (incorporated **December 27**). Freud supports this, warning WR, "you'll be stepping into a hornet's nest."

1929 First of nine volumes of an encyclopedia on interplanetary travel by Prof. Nikolai A. Rynin is published in the Soviet Union. Wiley writes *History of a Crime,* detailing perversion of US food and drug laws by monopolies. Malinowski's *Sexual Life of Savages.* Trotsky is expelled from USSR to Turkey. **June:** all inner-party opposition destroyed within KPD, completing Stalinization of German Communism. Karl Frank and William S. Schlamm (future cofounder of the *National Review*, with William F. Buckley) break with the Communist Party. **September:** WR and Annie Reich travel to Moscow, meeting "pedologist" Vera Schmidt, who, with Trotsky's patronage, had incorporated psychoanalytic teachings in her work at a childcare center (where Stalin's son, along with the children of other top Party functionaries, attends.) **August 23-October 31:** Russian plane, *Land of the Soviets*, flown on goodwill tour of the United States from Moscow to Seattle, and on to New York. **October:** New York Stock Market crash. **October 15:** Premier of German movie film, *Frau im Mond* (The Girl in the Moon) directed by Fritz Lang, which assisted popular awareness of rocket potential in Germany. NKVD agent Arnold Deutsch recruits his friend Edith Tudor-Hart for Soviet espionage in Vienna. WR's articles, "Dialectical Materialism and Psychoanalysis" and "Psychoanalysis in the Soviet Union" are published. Arnold Deutsch publishes WR's *Sexual Excitation and Sexual Satisfaction* in Vienna. **December 13:** WR founds Committee of Revolutionary Social Democrats, trying to forge an alliance between Communists and Social Democratic paramilitaries (Schutzbund). 1500 workers meet at Stalehner restaurant. WR is outraged at the mute response of the left to nascent fascism.

1930 Robert Goddard tests rockets in Roswell, New Mexico. Last meeting between WR and Freud. **January:** WR publishes a political weekly journal, *der Revolutionäre Sozialdemokrat* (The Revolutionary Social-Democrat) that decries the timidity of the SPÖ leadership in the face of fascism. **January 7:** Josef Frey tells Trotsky, regarding the Revolutionary Social-Democrats, that "Reich is in reality a tool of the Stalin Central Committee." **January 16:** in an arbitration hearing, WR confesses to cooperating with the KPÖ and chastises the SPÖ for cooperating with the fascist Heimwehr. SPÖ expels WR from its party. **February 27:** WR speaks on the crusade by the Catholic Church against the Soviet Union. **March 6:** WR, in the name of the Revolutionary Social-Democrats, speaks to 10,000 demonstrators against the Austrian Social Democrats' policy of "social fascism" and capitulation to the right. **March 14 & 21:** WR speaks to large crowds in Vienna. **April 15:** WR disbands Committee of Revolutionary Social Democrats, declares his support for the KPÖ. **September:** WR moves from Vienna to Berlin, living in an apartment on Schwäbische Strasse, where he shares his ideas on political psychology with Barbara Lantos, Edith Jacobson, Otto Fenichel, and Erich Fromm, all young psychoanalysts. Siegfried Bernfeld, an older analyst, also participates, without much comprehension. WR organizes a Berlin Technical Seminar. WR enters a brief psychoanalysis with Sandor Rado, husband of Emmy Rado. **September 16-23:** 4th Congress of the World League for Sex-Reform, Vienna. WR speaks, for one of the last times in Vienna, on "Sexual misery of the working masses," and argues for recognition of the political implications of the sex-reform movement. In the Social-Democrat journal, *Freiheit*, Karl Kautsky, Jr. denounces WR, accusing him of taking away all "ideals" from the worker with sexual propaganda. **November:** WR reads Malinowski's *Sexual Life of Savages*.

1931 American economist Robert A. Brady makes first visit to Moscow. WR's articles on youth sexuality are published by the KPD. **Spring:** WR lectures at Berlin's Marxist Workers' School (MASCH). **May 1:** WR participates in May Day demonstrations in Berlin. **May:** successful regional unification of sex-reform organizations in Düsseldorf, Germany; WR is keynote speaker. **June:** Reichsverband für Sexualpolitik founded in Berlin. **June 20-21:** failed attempt at a national unification of sex-reform organizations. **July:** collapse of German banking system. Nazi Storm troopers stage frequent marches on Berlin. WR's KPD cell prepares to defend its "Red Block" of apartments against the Nazis. **Summer:** Sandor Rado tells Annie Reich that WR is psychotic, that she should leave him, and leaves Berlin himself for the US. WR submits *Sexual Struggle of Youth* to KPD, who forward it to Moscow for final approval. Eva and Lore Reich placed in a Communist youth commune, a nightmarish experience for the two girls. Karl Frank rejoins the SPD, along with a group of younger people, to fight Nazism. Sexpol Verlag founded. **December 1:** WR becomes a full member of the German Psychoanalytic Society.

1932 Hitler becomes a German citizen. Six million unemployed in Germany. Zernike develops phase contrast microscope for the study of living, unstained cells. WR publishes *Invasion of Sex-Morality*, and *The Sexual Struggle of Youth*. Annie Reich publishes *When Your Child Asks You*, and *The Chalk Triangle*. The Norwegian psychoanalyst Nic Hoel studies in Berlin and calls the attention of other Norwegian psychoanalysts to Reich. Nathan Gregory Silvermaster receives Ph.D. in Economics at the University of California at Berkeley, a dissertation on "Lenin's Economic Thought Prior to the October Revolution" - sponsored by Robert A. Brady. **January:** Arnold Deutsch is summoned to Moscow. **January:** Freud attaches note to WR's article on masochism, stating that the author is a Bolshevik, and that like "the Society of Jesuits," the Reich was a parrot of the Party. In Berlin, Fenichel calls a meeting of Marxist analysts to discuss the situation, and asks WR to make changes to his masochism article, which WR refuses. The article is published without Freud's note, due to the influence of Socialist analysts Siegfried Bernfeld and Max Eitingon. Instead, Bernfeld writes a rebuttal to WR's unchanged article on masochism, and it is published although it did not address the points made by WR. Meanwhile, Bernfeld is engaged in the clandestine group, the Org, and takes a prominent role in its internal debates over infiltration of other left groups;

Wilhelm Reich Chronology

Bernfeld argues for sending Org cadres into fascist organizations to work as moles after the Nazi takeover. Freud, concerned with making psychoanalysis appear conservative, decries Otto Fenichel and WR as "Bolshevist attackers," and strips Fenichel of his editorial position. Freud recalls the publications *Imago* and *Internationale Zeitschrift für Psychoanalyse* from Berlin to Vienna to reduce Marxist influences. **May 1:** WR meets second wife, Elsa Lindenberg in Berlin. Due to difficulties with KPD publishing his work, WR founds Verlag für Sexualpolitik. **June 17:** "Bonus Marchers" - 10,000 WWI veterans encamp in Washington DC. **July 20:** von Papen coup in Prussia replaces legal government. **September:** WR's participation with "United Association of Proletarian Sex-Reform." **October 1:** Wernher von Braun joins the German Army Ordnance Office rocket program at Kummersdorf. **October:** Max Eitingon vetoes WR's election to the Berlin Training Institute. **November 27:** Trotsky gives speech in Copenhagen, endorses a fusion between Marxism and psychoanalysis. **October-December:** KPD describes Sexpol as "counter-revolutionary". **December 5:** German Communist youth-sport journal announces the prohibition of distribution of WR's tracts. **December:** National Socialists and Communists jointly call for a strike of transportation workers in Berlin, against the Social Democratic Administration. **December 13:** *Die Rote Fahne*, the leading Communist journal in Austria, commends WR's pamphlet, *The Sexual Struggle of Youth*, embarrassing Party leadership.

1933 Forty million unemployed in the US. British Interplanetary Society organized. Einstein arrives in the US from Germany; his second wife, in an interview, says, "Prof. Einstein is no eccentric. He is not absent minded. He dislikes dirt and confusion. Politically we are Socialists." Walter Löwenheim's *Neu Beginnen!* pamphlet appears, calling for a clandestine period for Leninist cadres. Edith Jacobson joins this Org. **January 29:** In Berlin, the United Association of Proletarian Sex-Reform passes resolution against WR, accusing him of seducing the youth of the working class and creating "fuck organizations." **January 30:** President Hindenburg appoints Adolf Hitler as German Chancellor. **February 1:** German Reichstag dissolved. **February 18:** Dutch anarchist, former communist, van der Lubbe, arrives in Berlin from Holland. **February 18:** United Association of Proletarian Sex-Reform again resolves to dissociate itself from WR, effectively ending his activity within the KPD. **February 24:** Erik Jan Hanussen, a performing occult magician and hypnotist with ties to Nazi officials, publishes "prophecies" of the burning of the Reichstag. **February 24:** WR travels to Copenhagen, Denmark, to address a student organization "Clarté, "on the problems of Fascism and race. **February 27:** van der Lubbe sets fire to the Reichstag. Emergency laws allow Hitler's dictatorship to begin. Nazis begin arresting leftists and intellectuals. **February 28:** Danish-Communist newspaper *Arbejderbladet* reviews WR's address to students in Copenhagen positively: an example of "how intellectuals can really stand by the working class." **February 27-28:** WR returns to Berlin the evening of the Reichstag Fire; he escapes immediate arrest only because he never held an official position in the KPD. **March 1:** Communist deputy in the Reichstag assures WR personally that 40,000 fighting workers are prepared to take the streets in the event of a Nazi victory in the national election. That day, WR lends his car to armed workers smuggling weapons into Berlin. **March 2:** Nazi paper *Völkischer Beobachter* attacks WR's *Sexual Struggle of Youth*. WR prepares to leave Berlin that night. **March 4:** Dolfuss regime restricts civil liberties in Austria. **March 5:** German national elections solidify Nazi control of the government. **March 5-12(?):** WR and Annie Reich cross the Bavarian border into Austria, Annie Reich immediately returns to Berlin, to find it is unsafe for her husband to return. During his flight, some of WR's private archives are lost. Their children, Eva and Lore, are already in Vienna living with Annie's parents. **March 12:** Karl Frank goes to the Berlin flat of Max Seydewitz, where he finds the SA are there, to arrest Max and his brother-in-law, Fritz Levy. Frank is detained but released because he is a "foreigner." He begins his illegal life and lives for several months at the home of Siegfried Bernfeld. **March 17:** Freud informs WR that International Psychoanalytic Publishers will not honor its contract to publish *Charakteranalyse* (Character Analysis.) **March 23:** nearly 5,000 communists and social democrats interned at Dachau, near Munich. **March-April:** WR re-enters Germany and returns to Berlin under his own name, having learned that the border guards do not have lists of people yet. Retrieves some clothes and personal possessions (his card catalogues and manuscripts having been previously hidden at various locations throughout Germany.) After a few days in Berlin, WR returns briefly to Vienna; speaks before a large group of university students about his new book and ideas. **April 17:** historic meeting between Felix Boehm, Paul Federn, and Sigmund Freud to determine the future of psychoanalysis in Germany: Freud demands: "free me from Reich." **April 23:** WR meets with IPA executive council over their prohibition against his speaking to socialist groups. Tage Philipson, a Danish analyst, invites WR to come to Copenhagen and train students; growing hostility from Viennese colleagues forces WR to accept. **April 30:** WR leaves Vienna for Denmark via Poland on a freighter. **May 1:** WR arrives in Copenhagen. Danish Communist party refuses WR's request for aid to German refugees, instead asking to see his immigration papers. **May:** Gertrude Gaasland (WR's assistant in Norway, and later Willy Brandt's fiancé) is arrested by Nazis, and freed five weeks later. **May 3:** WR and Karl Motesiczky (a Sexpol activist) are named with 13 other people as officially expelled from Prussia because of political activity. WR rents an apartment in Copenhagen and begins work. **June:** WR publishes *Charakteranalyse* in Copenhagen. **Summer:** Kim Philby arrives in Vienna, Austria. **August:** WR secretly expelled from the German Psychoanalytic Society, but is not informed until the 1934 meeting of the International Psychoanalytic Congress in Lucerne, Switzerland. **September:** ACLU Lawyer Arthur Garfield Hays meets with Nazi officials in Leipzig. He attends the trial of Georgi Dimitrov, who was accused of plotting the Reichstag afire. **October:** WR self-publishes *Mass Psychology of Fascism*. The German Communist Party in exile in Denmark objects to the very first sentence of the book: "The German working class has suffered severe defeat." **October:** A female, former Danish patient of WR's attempts suicide after Danish authorities would not permit him treat her. Danish authorities blame the suicide attempt on the treatment, rather than the lack of it, and WR is forced to leave Denmark at the end of the year. **October:** WR writes to Trotsky suggesting "close contact" and cooperation. Trotsky's mail is reviewed by Soviet intelligence. **November 7:** Trotsky replies to

WR's previous letter, indicating he would like to learn more about WR's views. **October 27:** Danish Communist newspaper *Arbejderbladet* publishes without commentary an announcement for a meeting in Copenhagen to discuss "the fight for psychoanalysis" and WR's ideas. **November 21:** Danish Communist Party, via the newspaper *Arbejderbladet*, announces the expulsion of WR from the German Communist Party, for his mixture of Marx and Freud, his book, *Mass Psychology of Fascism*, and his association with Brandler, the exiled former leader of the German Communist Party. **December 1:** *Arbejderbladet* publishes a large, anonymous feature review of WR's book, now viewed as "dangerous for our revolutionary work." **December:** WR meets Bronislaw Malinowski in London. In Paris, WR meets with members of Trotsky's Fourth International, as well as with German political refugees in the SAP, or Socialist Workers Party. In response to his disappointment with these exiled politicians, WR outlines "What is Class Consciousness?" After a few days in Paris, WR meets Fritz Brupbacher in Zurich, a veteran, radical critic of the anti-sex workers's movement and the establishment "sex-reform" organizations based on "race-hygiene." WR meets Max Hodann in Basel, Switzerland, after Hodann's escape from a concentration camp. Hodann and WR agree on future collaboration but a proposed journal never materializes. **December 24:** WR moves on to Tyrol, Austria, visiting Annie and their daughters. **December 26-27?:** Then he goes to Prague, and seeks out old friends, but finds only confusion. **New Years:** WR returns to Berlin after learning border authorities still have no list of names, stays with Elsa Lindenberg and her family.

1934 Birobidzhan, USSR, becomes "autonomous Jewish state." Lewis W. Douglas' Germanic, Rockefeller-funded father-in-law, Hans Zinsser, publishes *Rats, Lice and History*. Gloria Steinem is born. Hindenburg dies, Hitler becomes *der Führer*. US diplomatically recognizes the USSR. FBI kills John Dillinger. *New Frontiers* by US Secretary of Agriculture Henry A. Wallace is published. John F. Kennedy later borrowed this title. **January:** Neu Beginnen leader Karl Frank opens a "foreign bureau" in Prague to raise money for underground resistance to the Nazis. Elsa Lindenberg accompanies WR to Sweden, and they take up residence in Malmö. They remain there for six months; "at least," WR wrote of Malmö, "it was better than a concentration camp." **February 12:** Austrian government arrests Social Democratic leadership, begins artillery shelling of worker's council estates in Vienna suburbs. **February 24:** Kim Philby marries his first wife, Sexpol activist Alice Kohlman "Litzi" Friedman in Vienna. An underground wing of the SPÖ, called Revolutionary Socialists, mounts an armed defense of the workers; Kim Philby joins them and stands by a machine gun. Karl Frank is invited by the SPÖ to be an instructor for "Revolutionary Socialist" underground groups in Austria. **March:** Otto Fenichel begins Rundbriefe Circle in Olso. **April:** WR publishes *Zeitschrift für Politische Psychologie und Sexualökonomie* (Journal for Political Psychology and Sexual-Economy.) WR coins the term Sexpol. Danish and Swedish police launch coordinated raids on WR's premises and those of his associates in both countries. WR's colleague, Jonathan Hoegh ("Joyce") Leunbach runs as a candidate for the Danish parliament on a Sexpol platform. **April 13:** Gestapo bans WR's pamphlets "What is Class Consciousness," "Dialectical Materialism and Psychoanalysis," and all existing and future editions of the "*Zeitschrift für Sexualpolitik* (Copenhagen, Prague, Zurich.)" **April:** Kim Philby and his wife flee Austria. Litzi Friedman and Kim Philby had been acting as couriers for the Comintern underground, and as a liaison between the KPÖ and the Revolutionary Socialists. **Easter:** WR meets Ola Raknes for the first time. WR rejoins Otto Fenichel in Oslo, Norway, along with Edith Jacobson and Harald Schjelderup, as members for the "Psychoanalytic Opposition Congress." Only one member of this group, Edith Jacobson, is aware that WR has already been secretly excluded from the International Psychoanalytic Association, but she fails to inform him. WR visits the University of Oslo, Norway, where he is offered a position teaching character analysis. 7th Comintern Congress in Moscow shifts party policy to a "United Front" against fascism, members encouraged to join Social Democrats, trade unionists, and other non-Communist progressive organizations. Michael Straight attends London School of Economics, marches in **May Day** parade in Hyde Park. **June:** Sweden revokes WR's permit to reside and work in that country. Willy Brandt joins the board of Mot Dag, a Norwegian left-cult based in Oslo. **June 4:** WR illegally re-enters Denmark and lives in Sletten under the alias "Peter Stein." **June:** Fenichel lectures WR's group in Copenhagen, and comes under heavy criticism. **June 30:** "The Night of Long Knives" in Germany; Stormtroopers' leaders (Röhm, Heines, etc.) arrested and shot on the grounds that these "left-leaning" elements were conspiring to bring about a "second revolution" with the support of a foreign power. **July 14:** WR receives a request from Soviet filmmaker, Sergei Eisenstein, to send Sexpol literature. **August 1:** WR informed that his name will not appear on the list of members of the German Society when the IPA meets in Lucerne, but that this is only a formality to avoid embarrassment. Ex-Sexpol activist Arnold Deutsch (aka OTTO) recruits Kim Philby as an illegal agent of the NKVD. Guy Burgess leaves Cambridge to become a recruiter of "student moles" for Soviet intelligence. (Friends note his "bizarre break" with former Communist associates.) **August 26:** in Lucerne, Switzerland, at the Congress of the International Psychoanalytic Association, WR formally expelled from the IPA. Delivers his paper, "Vegetative Streaming and Psychic Contact," as a guest of the Association. None of the German-Austrian members of the "Opposition Group" offer any support. Each one presents papers that depend on Reich's prior contributions without mentioning his name. **September:** WR and Elsa Lindenberg return to Copenhagen, via France, then move permanently to Olso, Norway. **October:** Karl Frank returns to Berlin from Prague, to assume leadership of the Org, and keep it functioning as an underground resistance against the Nazis. **November:** Guy Burgess recruits Anthony Blunt, teaching Fellow of Trinity College. Michael Straight begins study at Trinity College, Cambridge, England. Blunt, Burgess and James Klugman recruit John Cairncross, introducing him to Arnold Deutsch (NKVD).

1935 Soviet "show trials" and massive purges begin. Robert A. Brady makes second trip to Moscow. Stakhanov "super-worker" campaign in USSR. Arthur Kallet leads a violent strike, with the support of the CPUSA, against Consumers Research, and later establishes the Consumers Union with Robert A. Brady and Mildred Edie. Alger Hiss brought in by Frank and Tugwell at AAA (Agricultural Adjustment Administration) in Department

Wilhelm Reich Chronology

of Agriculture. Anthony Blunt joins Cambridge secret society "Apostles" and Michael Straight joins Communist cell at Cambridge. Guy Burgess' cover is to work for Captain Jack Macnamara, homosexual MP and Nazi sympathizer. **January:** Mao leads Chinese Communist Party. **January:** WR begins lectures at Oslo University. **Spring:** Willy Brandt leaves Danish left-cult, Mot Dag. Norwegian Labor Party elected to power. March: WR obtains oscillograph. **May-December:** bioelectrical investigations: WR begins research into the nervous system, searching for a physical basis of "libido" at the Psychological Research Institute laboratory at the University of Oslo. Future German Chancellor Willy Brandt participates as a volunteer in the experiment. **June:** Trotsky moves to Norway. **Summer:** Eva Reich visits WR, and together they go to Grundlsee to visit Annie Reich and Thomas Rubinstein (Yakov Reich), to ask if Eva can live with her father. They refuse until she completes her treatment with Berta Bornstein, who, Eva says, "brainwashed" her and poisoned her childhood relations with her father. **June 25:** series of meetings in underground Berlin, resulting in the reformation of the Org as "Neu Beginnen," with Walter Löwenheim ("Miles") and a few others emigrating and leaving the group, the rest remaining underground. Karl Frank and younger members of various labor and political organizations anticipate a foreign military liberation from the Nazis. **August:** a group of Cambridge students, including "Apostles" society members Blunt, Straight, and future WR biographer Charles Rycroft travel to Leningrad, USSR. **September 10:** WR writes to Trotsky on behalf of Sexpol. Proposes a meeting. **September 18:** Trotsky replies to WR, his medical problems delays a meeting, but he agrees to meet a future date. **October 3:** Mussolini invades Ethiopia. **October 24:** Psychoanalyst and Org member Edith Jacobson arrested in Berlin. **December:** Wilhelm Hoffmann of the Kaiser Wilhelm Institute, Berlin, collaborates with WR's bioelectrical experiments in Oslo. **December 14:** final meeting of the "Opposition Group" of analysts within the IPA to discuss WR's expulsion. WR'S break with Otto Fenichel - who moves from Oslo to Prague. **Fall 1935-Jan 1936:** Karl Frank travels throughout the US, meets Eleanor Roosevelt; he raises money and awareness of the Neu Beginnen group's underground resistance. **Winter:** WR publishes about 100 copies of *Menschen im Staat*, later reworked into in the English edition of *Mass Psychology of Fascism* as the ninth chapter); he sends this to selected comrades and party officials, including Stalin.

1936 US Federal Court of Appeals rules that physicians may prescribe contraceptives to patients. One of WR's early influences, the Viennese editor, pacifist and epigrammatist Karl Kraus dies. "They pick our dreams as if they were our pockets," Kraus once wrote of psychoanalysts. **January:** Karl Frank visits Robert and Mildred Edie Brady in Berkeley, California. **February:** WR founds Institute for Sex-Economic Bio-Research with nine members. Works with Ola Raknes, Ph.D. Meets Roger du Teil in Paris, who will collaborate in bion research. **February:** Spanish voters elect Popular Front coalition of left-liberal parties. **March 7:** German soldiers occupy the Rhineland in violation of the Treaty of Versailles. **May:** WR begins microscopic study of protozoa, discovering "the preliminary stages of life" in earth preparations, leading to the discovery of bions. **Spring:** "Marxist-Psychoanalytic Workgroup" formed in Prague. Karl Frank, under the alias "W. M." (Willi Müller), publishes "Sigmund Freud and Revolutionary Marxism" in Der Kampf, published by Otto Bauer. **July 17:** Kim Philby and Litzi Friedman travel to Spain after outbreak of Civil War. Michael Straight visits Paris, meets Austrian students travelling to Spain. **August 1:** Olympic Games in Munich. **August:** WR travels by car, alone, through Norway, Denmark, Poland, Czechoslovakia, Austria, Switzerland, France, England, Denmark, and returns to Norway. **September 4:** WR's name appears on an NKVD death list (Document 20) of suspected "Trotskyists." **Fall:** Straight tapped for membership in "Apostles" secret society by David Champernowne. In Copenhagen, a Sexpol clinic under the direction of Leunbach and Tage Philipson is established - among other accomplishments, they trained 80-90 % of Danish medical students to fit diaphragms. The Copenhagen Sexpol chapter is closed down by Danish authorities; Leunbach is jailed for three months for performing an illegal abortion. Philipson jailed for "neglect of patients" for three months also. Both MD's have their licenses revoked for several years. Because Sexpol members do nothing to support the two doctors, WR considers disbanding the

Wilhelm Reich Chronology

organization. **December:** Trotsky exiled from Norway to Mexico. Karl Frank meets Willy Brandt for the first time in Prague. **December 10:** WR demonstrates bion cultures to Albert Fischer at the Rockefeller Institute. Meets A. S. Neill, founder of alternative Summerhill School.

1937 Much of Red Army command, accused of "Trotskyism," is executed in Russia. Robert A. Brady publishes *The Spirit and Structure of German Fascism*. WR observes that bion cultures immobilize bacteria, including "T-bacilli" cultured from animal tissue. WR publishes results of his experiments on bioelectricity and sexuality. Pastor Reinhold Neibuhr marries American Anna Caples to Karl Frank in New York. **January:** Gerhart Eisler starts illegal German-language radio station for the Comintern ("Deutscher Freiheitssender 29.8") near Madrid, Spain. **January:** Michael Straight learns that John Cornford was killed in Catalonia in the Spanish Civil war. Straight agrees to go underground for the secret work of the Comintern, and to return to the US and find work at the Morgan Bank. Friends notice Straight's sudden distance from the communist student organizations. **February 2:** Jørgen Neergard, WR's close friend, dies in Oslo hospital after a long illness. **March 2:** WR requests a grant from Frank Blair Hansen, H. M. Miller, and Tracy B. Kitteridge, representatives of the Rockefeller Foundation in Paris. **March 8:** bion cultures received by Prof. Theodor Thjötta, who, unbeknownst to WR, is a Rockefeller-funded bacteriologist. **March 28:** Rockefeller Foundation rejects WR's grant request. **April 6:** Du Teil successfully cultures bions in Paris. **April 16:** Willy Brandt writes to WR to inform him about the situation in Spain. **April 26:** German planes drop incendiary bombs on civilian population in Guernica, Spain. **May 20:** Prof. Theodor Thjötta refuses WR's request for permission to use animals in his studies. WR injects lab mice with T-bacilli in a private lab, showing the relationship between cancer and "bionous breakdown". **May 26:** Roger du Teil works with WR in Oslo. **June 4:** Norwegian Communist journal publishes criticism of WR's scientific work. **June:** Karl Frank visits Spain, where several friends had disappeared. Willy Brandt returns from Spain, where he advocated a Stalinist "people's front" against the POUM. **July:** Michael Straight returns to US. Tours the Midwest with ACLU founder

Roger Baldwin, meets with labor leaders. Tells no one of his agreement with Blunt. Meets with FDR about government job possibilities. Tours the South with Mrs. Eleanor Roosevelt, a childhood friend of his mother's. She comments to him: "I would be a communist if I thought Russia was like America." **September 1:** WR begins preliminary work on cancer. **September 22:** Norwegian daily paper *Aftenposten* publishes critical review of WR's bion experiments. Biologist Klaus Hansen dismisses bions as "Brownian movement" (a process earlier identified by Bastian, the leading Darwinian advocate of spontaneous generation.) Beginning of "Scandinavian Press Campaign" against WR. **December 12-13:** Thirty-one year old Herbert Wehner, the future "gray eminence" of post-war German national politics, is interrogated at Lubyanka Prison, Moscow, where he denounces WR as a "Trotskyite." **December 28:** the Adlerian psychologist Ingjald Nissen notes in *Arbejderbladet* that WR's therapy is quackery. Summerhill School founder A. S. Neill becomes WR's patient at the end of the year.

1938 Stalinist purges: Bukharin shot. Cerletti and Bini introduce electroshocking as psychiatric treatment. Orson Welles broadcasts radio version of "War of the Worlds." Anschluss unites Germany and Austria. Thomas Wolfe dies. Albert Einstein & Leopold Infeld publish *The Evolution of Physics*. Richard Whitney, former president of the New York Stock Exchange, and a partner in J.P. Morgan & Co., is sentenced to Sing Sing prison for grand larceny. His cousin, Michael Straight, now a Comintern agent, volunteers at the State Department, European Affairs division, and has his first meeting with "Michael Green," an NKVD agent. Donald MacLean enters British Foreign Service, posted to Paris. Willi Schlamm emigrates to the US. **February 1:** WR's *Die Bione* is published. **March:** Hitler annexes Austria. Sigmund Freud arrested. Norwegian press begins virulent attack on WR; Scharffenberg accuses WR of arranging sexual intercourse between mental patients. **April 21:** Thjötta tells newspapers that bion cultures are contaminated with "air germs." **April 22:** Norwegian fascist newspaper calls for WR's internment in a concentration camp. Prof. Schreiner reports WR to the authorities for animal experimentation. **April 23:** German Embassy in Norway writes to German State Depart-

Wilhelm Reich Chronology

ment concerning WR, arguing for an accelerated expatriation from Germany. **April 27:** Reich answers his critics in the *Aftenposten* and demands controlled experiments. The medical department of the university reacts positively, but later this offer is withdrawn. **May 16:** upon invitation of the local physician association WR lectures to hundreds of physicians about the bion experiments. **May:** Michael Straight spots Alger Hiss as a possible recruit and recommends him to his control agent, "Michael Green." **May 10:** One hundred microphotos from WR's labwork stolen in a break-in at a photo-developing shop. Roger du Teil suspended from university post because of private association with WR, but reinstated on **June 30. July:** WR camps out in a tent for the month. **August 2:** Einstein and other physicists urge FDR to develop an atomic bomb. **September 1:** WR posits spinning wave hypothesis. Michael Straight joins campaign of Texas Congressman Maury Maverick, an ally of Lyndon Baines Johnson. Maverick spreads rumor that Straight is an FBI agent sent to ensure a legitimate voting process. **September:** Straight returns to State Department. Alger Hiss praises Straight's reports in a personal meeting. Straight, offered a post in the Office of the Economic Advisor, declines. Straight passes his reports and trade summaries to an illegal Soviet agent. **Fall:** American Theodore Wolfe visits Oslo for training, and suggests that WR relocate to the US. **November 11:** *Kristallnacht.* **December:** WR attends Willy Brandt's 25th birthday party and gets into an argument with Brandt about work democracy. WR applies for an American visa.

1939 Ralph Nader is born. Whitney Straight (Michael Straight's older brother) called to active duty in the RAF. Kim Philby assigned to kill Francisco Franco on Stalin's orders. Stalin named "Man of the Year" by *Time*. Robert Brady and Haakon Chevalier edit the journal *Black and White*. **January 10:** NKVD officer Walter Krivitsky defects and is debriefed by the spy-packed US Department of State. **February:** Guy Burgess begins working for British Secret Service. **February 16:** Isolation of bions from beach sand (SAPA bions), leading to WR's accidental discovery of orgone radiation. **February 20:** WR invents "Kirlian" photography by placing SAPA bions on a one-kroner coin and covering it with X-ray film for 20 hours. **March:**

Germany overruns Czechoslovakia, gaining some of the richest uranium deposits in the world. **March 14:** WR calls radiation from SAPA bion cultures "orgonicity." **March 15:** Hitler takes over Prague. **March 22:** exiled KPD and KPÖ meet in Paris and propose a united front with German Socialists and the Revolutionary Socialists in Austria. **March 28:** definition of "orgone," "orgonotic," etc. **April:** Karl Frank cautions Org members in several countries not to cooperate with the Communist united front. **April 13:** WR decides to move permanently to the US. **May:** Gertrud Gaasland, Willy Brandt's wife, brings WR's archives, laboratory instruments and furniture to the US in advance of his emigration. **August 19:** WR departs Olso on the *Stavenger*. **August 23:** Stalin and Hitler sign non-aggression pact. **August 27:** WR arrives in New York. **September 1:** Germany invades Poland, beginning World War II. **September 3:** Britain & France declare war on Germany. **September 24:** Freud dies. **September & October:** Soviet agent "Michael Green" meets Michael Straight in Washington, DC. **October 11:** Albert Einstein writes to FDR urging the development of the atomic bomb. **October 18:** Lee Harvey Oswald is born in New Orleans. **Christmas Day:** WR and Ilse Ollendorff decide to live together.

1940 Eleanor Roosevelt uses her influence to help Karl Frank and the German underground movement against Hitler. Mrs. Roosevelt helps relocate to the US a number of Labor and Socialists deputies stranded in Europe. Henry Wallace becomes US Vice President. Erling Falk, founder of Norway's Mot Dag socialist sect, dies. Development of the Reich Blood Test, anticipating PAP smear technology by 15 years. NKVD agent Arnold Deutsch dies on a ship torpedoed by a U-boat, en route to the US, where he was to re-establish contact with Michael Straight. **February:** American Youth Congress in Washington, DC organized to oppose US intervention in WWII - Michael Straight's "staged break with the CP." Straight is Eleanor Roosevelt's personal advisor in matters dealing with the CPUSA. Eva Reich attends these "pacifist" demonstrations. **April:** WR invents orgone accumulator in an attempt to isolate orgone radiation from SAPA bions. Discovery of the temperature difference between orgone accumulators and surrounding atmosphere. (To-

Wilhelm Reich Chronology

T). **April:** Hitler invades Norway, with the approval of the Soviet Union. Whitney Straight, stationed in Norway, is wounded when the Luftwaffe strafes his fighter squadron. **May 10:** Winston Churchill becomes Prime Minister. **May 14:** German Luftwaffe bombs Rotterdam, Holland. **June 9:** Norway falls to Germany. **June 16:** Petain becomes French Prime Minister. **June 22:** Petain signs armistice with Germany. **June 22:** Britain's Communist organ *Daily Worker* calls for a "People's Convention" for a "People's Peace," to oppose the war against the Nazis. **July:** Discovery of atmospheric orgone. NKVD agent Guy Burgess visits with Michael Straight in Washington, DC. **August 2:** Beginning of the Battle of Britain. **August 21:** Trotsky is murdered in Mexico by an agent of the NKVD, commanded on-scene by Leonid Eitingon. **August 25-26:** First RAF bombing of Berlin. NKVD agent Anthony Blunt joins British Secret Service. **November:** US forms Foreign Activity Correlation Division to monitor European exile groups. **December 30:** WR writes to Albert Einstein requesting a meeting to discuss orgone energy research.

1941 Charles A. Willoughby becomes head of General Douglas MacArthur's G-2 intelligence section. Kim Philby assigned to British Secret Service. **January:** Comintern-inspired "People's Convention" for a "People's Peace" to end Britain's war with Germany. Stalin named "Man of the Year" by *Time* again. **January 13:** WR meets Einstein in Princeton for a five-hour discussion. **January 27:** WR visits Einstein again, bringing an orgone accumulator and an orgonoscope for objective atmospheric observation. Einstein asks to keep it for further study. **February 7:** Einstein writes to WR explaining away his findings; Einstein's office ceases further discussion. **February 10:** NKVD defector Walter Krivitsky is found dead from a gunshot wound in Washington, DC - police rule it a suicide. **Spring:** WR teaches his last term at the New School for Social Research. Fromm's *Escape from Freedom*. **March 8:** WR begins treating 15 human cancer patients with orgone accumulators. **May 15:** first jet plane flown in Britain. **June 22:** German tank divisions enter Soviet territory, ending Soviet collaboration with the Nazis. **July 1:** First commercial television broadcast over WNBT, New York. **September 7:** Dies Committee demands the removal of Robert A. Brady and Mildred Edie Brady from their federal government jobs on the basis of their known allegiance to the Soviet Union. **September 21:** Emmy Rado, wife of psychoanalyst Sandor Rado, joins the forerunner of the OSS in New York, working with the Oral Intelligence section interviewing exiles, and developing exile networks of agents and sources. **September 23:** WR writes last of a series of letters to Einstein countering Einstein's objections, point by point. Einstein's office does not return WR's loaned equipment until **November 14th**, after several requests are made. **September 25:** Moscow is informed about British-American plans to build the atomic bomb. **October:** WR's study group, drawn mainly from his seminar at the New School for Social Research, buys him a house in Forest Hills for a home and lab. WR founds Orgone Research Institute. **December 7:** Pearl Harbor. **December 8:** US declares war on Japan. **December 8:** first Jews die in Nazi gas chamber, in Poland. **December 11:** Germany and Italy declare war on the US. **December 12:** On orders of the Attorney General, WR arrested at 2 a.m. by the FBI as an "enemy alien" and taken to Ellis Island and held in "custody detention" for three weeks. Joe Kennedy, Sr. announces the "inevitable" defeat of Britain, and FDR fires him as Ambassador to Britain.

1942 "Casablanca" wins the Academy Award for Best Picture. *Function of the Orgasm* and the *Int. Journal of Sex-Economy and Orgone Research* are published by WR. **January 1-2:** Enigma machine - British breakthrough in decoding German communications. **January 5:** WR released from Ellis Island. Purchases 280 acres in Rangeley, Maine. "Michael Green," Michael Straight's control agent, is recalled to Moscow, where he is shot as a "British spy." **February 25:** Air defense fires 1,430 rounds of antiaircraft shells without effect at UFOs hovering over Los Angeles, California. Napalm developed in the US. **April 14:** The FBI interviews Robert A. Brady about subversive activities. **May 16:** Malinowski dies. **June 13:** Office of Strategic Services (OSS) is formed out of the COI to coordinate US intelligence operations. **June 10-20:** FDR and Churchill approve cooperation on "Tube Alloys" i.e., the atomic bomb. **July:** Karl Frank offers his services and underground contacts within Germany, the Neu Beginnen organization, to Calvin Hoover of the OSS.

Wilhelm Reich Chronology

His efforts are opposed by the State Department, Ruth Fischer and Emmy Rado, among others. **December 2:** First controlled chain reaction, producing heat, in a layered compile of graphite and uranium in a secret room beneath University of Chicago's football field.

1943 WR publishes initial work on treatment of cancer patients. Nikola Tesla dies. Max Eitingon dies. Swiss chemist Hoffmann discovers hallucinogenic properties of LSD-25 first-hand by accident. D. Ewen Cameron becomes chair of McGill's department of Psychiatry, in Montreal, Canada. **January:** Vannevar Bush's Manhattan Project is approved. **January 14:** FDR and Churchill meet in Casablanca. **February 16:** Karl Motesiczky is deported to Auschwitz. **April 10:** Steve Nelson, of the CPUSA, discusses espionage with San Francisco rezident Vassili Zarubin. **April 17:** liquidation of over four thousand Polish officers in Katyn forest by Red Army on orders from Stalin. **April 19:** Warsaw ghetto uprising begins. **May 15:** Comintern is disbanded by Stalin. **July 7:** Adolf Hitler gives the German V-2 program highest military priority. **August 7:** German turbojet fighter, a Messerschmitt Me-262, is demonstrated before Adolf Hitler in East Prussia. **August 29:** Germans re-occupy Copenhagen after Danish resistance steps up sabotage. **Fall:** first sightings of "foo fighters" (UFOs) in Northern Europe. **September 3:** Italians secretly surrender to Allies. **September 12:** Colonel Otto Skorzeny's airborne rescue of Mussolini from Allied prison.

1944 Using a new discovery - the atmospheric jetstream - Japan launches approximately 10,000 Fugo balloons (30-foot diameter) carrying incendiaries and aimed at the North American continent. V-2 rockets bombard England. Uranium reactor developed in Germany. Soviet spy Donald MacLean posted to Britain's Washington Embassy. German exile psychoanalyst Paul Federn sees novelist William S. Burroughs as a patient. Eighteen year-old Myron Sharaf meets WR. Churchill orders all Communists out of Secret Service. Soviet spy Kim Philby becomes head of Section Nine (British counterintelligence). WR's old socialist youth friend, Willi S. Schlamm joins Henry Luce's magazine empire as assistant to the director. Britain's Ministry of Information bans George Orwell's *Animal Farm*.

FDR and Democratic Party wheelhorses remove Vice President Henry Wallace from the ticket. **April 7:** Peter Reich is born to WR and Ilse Ollendorff Reich. **July 20:** failed von Stauffenberg assassination attempt on Hitler. **October:** Dumbarton Oaks conference agrees to form United Nations. **December:** OSS purchases Soviet codes and ciphers from Finnish contacts; the Roosevelt administration orders the materials returned to the Soviets.

1945 WR invents orgone field meter. Attorney Arthur Garfield Hays advises WR not to file a libel suit against the head physician of the Norwegian Army, Karl Evang, who claimed that WR had been locked away in an insane asylum in America. Evang's brother, Vilhelm, is head of Norway's military intelligence, and both had belonged to Mot Dag. Koestler's *The Yogi and the Commissar*. Willy Brandt returns to Germany this year. **January 24:** Germans successfully launch A-9, a winged prototype of the first intercontinental ballistic missile. **January 26:** journal *Amerasia*, funded by Nelson Rockefeller and by Michael Straight, publishes a report on "British Imperial Policy in Asia" taken verbatim from a highly classified State Department report. **February 11:** FDR, Churchill, and Stalin confer in Yalta and agree on postwar division of "spheres of influence" in Europe. Alger Hiss, of the US State Department, leaves Yalta, goes to Moscow and is secretly given a Soviet medal for services rendered. **February 20:** US Secretary of War approves ordnance plans for White Sands Proving Ground. **March 9:** US General Curtis LeMay firebombs Tokyo with napalm; that night, an estimated 80,000 people die. **April 12:** FDR dies. **April 13:** Red Army occupies Vienna. **April 24:** Himmler proposes separate peace with US and Britain, who refuse. **April 29:** Dachau concentration camp is liberated by US soldiers. **April 30:** non-profit Orgone Institute Research Laboratory is formed. **April 30:** Hitler commits suicide. **May 8:** World War II ends in Europe. **June 6:** *Amerasia* Case: Andrew Roth (Office of Naval Intelligence), Philip Jaffe (publisher of *Amerasia*), Mark Gayn (free-lance journalist and Soviet agent), John Stewart Service (State Department official), Emmanuel Larson (State Department), and Kate Mitchell (Jaffe's secretary) are all arrested by the FBI for espionage for the USSR. US Attorney General Tommy "The

Wilhelm Reich Chronology

Cork" Corcoran drops all charges. Paul Goodman publishes "The Political Meaning of some Recent Revisions of Freud," a positive review of WR's work, in the leftist journal *Politics*. C. Wright Mills and Patricia J. Salter rebuke Goodman in a following issue. **July 16:** Trinity: first test of atomic device in Alamogordo, New Mexico. **July 17:** Potsdam conference. **July 23:** *Life* publishes drawings of a manned space station as envisioned by the German rocket scientists. **August 6:** US drops atom bomb on Hiroshima, Japan. **August 9:** atom bomb dropped on Nagasaki, Japan. **August 14:** Japan surrenders, WWII ends. UN is formed in San Francisco, Alger Hiss selected as first United Nations Secretary General for the San Francisco Conference. **September:** Branding the OSS as British-controlled, Truman disbands it. During **October:** Secretary of War Patterson approves plan to bring top German scientists to United States to aid military research and development. More than 100 German rocket scientists and engineers, who had agreed to come to the United States to work on missile development under Project Paperclip, arrive at Fort Bliss, Texas, and White Sands Proving Ground, New Mexico. **October 20:** Nuremberg war crimes trials begin. **December**: Office of Deputy Chief of Air Staff for Research and Development created in US Army Air Force, headed by Maj. Gen. Curtis E. LeMay. Last issue of the *Int. Journal of Sex-Economy and Orgone Research* is published.

1946 Revised, English-language edition of *Mass Psychology of Fascism* is published. **May:** WR becomes US citizen, begins student training and research at his Orgonon estate in Rangeley, Maine. Elsworth Baker, chief of the Women's Department of the Marlboro State Hospital in New Jersey, comes to WR for therapy and training. **June:** first National Congress of the American Veterans Committee (AVC) formed to counter the influence of conservative veteran's groups. **June 6**: Joint Army-Navy Research and Development Board is created for purpose of coordinating all activities of joint interest in fields of aeronautics, atomic energy, electronics, geographical exploration, geophysical sciences, and guided missiles. **July 17:** Attorney General Tom Clark persuades President Truman to expand peacetime electronic surveillance of "persons suspected of subversive activities." **August 1:** Atomic Energy Commission formed.

November 13: Irving Langmuir and Vincent Schaefer employ "cloud-seeding" with frozen carbon dioxide particles. First snow from a natural cloud is produced by V. Schaefer of General Electric, the experiment carried out by means of dry-ice pellets dropped from a plane over Greylock Mountain, Mass. **Autumn:** Mildred Edie Brady "interviews" WR at his home in Forest Hills, NY. **December 2:** *The New Republic*, under editorship of Henry Wallace, publishes Frederic Wertham's review of *Mass Psychology*, condemning WR's "utter contempt for the masses." (First US mass media attack on WR.) Lavrenty Beria assumes control of USSR's entire nuclear program, from foreign spies to German scientists, as well as domestic research and production. Alger Hiss meets several times with Donald MacLean. US General Charles Willoughby (military intelligence) exchanges freedom for Japanese war criminals for biowarfare data from Unit 731. **October 16:** Ten Nazis hang in Nüremburg. **October:** Stalin targets activists in the Jewish Antifascist Committee, a campaign against Jews in the Party. Builds up for the Doctor's Plot. By the end of this year, in a project which would become known as VENONA, Meredith Gardner of the US Army Signal Intelligence Corps will have decoded enough wartime KGB ciphers to determine deep Soviet involvement with atomic espionage. Karl Frank, barred from returning to Germany, contemplates writing a book about "political psychology."

1947 *Annals of the Orgone Institute* appear. Eleanor Roosevelt spearheads a drive by American supporters of New Beginning to create Americans for Democratic Action (ADA). The ADA espouses domestic social reform and supports President Truman's toughening stance against Russia in the developing Cold War. Cominform is founded in Warsaw. Soviet agent Donald MacLean becomes joint secretary of the Anglo-American Combined Policy Committee on atomic energy. Alger Hiss is president of Carnegie Endowment for International Peace. **February 5:** President Truman directs that production of nuclear weapons continue, following the recommendations of the AEC and the Secretaries of War and Navy. **March:** Progressive Citizens of America launches Henry Wallace's 3rd party candidacy for the Presidency. Myron Sharaf introduces William Washington to WR. Washington works for WR as a mathematical assistant. **March**

Wilhelm Reich Chronology

17: Vladimir Androsov, later a Russian orgone energy researcher in the 1990s, is born in Moscow. **April:** *Harper's* Magazine publishes Mildred Edie Brady's "The New Cult of Sex and Anarchy" describing WR's works as "the guiding light" of a new bohemian culture on the West Coast. Michael Straight and Wallace travel to Europe: London, Copenhagen, Paris. Churchill calls Wallace a "crypto-communist" and then retracts it. Wallace and Straight meet Ole Calvring, publisher of the newspaper *Extrabladet* that published some of the first articles against WR in Copenhagen. In Paris, Wallace meets with Pierre Cot, NKVD agent. **May:** Michael Straight's *New Republic* magazine publishes "The Strange Case Of Wilhelm Reich" by Mildred Edie Brady, who calls for government action against WR. **June 24:** private pilot Kenneth Arnold sights nine "flying saucers" in formation over Mount Rainier, Washington. **July 8:** Col. Blanchard at Roswell Air Force Base orders public information officer Walter Haut to issue press release on the recovery of a crashed "flying saucer" in New Mexico. **July 17:** Raoul Wallenberg dies, probably of a lethal injection, in Leforto prison, Moscow. **July 23:** Federal Trade Commission official (J. J. Durrett) sends Brady article to the Food and Drug Administration (FDA). R.M. Wharton, chief of the Eastern Division of the FDA, directs Charles A. Wood, Maine's FDA inspector, to launch a preliminary investigation. **July 26:** President Truman signs the National Security Act, birth of the Central Intelligence Agency (CIA). **August 8-9:** WR demonstrates the motor force in orgone energy. **August 27:** Charles A. Wood visits Orgonon. **September 24:** Wood visits Orgonon again. At first, WR cooperates with the FDA. **September 30:** Research and Development Board (RDB) of DOD superseded Joint Research and Development Board, with Vannevar Bush named as Chairman. **October:** Wallace and Straight meet with leaders of the Stern gang in Palestine, and then have an audience with the Pope in Rome. **November 19:** Julian Culver, WR's lawyer, telephones Wood and asks about FDA agents recent inquiries to WR's coworkers about a "sex racket". Wood denies any preconceived attitude, and invites WR to demonstrate the accumulator. WR refuses to hand the FDA a list of orgone accumulator users. The FDA already had that list. **December:** *Everybody's Digest* regurgitates Brady with "Is the World Sexually Sick?" *Collier's* also retransmits Brady libels. **December 23:** Shockley invents transistor. **December 15:** Karl Frank confirms Mildred Brady's ties to the Communists in a letter to WR. **December 20:** WR transfers all rights for the medical use of the orgone accumulator to the non-profit Orgone Institute Research Laboratory. **December 21:** New Soviet Ambassador Panyushkin arrives Washington D.C. (Former head of Division Secret Police Moscow.) **December 29:** Progressive Party formed, with Henry Wallace as Presidential candidate. During a Politburo meeting reviewing the problem of developing an intercontinental ballistic missile, Premier Joseph Stalin reportedly states that a transatlantic rocket capable of hitting New York City "would make it easier to talk with the gentleman-shopkeeper, Harry Truman, and keep him pinned down where we want him."

1948 Richard Nixon forces confrontation between Whittaker Chambers and Alger Hiss, leading to Hiss's perjury charge and conviction. Norman Mailer's *The Naked and the Dead*. Kinsey Report is issued. Norbert Wiener founds cybernetics. WR writes *Ether, God & Devil*. *Cancer Biopathy* and *Listen Little Man* are published. Formation of American Association for Medical Orgonomy (AAMO), consisting of Drs. Baker, Cott, and Willie as board members. Wolfe issues *Emotional Plague versus Orgone Biophysics*. **January 30:** Orville Wright dies in Dayton, Ohio, at the age of 76. During his lifetime, the speed of aircraft had increased from 0 mph to nearly 1,000 mph. **April 30:** Meeting between OIRL and AEC representatives to report Geiger-counter effect of Orgone Energy and Vacor Phenomena. WR told that orgone research is outside scope of AEC. **May 28:** Local FDA agent Wood hands over investigation to Washington: recommends that the FDA obtain an injunction, but says "no dissatisfied users were located." **Summer:** William Washington arrives at Orgonon to develop the orgone energy motor. **June 26:** Berlin airlift begins. WR demonstrates orgone energy motor. **July 20**: Twelve CPUSA leaders are arrested under the Smith Act. **July 31:** Elizabeth Bentley accuses Treasury Department's Harry Dexter White and State Department's Lauchlin Currie of being Soviet agents. **August 3:** Whittaker Chambers publicly accuses Alger Hiss of being a Soviet Agent, repeating earlier

Wilhelm Reich Chronology

information given to the State Department in 1939. Wernher von Braun is named an Honorary Fellow of the British Interplanetary Society. **August 28:** WR's colleague Walter Hoppe is detained at Ellis Island. **August 30:** First International Orgonomic Conference at Orgonon. WR intervenes with immigration officials and Hoppe arrives at the Conference **August 31. October 14:** First test of silver iodide cloudseeding is declared successful. **November 4:** Harry Truman elected over Thomas Dewey, Henry Wallace and Strom Thurmond. **December 29:** The first report of the Secretary of Defense, James Forrestal, reported that the United States had been engaged in research on an earth satellite.

1949 *The Orgone Energy Bulletin* appears. Orwell's *1984* is published. Mao establishes People's Republic of China. **February 16:** secret conference of American scientists, including Edward Teller, is convened at Los Alamos, New Mexico, to investigate sightings of "green fireballs" over sensitive AEC installations. **May:** W. S. Burroughs builds his first orgone accumulator in Texas. **May 12:** USSR's blockade of Berlin ends. **May 22:** former Secretary of Defense James Forrestal plunges to his death from a six-story window of Bethesda Naval Hospital. **August:** Kim Philby is appointed as CIA-FBI-SIS liaison to Washington. Michael Straight bumps into Guy Burgess at an Apostles' gathering in London. Straight also meets with Anthony Blunt. Back home, Straight is elected chairman of the American Veterans Committee. **September 23:** Truman announces the detonation of the first Soviet atomic bomb. **October 29:** Oppenheimer advises AEC not to develop hydrogen bomb. **November:** Orgonomic Infant Research Foundation is founded. **December 9:** Wilhelm Reich Foundation is incorporated as a non-profit umbrella for research activities. **December 12:** Last monkey, Albert IV, launches in space tests at White Sands Proving Ground, a successful flight with no ill effects on monkey - until final impact of the vehicle. **December 28:** USAF reports that their two-year investigation had found that there is no such thing as a "flying saucer" and that Project Saucer at Wright-Patterson Air Force Base had been discontinued.

1950 **January 21:** Alger Hiss is sentenced to a jail term in Lewisburg federal prison for perjury. **January 24:** Klaus Fuchs confesses to atomic spying. **January 30:** President Truman announces his decision to go ahead on the hydrogen bomb development program. **February 9:** McCarthy's speech at Wheeling, WV announcing his list of 205 communists in the State Department. Communist double-agent Harvey Matusow begins working for FBI and is paid $325 as an undercover informant on the Communist Party for the FBI. **May:** WR moves to Orgonon, in Rangeley, Maine permanently. Sharaf moves to Orgonon as a research assistant. **May 22:** FBI arrests Soviet spy Harry Gold. **June 25-27:** Korean War begins. Within 48 hours of capture, some American POWs are issuing broadcasts favoring communism; American intelligence officials suspect Soviets have developed a form of mind control. Eleanor Roosevelt teams with her son Elliott and NBC on a television and radio show featuring famous guests, such as Albert Einstein and the Duke and Duchess of Windsor. **July 17:** Julius Rosenberg is arrested. **July 30:** first of several UFO sightings over the Hanford AEC plant is recorded in Air Force memoranda. Einstein's coauthor Leopold Infeld defects to Poland, working directly for the NKVD chief Lavrenty Beria. Develops Poland's Institute of Theoretical Physics in Warsaw. **September:** *Miami Herald* journalist on CIA payroll introduces the term "brain washing" to describe the defections of POWs in Korea. In Jerusalem, Teddy Kolleck tells the CIA's James Jesus Angleton that Kolleck had been present at the marriage of Kim Philby and Litzi Friedman, an open Communist, in Vienna in 1934, and wonders why Philby is now a British intelligence officer. (Philby had trained Angleton in Rome during the war.) **December:** WR publishes findings on healing X-ray burns with an orgone accumulator.

1951   Fritz Perls & Paul Goodman publish *Gestalt Therapy*. William F. Buckley and William S. Schlamm found the *National Review*. **January:** James Forrestal Center is established at Princeton University for jet-propulsion research. **January 5-12:** Oranur Experiment: WR places one milligram of radium inside an orgone accumulator. Geiger counters jam with unusually high readings. WR warns that no shielding is possible from the noxious effects of nuclear radiation. Sends Oranur Report to various government agencies, including the Atomic Energy Commission. **January 15:** J. Robert Oppenheimer writes to Mrs.

Wilhelm Reich Chronology

Roosevelt that Oranur experiment is a "hoax." WR takes up oil painting. Confronted with evidence of infidelity by his wife, Mildred, Robert A. Brady attempts suicide by swallowing 500 Nembutals. **April 2:** USAF Air Research and Development Command becomes operational, to which is assigned: Air Development Force at Wright Field and AF Cambridge Research Division. **April 12:** Truman relieves General Douglas MacArthur of command in Korea. Returns with Willoughby to the US and considers running for president. **May 27:** Donald MacLean defects to Moscow with Guy Burgess. Kim Philby is recalled from US, where he served as Anglo-American intelligence liaison. CIA's James Jesus Angleton is asked to describe relationship with Philby. WR writes *Murder of Christ*. Eva Reich and Bill Moïse move into Orgonon. **July:** CPUSA announces reorganization as a underground "cadre" group. **August:** FDA renews investigation. **September:** Ilse Ollendorff obtains legal separation from WR in Hot Springs, Arkansas. **September 5:** USAF awards contract to Consolidated-Vultee to fly a B-36 with a nuclear reactor aboard, to be built by General Electric, for added boost. **September 12:** Sergei Kolokoltsev, Russian orgone energy researcher in the 1990s, is born near Moscow. **October:** WR suffers major heart attack. Recovers by **November**. **November:** WR publishes *Cosmic Superimposition*. Sharaf begins assisting WR in organizing his archives, and translates *People in Trouble*.

1952 Martin Gardner's *In the Name of Science* appears with understated and underhanded attack on WR. **March:** Orgonon is evacuated due to continuing Oranur effects. **May-June:** WR discovers cloud-buster effect of metal pipes grounded in water. Discovery of Melanor, or "blackening of the rocks" at Orgonon. Hypothesis of DOR relationship to desert development. **May:** Sharaf leaves Orgonon after WR finds him searching through his personal files. **June 7:** WR sends samples of changing minerals to President Truman's office. **June 8:** Secret Norwegian investigation: US State Department inquires about press campaign in Norway against WR (1937 - 1938.) **June 9:** Johns Hopkins Hospital, on orders from the FDA, treats a patient four times with an orgone accumulator. **July 19:** Major UFO "flap" in Washington, D.C., with photographs and many witnesses. **July 29:** Three

FDA agents arrive, unannounced and without warrants, at Orgonon for "inspection." **August:** Mayo Clinic tests orgone accumulator for FDA. **September 18:** Construction begins on Thule Air Force Base in northwestern Greenland, 930 miles from the North Pole. **October 18,19:** Kurt Eissler, secretary of the Freud Archives, conducts interviews with WR on his reminiscences of Freud. **November 4:** Eisenhower is elected president. Truman creates the National Security Agency (NSA). Rangeley's local druggist organizes a demonstration by children at Orgonon, shouting "Orgy, Orgy, Commie, Commie!" **November 18:** date of "Majestic-12 Briefing Document," an alleged report to President-elect Eisenhower, by Admiral Roscoe Hillenkoetter, reporting on the Roswell UFO incident, and discussing the retrieval of humanoid occupants from a crashed "flying disc-shaped aircraft." **December 5:** Straight testifies before the House Select committee, says, "I regard the Communist Party as a source of espionage and of subversion".

1953 *People in Trouble* and *Murder of Christ* are published. Ray Bradbury publishes *Fahrenheit 451* about book burning in the future. E. G. Bowen of the Australian Radio and Physics Division of the Commonwealth Scientific and Research Organization first propounds the theory that meteoric dust provides the nuclei for heavy rainfall. (In 1960, additional data were acquired with U-2 aircraft.) **January 13:** "Doctors Plot" against Stalin's life is announced in Moscow. **January 14:** Closed meeting of meteorologists and Eisenhower's advisor Lew Douglas at University of Arizona, at Tucson. **January 14-16:** USAF scientific advisory panel concludes that unidentified flying objects: (1) hold no physical threat; (2) are not foreign weapons systems; (3) are known phenomena consistent with current scientific mechanisms; and (4) rashes of sightings offer a threat from hostile propagandists. **January 20:** Eisenhower assumes presidency. **March 6:** Stalin's death is announced. Beria's state security apparatus is mobilized, but forced down by military units loyal to Khrushchev. **Spring:** WR publishes *The Einstein Affair*, documenting his contacts with Albert Einstein at Princeton. **April 13:** CIA Director Allen Dulles authorizes MKULTRA mind control experimentation. Dr. Ewen Cameron, president of the American Psychiatric Association (APA) and MKULTRA fundee, tells a patient that WR is a "charlatan" and that WR's therapy is a "pure fake" - the APA will bring charges. **May:** Robert McCullough is hired on at Orgonon as biologist. **June 19:** Ethel and Julius Rosenberg are executed. **June 28:** Head of Soviet security apparatus, Beria is arrested and executed. General Noam Eitingon (the director of Trotsky's 1940 murder) is arrested and sentenced to 12 years. **Summer:** WR develops medical DOR-buster. **July 5-6:** Ellsworth OROP: first public demonstration of the cloudbuster at the request of Maine blueberry growers during a prolonged drought. Rain follows the next day and through the night. **July 7:** WR observes Air Force planes over Orgonon, dragging measuring instruments behind them. **July 27:** Korean War ends. **August:** Lew Douglas and his brother-in-law John Jay McCloy (chairman of Rockefeller's Chase Manhattan Bank and political advisor to Eisenhower) spend time fishing in Arizona. **August 3:** Eisenhower signs Weather Control Act into law. **August 7:** Kerr-McGee signs with the AEC to build uranium processing mill at Shiprock, New Mexico. **August 12:** U.S.S.R. announces H-bomb explosion. **August 22:** Operation Ajax: CIA-British restoration of the shah of Iran. **August 26-27:** WR holds seminar on DOR. **September 18:** The Orgone Energy Corporation incorporated. **November 18:** WR reads Keyhoe's book on UFOs. **November 19:** CIA chemist Sidney Gottlieb, associate of Ewen Cameron, spikes CIA agent Frank Olson's drink with LSD-25. After several days of "depression," Olson jumps, or is thrown, to his death from an upper story window of the Statler Hotel in Washington. **November 24:** Immigration and Naturalization Service agent interviews WR, who is outraged by the agent's request to be an informer against individual Communists, including Lion Feuchtwanger and Hannah Peters. **December 9:** Wilhelm Reich Foundation incorporated.

1954 WR operates cloudbusters to break drought in the Northeastern US. ADA leader and US Senator Hubert Humphrey (D-Minn.) sponsors a bill making it a crime to be a member of the Communist Party. **January 13:** WR observes an "Ea" (Energy Alpha, or UFO). FDA applies for an injunction against WR. **January 23:** Lew Douglas obtains $150,000 from the Sloan Foundation, where he is a board member, to start the University of Arizona's

"Institute of Atmospheric Physics," the first well-funded scientific institute to study weather modification. **February 5:** *New Republic* editor Michael Straight testifies before the Reece Commission on the influence of his family's foundation. **February 10:** Peter Mills, formerly WR's attorney, now US Attorney in Maine, files a federal civil injunction against WR, Ilse Ollendorff, and the WRF. **February 25:** WR's "Response to the Court," refusing to appear. **March 15:** AEC releases report on health conditions in uranium mines: "Control of Radon and Its Daughters in Mines by Ventilation." **March 16:** US Senate votes to hold "Army-McCarthy" hearings, televised nationwide for 36 days. **March 17:** WR sends "Survey on Ea" (on possible orgone-propulsion of UFOs) to USAF. **March 19:** WR's preliminary hearing. Injunction prohibits WR from transporting orgone accumulators across state line and cites WR's books as "mislabeled." **April:** Bell Laboratory announces invention of the silicon solar battery. **April 12:** Security Board of the AEC convenes an investigation of Oppenheimer as a possible Soviet agent. **April 22:** Army-McCarthy hearings commence with James D. St. Clair representing the Army. (Later, St. Clair's firm, Hale and Dorr, represent WR in his contempt trial.) **May:** Michael Straight meets the political counselor of the Soviet embassy, Striganov. **May 6:** Revocation of Oppenheimer's security clearance. **May 12:** WR aims cloudbuster at a UFO and it speeds away. **May 15:** Medical orgonomists file "amicus curae" brief in WR's court case, to protect their right to practice medicine. Court rejects their appeal: injunction is only against WR "ad personam." **May 26:** The Orgone Energy Corporation is dissolved. **June:** Theodore Wolfe, WR's former colleague and translator, dies in New Mexico, an apparent suicide. **July 9:** Psychoanalyst Carl Jung mentions his interest in "flying saucer" phenomena, and reports that he has been collecting a "voluminous dossier" on sightings since 1946. **July 14:** date of the famous "Cutler/Twining Memo" from the National Security Council; only authenticated reference to "MJ-12" now held at the National Archives; the memo refers to a highly-classified "Special Studies Project." Allegedly, "Majestic 12" consisted of the President and a group of government officials convened to study human interaction with "extraterrestrial biological entities."

Wilhelm Reich Chronology

This NSC memo, addressed to Air Force General Nathan Twining, and signed by Robert Cutler, Special Assistant for National Security to Eisenhower, notes that a special projects briefing would be rescheduled to take place "during the already scheduled White House Meeting of **July 16**." **July 27:** Office of Lewis W. Douglas telegrams WR's assistant, William Moïse in Washington DC, regarding weather modification, invites further written correspondence. **July:** *Cosmic Orgone Engineering (CORE)* journal appears with "Space-Ships, DOR and Drought." **August:** Ilse Ollendorff separates from WR permanently. **August 3:** Eisenhower appoints Lew Douglas to President's Advisory Commission on Weather Control. Douglas, former ambassador to Great Britain, writes to WR regarding rainmaking. **August 8:** First international conference on weather control at the Institute of Atmospheric Physics in Tucson. President's Advisory Commission on Weather Control inspects time-lapse cameras used to record operations and cloud structure. **August 12:** Air Force establishes AFR 200-2, regulations on reporting UFO sightings to Air Technical Intelligence Command (ATIC). **September 5:** *New York Post* publishes Irwin Ross' "The Strange Case of Wilhelm Reich." **September 21:** WR discovers that the radium used in the Oranur Experiment has become de-ionized, a transmutation he calls ORUR. **September 24:** U.S.S.R. establishes Ziolkovsky Gold Medal for outstanding contribution to interplanetary communications (Russian term for space travel). **Fall:** U.S.S.R. creates the Soviet Interdepartmental Commission on Interplanetary Communications. **September 26:** Moscow radio reports rockets have been sent to 240 in miles height; claims rocket for interplanetary travel designed and flight principles worked out. Sharaf's estranged wife, Grete Hoff, becomes WR's lover. **October 10:** In Maine, WR disables two UFOs using the cloudbuster above Bald Mountain. **October 13:** Robert McCullough and his daughter, en route to Tucson, sight a UFO in connection with a terrific storm. **October 18:** McCullough observes a UFO in the Tucson area at 8 pm, and at midnight a B-47 stratojet bomber crashes and burns ("Planetary Valley Forge"). (This same year, James E. McDonald and three other physicists observe a UFO over Tucson, sending their precise and detailed observations to the Air Force.) **October 15:** Bill Moïse meets with Air Force Technical Intelligence Command (ATIC) officials in Dayton, Ohio, discussing UFOs, anti-gravity equations. **October 18:** WR leaves Orgonon for Tucson, Arizona, to continue atmospheric research with cloudbuster. **October 29:** WR stops overnight in Ruidoso Downs, near Roswell, New Mexico. **October 30:** WR arrives in Tucson, and with the help of Lew Douglas' Southern Arizona Bank, locates a property north of Tucson for a base of operations. **November 22:** Eisenhower's "Atoms for Peace" proposal is accepted by UN. **November 24:** UFO sighting over "Little Orgonon," Tucson. **December 7:** Silvert arrives in Tucson with "Orur," denatured radium from the Oranur Experiment. (Filmed by KEAN-TV, Phoenix). **December 20:** Kim Philby's pupil, James Jesus Angleton is appointed as chief of the CIA's Counterintelligence staff. **December 30:** US Marshal accompanies FDA agent in "inspection" of "Little Orgonon" in Tucson. **December:** Disney's *Man in Space*.

1955 Vaccine against polio. **January:** Michael Silvert ships orgone accumulators from Maine to New Jersey. **January 17:** WR uses cloudbuster to clear fog over the Tucson Municipal Airport, one of Lewis W. Douglas' local investments. **January 28:** WR meets with local farmers to discuss University of Arizona's cloudseeding. **February:** Straight sells *The New Republic* to Gil Harrison. **February 14:** Above-ground nuclear test, Nevada. **February 18:** televised appearance of WR's cloudbusting work in Tucson. **March 1:** Trevor Gardner becomes the first Assistant Secretary of the Air Force for Research and Development. **March 6:** USAF Chief of Staff, Nathan F. Twining, reports that ICBMs are receiving top priority in because of known Soviet progress. Navaho, Snark, and Atlas programs are accelerated. **March 8:** National Security Counsel approves the FBI's COINTELPRO program against domestic political groups. **April 18:** Einstein dies. **April:** WR leaves Tucson, ending operations there, and returns to Maine. He is alone save for Tom Ross. **June 16:** Peter Mills, WR's former attorney, institutes a complaint of contempt of the injunction. WR is ordered to appear **July 26. June:** Grete Hoff separates from WR and returns to Norway. **June 22-27:** Eisenhower's logs reveal a fishing trip to the Rangeley, Maine area. **July 20:** USAF aircraft carrying an atomic reactor makes its first flight; the reactor is not activated. **July:** WR and Michael

Wilhelm Reich Chronology

Silvert appear for jury trial on charges of contempt of court. **August:** last Orgonomic conference at Rangeley, Maine. WR meets his last wife, Aurora Karrer, a biologist for National Institute of Health. **November 2:** The Atomic Energy Commission approves, on the basis of a statement of interest by the Department of Defense, the proposed plans of the Los Alamos Scientific and the Radiation Laboratories of the University of California, for the study and development of nuclear power for rocket propulsion. **November 30:** Lew Douglas among investors opening Happy Jack Uranium Mine, Utah. **December 1:** Rosa Parks takes a seat at the front of a bus in Montgomery, Alabama.

1956 WR asks: "Am I a Spaceman?" FBI establishes COINTELPRO operations. Straight resigns as editor of *The New Republic*. Willy Brandt becomes Mayor of West Berlin. 20th Party Congress: Khrushchev's "Secret Speech" denounces Stalin's "cult of personality" and lists his crimes. **January:** civil marriage contract between WR and Aurora Karrer. **February 7:** WR submits to Judge Clifford a message he sent to CIA Director Allen Dulles, claiming that he is the unnamed "creative scientist" named in chapter five Ruppelt's book *Report on Flying Objects* (1956). **March 8:** Judge Clifford recuses himself from WR's case. **April 30:** WR's contempt trial commences; WR and Silvert do not appear, since the court order is unsigned. WR arrested in Washington **May 1. May 3-7:** Trial against WR. **May 22:** James A. MacKeen, US Probation Officer, files lengthy and detailed pre-sentencing report with the Court on the defendant WR's personal history. This document had been prepared with the assistance of an unnamed psychiatrist who had known WR in Vienna. "A highly regarded individual in the field of psychiatry estimated that the defendant's I.Q. might be as high as 150. It is believed by this individual who refuses to be quoted that the defendant is suffering a mental retrogression and has reached a stage of paranoia in which he suffers from delusions of grandeur and persecution." **May 25:** Found guilty of violating the Food and Drug Act, and of criminal contempt of court, WR is sentenced to two years, and Silvert to one year and one day. WR Foundation is fined $10,000. **June 5:** FDA agents supervise the destruction of orgone accumulators at Orgonon. **June 26:** FDA agents supervise burning of WR's books, including *The Oranur Experiment* and *Ether, God and Devil*. **July 9:** ACLU sends a protest letter regarding the injunction but no media print it. **July 23:** Destruction of fifty orgone accumulators in New York City. **August 23:** Incineration of six tons of orgonomic literature in New York City. **October:** WR submits petition to the Appellate Court, with an affidavit that Nelson Rockefeller of HEW was sent a copy of Vol. III of the *Suppressed Documentary Evidence*. **November:** WR goes to Washington, DC, stays at the Hotel Alban Towers under the alias "Walter Roner," to pursue an appeal to the Supreme Court. He informs the FBI of his true identity. **December 11:** Appellate Court denies WR's appeal. WR appeals to the Supreme Court.

1957 WR's last book, *Contact With Space* is published. Eleanor Roosevelt travels to the Soviet Union for the New York Post, meets Nikita Khrushchev. Mao: "Let a thousand flowers bloom!" Lee Harvey Oswald is assigned to Atsugi U-2 Spy plane base, Japan. **January:** WR speaks to medical orgonomists, individually, for the last time. **February 25:** CIA director Allen Dulles approves funding for D. Ewen Cameron's mind control lab. **February 25:** Supreme Court denies WR writ of *a certiorari*. **March 10:** FBI informs Peter Mills (at home) that they have learned that William Moïse and William Steig had gone to the Portland Police Department to inquire if they had a cell ready, and that they intend to pursue a citizen's arrest (presumably of US Attorneys Peter Mills and Joseph Maguire.) Mills calls Maguire and tells him to come immediately to his house until the next day. **March 11:** final sentencing hearing, WR and Silvert are jailed in Federal Prison at Danbury, CT. Special agents of the FBI prevent any citizen's arrest and escort Joseph Maguire out of town. **March 9:** Court orders WR to undergo psychiatric evaluation before considering his plea to reduce sentence. **March 18:** Psychiatrist Richmond C. Hubbard, MD, concurs with the medical diagnosis of a retired surgeon, Chief Medical Officer of the US Prison Health Service, Clifford G. Loew, MD, that WR "*probably* suffers from Paranoid Schizophrenia." That same day, US Attorney Peter Mills sends letter to the Chief of the Justice Department's Administrative Regulations Section, informing him that "I had no reason to doubt the mental compe-

Wilhelm Reich Chronology

tency of these defendants to understand the proceedings against them..." and sends copies to Joseph L. Maguire, Assistant General Council for the FDA, and the Warden at Danbury. Officially, now, Reich is no longer a charlatan but a psychotic. **March 22:** After prison psychiatrist diagnoses WR as a paranoiac, WR is transferred to Lewisburg Federal Penitentiary, C Cell Block, near Harrisburg, Pennsylvania. Silvert remains at Danbury. **March 24:** WR passes sixtieth birthday, alone in prison. He is sent dozens of written greetings, but these are not passed to him by prison officials because of regulations about quarantine. **April 18:** In Lewisburg, Dr. Dominick J. Lacovara, Dr. L.A. Witkin, Dr. R. Gatski reverse earlier diagnosis of WR as "paranoiac' and declare him mentally competent, although "Reich could easily have a frank break with reality, and become psychotic, particularly if the stresses and environmental pressures become overwhelming." To find out, they release WR to general population in Lewisburg Prison. **April 16:** On the first day he is able, WR requests to be married to Aurora Karrer in prison, and is told his "chances are slight." **April 19:** WR is assigned to prison library, where he checks out books for prisoners. **June:** Attorney Roy St. Lewis prepares a petition for clemency from the President. **June 6:** WR seeks counsel from Rev. Silber, Protestant prison chaplain on how to "revive the experience of God (=Cosmic Energy") within the devilish organism. This would be far more beneficial than to eternally have to expel the devil into a hell as lasting as heaven." (Silber reports that WR attends prison chapel services "without fail.") **July 11:** WR grants Aurora Karrer power-of-attorney. **July 12:** WR transfers $4400 in traveler's checks to Aurora Karrer. **August 17:** sale of 1952 Ford pickup truck used in WR's mobile cloudbusting operations in Tucson. **September 2:** National Guard seize Central High School in Little Rock, Arkansas to enforce school desegregation. **September 14:** WR contacts Rev. Silber on the danger of "Emotional Poisoning... a la Little Rock racism." **September 25:** WR's prison guard rates his work in the prison library as "fair" and laments, "because of his legal work, he seems preoccupied." **October 1:** WR files request to staff member form, asking that he be informed as the "source and kind" of money being deposited in his prison account. **October 9:** WR petition to President for clemency is denied. Warden discusses this with WR and later notes, "he expressed deep disappointment." **October:** Russians successfully launch Sputnik. **October 31:** WR's Parole Progress Report is filed. **November:** Sputnik II explodes on the launch pad. **November 3:** Wilhelm Reich is found dead in his cell at the 7 am head count in Lewisburg Federal Penitentiary, seven days short of his parole eligibility. Dr. Lacovara arrives at 7:15 am and takes WR's corpse to the morgue. At 9:20 am, Eva Reich is notified of WR's death. The Warden tells her that an autopsy should be performed. Eva opposes this, but will call back when she has had a chance to speak with her family. Interim, the Warden calls the Director of the Bureau of Prisons in Washington, DC, and Dr. Janney, who discuss the matter and agree that Warden Taylor will order an autopsy. In shock and sadness, WR's immediate family, according to Eva, decide not to have an autopsy performed. She phones the Warden, telling him "He probably died of a broken heart." At 2:15 pm, Dr. Lavocara observes Dr. Willard Christman, a pathologist, conducting the autopsy. In Lacovara's opinion, WR died of "coronary insufficiency with calcified aortic stenosis and generalized arteriosclerosis." All "specimans [sic] sent to the National Institute of Health at request of Dr. Janney." **November 4:** J.C. Taylor, Warden, files WR's autopsy report with the Director of the Bureau of Prisons. **November 5:** WR's corpse arrives home to Rangeley, Maine. **November 6:** Prison officials send FBI WR's death notice. **November 10:** WR's parole hearing date. **November 19:** L.L. Ashburn, MD (chief of HEW's Section on Pathological Anatomy) advises Lewisburg Prison's chief medical officer that he has no facilities to check the stomach contents of WR's remains. Microscopic studies "reveal marked myocardial scarring and a severe bronchiopneumonia." **December 12:** Michael Silvert released from Danbury Federal Prison.

1958 Soviets seed clouds with tracer bullets impregnated with silver iodide. Psychologist Carl Jung publishes *Flying Saucers; a Modern Myth of Things Seen in the Sky* (Zurich). Nelson Rockefeller becomes Governor of New York. Alaska becomes 49th state. **January 14:** Senator Lyndon B. Johnson in a CBS radio address urges the United States "to demonstrate its initiative before the United Nations by inviting all member nations to join

Wilhelm Reich Chronology

in this adventure into outer space together." **January 16:** Secretary of State Dulles proposes the formation of an international commission to insure the use of outer space exclusively for peaceful purposes. **February 7:** The Advanced Research Projects Agency (ARPA), placed in charge of the US outer space program. **February 27:** Willy Brandt, Mayor of Berlin, writes statement on behalf of his old friend Karl Frank, to obtain a travel visa. **March 26:** Military telephone and telegraph system using the troposphere to bounce radio signals over long distances, called "White Alice," is activated. **April 2:** In a message to Congress, President Eisenhower proposes the establishment of a National Aeronautics and Space Agency (NASA). **April 17:** Simulated seven-day trip to the moon made by six Navy men in chamber at Philadelphia Naval Base. **May 1:** Scientific findings from the two Explorer satellites discloses an unexpected band of high-intensity radiation extending from 600 miles above earth to possibly an 8,000-mile altitude. The radiation is described by James A. Van Allen as "1,000 times as intense as could be attributed to cosmic rays." **May:** Michael Silvert commits suicide in Cortland Park, New York City. **August 25:** first international colloquium on space law, in The Hague. **June 26:** Robert Donner, American industrialist, having offered Viktor Schauberger $15,000 contract to develop his theory of implosive propulsion (anti-gravity) in the US, brings the aged Schauberger to Sherman, Texas. **August 27:** The first Argus experiment is conducted by ARPA, in which a small A-bomb is detonated in space. Launched from the rocketship *Norton Sound*, the initial flash is followed by an auroral luminescence extending upward and downward along the magnetic lines where the burst occurred. **August 30:** The second Argus small A-bomb detonation beyond the atmosphere in the South Atlantic. **September 2:** U.N. Ambassador Henry Cabot Lodge announces that United States would propose a plan for international cooperation in the exploration of outer space to the United Nations. **September 4:** President Eisenhower appoints Detlev W. Bronk president of the National Academy of Sciences; William A. M. Burden; James H. Doolittle; and Alan T. Waterman, Director of the NSF, to the National Aeronautics and Space Council. Additionally, the Space Council including the Administrator of NASA, the Secretary of

Defense, the Secretary of State, and the Chairman of the AEC as statutory members. **September 6:** The third of the Argus small A-bomb detonations beyond the atmosphere. **September 25:** Victor Schauberger, discoverer of "living water," dies in Linz, Austria, his entire life work stolen by American industrialist Robert Donner. **December 19:** President Eisenhower's Christmas message from PROJECT SCORE satellite in orbit, the first voice beamed in from space.

1959   W. S. Burroughs publishes *Naked Lunch*. Saul Bellow publishes *Henderson the Rain King*. Jerome Eden's *Suffer the Children*. **January 1:** Fidel Castro takes power in Cuba. **February 17:** USAF Committee, lead by J. Allen Hynek, Associate Director of the Smithsonian Astrophysical Observatory at Cambridge, Massachusetts, recommends that the USAF continue to take a positive approach to UFO's, investigate reported sightings by all scientific means, and keep the public fully informed of existing policy. Of the unknown objects sighted, it reports, no scientific evidence supports the conclusion that the objects were spacecraft. **March 31:** *True* magazine writer Carlton Brown writes to Warden of Lewisburg Penitentiary, inquiring if WR's "diagnosis" as a psychotic had been verified and/or treated at the prison. Brown later pens "The Marvelous Sex Box of Dr. Reich" for *True*. **May 4:** National Bureau of Standards releases details on the effect on the ionosphere of high-altitude nuclear shots. **September:** Khrushchev visits US and is turned away from Disneyland for security reasons. **September 18(?):** Eva Reich transfers trusteeship of the Wilhelm Reich Infant Trust Fund to Mary Boyd Higgins. **October 1:** Attorney for Mary Boyd Higgins, Leonard Kolleeny, seeks information from Warden at Lewisburg about materials sent to Aurora Karrer by WR before his death in prison. **October 16:** Lee Harvey Oswald defects to Moscow. **November 9:** Entire outer Van Allen radiation belt broken up, and disappeared for several days. **November 28:** During severe geomagnetic storm, two Geiger tubes on EXPLORER VII found anomalies in the outer radiation zone at about 1,000-km altitude, which appeared to be correlated in space and time with optical emissions from the atmosphere below. Very intense narrow zones of radiation were detected over a visible aurora during one orbit. **December 20:** Melvin

Calvin reports that molecules in meteorites resembled basic constituents of genetic material found on earth.

1960 First American medical orgonomist, William F. Thorburn, dies. Charles Kelley's "Interscience Research Institute" and *A New Method of Weather Control* appear. Ralph Markson's CORE operations in Los Angeles with Baker and Duvall. Despite the federal injunction, WR's posthumous *Selected Writings* appears, a compendium of his scientific work compiled by Baker, Raphael, and Higgins. Oral birth control pill is introduced. **May 1:** Gary Powers goes down over Russia in U-2 spyplane. **November:** JFK defeats Richard Nixon.

1961 *Function of the Orgasm* republished. *The Creative Process* appears. **January 20:** JFK inauguration. **April 6:** Corliss Lamont, son of Tom Lamont, senior Morgan Bank partner, founds Fair Play for Cuba Committee founded. **April 15-17:** Bay of Pigs. **July 4:** General Douglas MacArthur speaks of "harnessing the cosmic energy" and "of ultimate conflict between a united human race and the sinister force of some other planetary galaxy." **August:** Soviets erect Berlin Wall. **November 29:** John McCone succeeds Allen Dulles, who was fired by JFK as CIA director.

1962 Thalidomide is discovered to have caused birth defects. Russia and US renew atmospheric nuclear testing; US explodes 36 nuclear bombs in this year alone. Oranur Research Laboratory founded by Richard Blasband, MD. *The Sexual Revolution* is republished. **July(?):** James McDonald of the University of Arizona confirms his own sighting of a UFO over Tucson, to the press: "just another unidentified object." From the observatory atop the Institute of Atmospheric Physics building, in Tucson, Arizona, he and his family sees "an orange-yellow luminous object. It is round, and also, for a period, it appeared to be elongated downward in sort of a conical shape." **September:** Rachel Carson's *Silent Spring* appears. **October 22:** Cuban Missile Crisis.

1963 Medgar Evers shot. *Character Analysis* republished. Orson Bean starts orgone therapy with Elsworth Baker. **January 10:** Nicholas Elliot confronts Kim Philby as a KGB agent in Beirut, Lebanon. Philby is allowed to escape and defect to Moscow. **February:** Tracy Barnes (Michael Straight's first cousin) becomes head of CIA's Domestic Operations Division. **April 4-6:** General Charles Willoughby's extremist group, Congress of Freedom, Inc., meets in New Orleans. **May:** JFK asks Straight to head the Fine Arts Commission, he declines. **April 27:** Robert A. Brady dies of complications from a stroke, suffered after an earlier suicide attempt. **June:** JFK asks Straight to chair a new agency, the Advisory Council on the Arts; he tentatively accepts a vague position. **June:** Straight learns that his new appointment, Chairman of the new National Endowment for the Arts, will require an FBI background check. After sleeping it over he tells the whole story to White House aide, Arthur Schlesinger. Straight wants to tell his story to the FBI, and asks Schlesinger to remove his name from consideration as chairman of NEA. Schlesinger calls Attorney General Robert F. Kennedy to make the arrangements. The next day, in his role as an art lover, Straight goes to the FBI and makes a statement to Cartha de Loach and William Sullivan, identifying Anthony Blunt as a Soviet agent. Blunt confesses in exchange for immunity - the deal is not made public until 1979. **July 1:** British announce that Kim Philby has defected to Moscow. **July 19:** Navy Yeoman Nelson Drummond is convicted of spying for the Soviets. **July:** Straight flies to London to confront Blunt. Straight informs MI5 that Leo Long was an agent. **November 22:** JFK assassination.

1964   Warren Commission publishes a novel. Publication of Paul Ritter's *Orgonomic Functionalism* ceases. **January 24:** *Time* covers sex and morality in the US with a gratuitous and pornographic defamation of WR. **December 21:** Dorothy Kilgallen mentions Pentagon "mind control and cloudbusting" in syndicated column. US begins official military action in Vietnam.

1965   Bombing of North Vietnam begins. *Listen Little Man* is republished. Orgonomic Research Foundation is incorporated. Massive power-outage along Northeastern seaboard, coincident with UFO sighting at power stations. Malcolm X assassination. **July 27:** Mildred Edie Brady dies. Henry Wallace dies.

1966   *Murder of Christ* is republished. *Die Sexuelle Revolution* appears in Germany. **June 6:** University of Arizona Professor of Atmospheric Physics, James E. McDonald, declares that official statements about UFO sightings being nothing more than well-known atmospheric discharges ("globular lightning," etc.) were "pure bullshit." William W. Vickers and James F. Church, working for the US Air Force, develop a dry ice seeding machine called the Cloudbuster. James E. McDonald becomes a member of the Advisory Panel for ESSA-Navy Project STORMFURY. CIA officer Tracy Barnes (Straight's first cousin) dies. Willy Brandt becomes Vice Chancellor and Foreign Minister of West German Christian/Socialist coalition government. Gajdusek transfers "slow virus" Kuru from humans to chimpanzees.

1967 US Department of Agriculture approves irradiation of food products. First heart transplant. Stalin's daughter, Svetlana, flees to the US. Robert Oppenheimer dies. Bone-thin Twiggy is top fashion model. *Journal of Orgonomy* begins publishing. Baker's *Man in the Trap,* and Reich's *Reich Speaks of Freud* are published. Students at Strasbourg University demand French publication of WR's "The Sexual Struggle of Youth" with university funds, causing a scandal. **Spring:** The student resident associations at French Universities at Strasbourg, Lyon, Nantes and Toulouse publish an explicitly "Reichian" tract, *La Misere de la Sexologie et la Sexology de la Misere.*

1968 Women toss cosmetics, bras, girdles into "freedom trash can" in Atlantic City to protest the Miss America Pageant. Jackie Kennedy marries Aristotle Onassis. **January:** Tet Offensive in Vietnam. **May:** occupations of Nanterre, Sorbonne Universities trigger a wildcat general strike in France; the crisis erupts throughout the world. MLK and RFK assassinations. Riots at Chicago Democratic Party Convention. American College of Orgonomy is founded.

1969 Richard Nixon becomes president. Eisenhower dies. Ilse Ollendorff's biography of WR is published. **July 20:** Armstrong walks on the moon. Willy Brandt becomes Chancellor of West Germany. **November:** 700,000 anti-war protesters converge on Washington DC. WR's works begin to be published in Germany in widely-distributed editions by Kiepenhauer & Witsch, then Fischer-Taschenbücher, with print runs over 100,000 copies.

1970 First Earth Day. **April 5:** Willi Schlamm publishes article on WR, "the Father of the Sexual Revolution." Solzyenitsin wins Nobel. Andropov orders Hitler's remains be destroyed. DeGaulle dies. *Mass Psychology of Fascism* is republished. Raknes' *Wilhelm Reich and Orgonomy* is published.

1971   Willy Brandt wins Nobel Prize for Peace, Pablo Neruda for Literature. Berta Bornstein dies. Kubrick's *Clockwork Orange* premieres. Makavejev's hippie film, *WR-The Mysteries of the Organism.* **Spring:** *Drive, He Said,* by Jack Nicholson, a "Reichian" film exploring youth rebellion and sexual politics. **January 4:** New York Times' Christopher Lehman-Haupt writes: "Perhaps it is time to reconsider all of Wilhelm Reich... and to reopen the question of cosmic orgone energy..." **June 13:** James E. McDonald dies in Tucson Arizona, apparent suicide. Orson Bean publishes *Me and the Orgone*." **July:** Henry Kissinger secretly visits China.

1972   Watergate arrests. Nixon defeats McGovern by a landslide. *The Invasion of Compulsory Sex-Morality* is republished. *Sex-Pol Essays* are published. Jerome Eden's *Orgone Energy*.

1973   Richard Nixon appoints Michael Straight to chair the National Endowment for the Arts (NEA). Straight's tenure lasts until 1981. *Roe v. Wade* legalizes abortion in US. Arab Oil embargo; "energy crisis." WR's son, Peter Reich, publishes *A Book of Dreams.* Jerome Eden publishes *Planet in Trouble; the UFO Assault on Earth.* The American Board of Medical Orgonomy is founded. *Ether, God and Devil/Cosmic Superimposition* is republished.

1974   Oil/Energy Crisis. Charles Bronson movie *Death Wish*. Willy Brandt resigns after a close aide is found to be an East German spy. Fort Lauderdale Conference on Weather Modification. *The Impulsive Character and Other Writings* is republished. *Eden Bulletin* appears (until 1981). **August 9:** Nixon resigns. Nelson Rockefeller becomes US Vice President. Claire Edward Petty suspects CIA's chief of counter-intelligence, James Jesus Angleton, of being a KGB mole. Angleton resigns Christmas Eve. **March 7:** Lewis Williams Douglas dies in Tucson, Arizona. **Fall:** Roger du Teil dies.

1975   Last American troops are evacuated from Vietnam. Pol Pot regime in Cambodia. Apollo-Soyuz Mission. Franco dies. Putsch in Portugal. Onassis dies. *Early Writings, Vol. 1* is republished. Bernd Laska's journal *Wilhelm-Reich-Blätter* appears in Germany.

1976   Trevor James Constable publishes *The Cosmic Pulse of Life; the Biological Power Behind UFOs.* England experiences worst

Wilhelm Reich Chronology

drought on record. Legionnaire's disease. Mao dies. Heisenberg dies. Soviet biologist Trofim Lyssenko dies. *People in Trouble* is republished. **May Day:** Committee for the Scientific Investigation of the Claims of the Paranormal (CSICOP) is founded by academic Old Leftists cloaked in the banner of "Humanism."

1977 First AIDS case diagnosed in New York City. US tests neutron bomb. Carter becomes president. Baader-Meinhof gang commits suicide. Red China eliminates Gang of Four, ending the Cultural Revolution. Wernher von Braun dies. *Newsletter for the Friends of the Wilhelm Reich Museum* appears.

1978 Jim Jones & People's Temple. Ramon Mercader, Trotsky's assassin, dies in Cuba. Nobel Prize for Sadat and Begin. Italian Prime Minister Aldo Moro is kidnapped and murdered by Red Brigades. Travolta's *Saturday Night Fever*. First issue of the Canadian *International Journal of Life Energy* (until 1981).

1979 Three Mile Island nuclear accident. **April 2:** Anthrax accident at "Compound 19" biowarfare unit in Sverdlosk, USSR kills a thousand people. **November 12:** Margaret Thatcher announces Anthony Blunt's confession in Parliament. Ronald Reagan, the first US president to have been divorced, is elected. *The Bion Experiments* is republished. Italian journal *Scienza Orgonomica* (until 1981). James DeMeo presents evidence from two years of field experiments to the Univeristy of Kansas, Geography-Meteorology Department, showing Reich's cloudbuster works to increase rains. DeMeo founds the Orgone Biophysical Research Laboratory.

1980 Experimental vaccine tested for hepatitis-B. Rise of Lech Walesa in Poland. Iraq-Iran war begins. Tito dies. Erich Fromm dies. *Early Writings, Vol. 2: Genitality* is republished. The journal *Offshoots of Orgonomy* appears (until 1987). German journal *Emotion* appears.

1981 Reagan assassination attempt by Manson Family. Space Shuttle launches. KGB General Naum Eitingon dies, Moscow. *Record of a Friendship* is published.

1982 FDA approves marketing of genetically engineered products. Arms race and peace movement. Brezhnev dies. Andropov successor. *The Bioelectrical Investigation of Sexuality and Anxiety* is republished. Institute for Orgonomic Science splits from the American College of Orgonomy.

1983 Myron Sharaf's biography of Wilhelm Reich, *Fury on Earth*, is published. Livermore Labs antinuclear/disarmament demonstrations. Peace Nobel Prize for Lech Walesa. Peace movement grows. *Children of the Future* is published.

1984 Flatland founded by James Martin. Peace Nobel Price for Desmond Tutu. Reagan announces Strategic Defense Initiative. Andropov dies. Red China attacks Vietnam. Indira Gandhi assassination. Genetic engineering gains focus of attention. The first volume of *Annals of the Institute for Orgonomic Science*. Jerome Eden's *The Orgonomic Education Newsletter* appears.

1985 March: Mikhail Gorbachev ascends to power in Moscow. Kohl and Reagan in Bitburg. AIDS identified as a plague. Bhagwan is arrested. Elsworth F. Baker dies.

1986 Olof Palme assassination. Glasnost. Chernobyl. Hole in the ozone layer of Antarctica. The Gebauer-Müschenich double-blind and controlled study on the orgone accumulator is submitted to the University of Marburg, Germany, showing it has strong influences on human physiology. The journal *Sciences Orgonomiques* appears in France. James DeMeo presents his *Saharasia* cross-cultural study of over 1000 human societies to the University of Kansas, indicating variations in human behavior and culture can be explained on the basis of Reich's sex-economy, and that desert climates influence human behavior.

1987 Mathias Rust, 19 years old, lands his Cessna 172 in Red Square; Air-defense commander Koldunov is sacked. Senator Gary Hart drops out of US presidential race after adulterous affair made public. Robert Anton Wilson's *Wilhelm Reich in Hell*.

1988 *Passion of Youth* is published. Ex-CIA chief George Bush becomes president.

1989 **June 4:** Poland freely elects Solidarity government. **Summer:** USAF General (ret.) Curtis LeMay provides the use of his Riverside, California, home to Trevor James Constable for cloudbusting smog-removal work. **October 7:** Hungary renounces communism. **October 18:** East German Prime Minister Erich Honecker resigns. **November 10:** Berlin Wall comes down. James DeMeo's *Orgone Accumulator Handbook* is published. First issue of *Pulse of the Planet*. Günter Hebenstreit presents a double-blind controlled study to the University of

Wilhelm Reich Chronology

Vienna, showing the orgone accumulator has a strong influence on human physiology.

1990 **August 19:** Yanayev, Pugo, Yazov and 3 others announce take-over in Moscow. The coup fails. Nelson Mandela is freed from South African prison. First volume *Orgonomic Functionalism* with Mary Boyd Higgins as editor.

1991 US and Allies bomb Iraq. James DeMeo leads a cloudbusting expedition in Israel, ending a major drought.

1992 **March 16:** WR's prison records released to researcher Kenn Thomas as per FOI/PA request. Andrew Collins's *The Circlemakers*. William Jefferson Clinton becomes President.

1993 **July:** Andrew Collins' Orgone 93, England. Richard Blasband and Barbara Koopman leave the ACO.

1994 **May:** *Jones v Clinton* files sexual harassment lawsuit. **June:** Nelson Mandela becomes president of South Africa. Rwandan genocide. Andrew Collins' Orgone 94, England. *Beyond Psychology* is published. Meyerowitz's *Before the Beginning of Time*.

1995 Yitzhak Rabin assassination. *Space Ships, DOR and Drought* is published in Germany. Müschenich's medical dissertation, *Der Gesundheitsbegriff im Werk des Arztes Wilhelm Reich (1897-1957)*. **July:** White House hires a new intern: Monica Lewinsky.

1996 **May:** First Greensprings Conference, Orgone Biophysical Lab. France declares last nuclear test. Clinton re-election.

1997 Plans for dealing with "Y2K" computer bug. WR's Centenary: talk by Lore Reich at the German Goethe Institute in Boston, Massachusetts, **March 21**, about Reich, Anna Freud, and Reich's expulsion from the IPA. Larry Warren and Peter Robbins publish *Left at East Gate*, documenting military use of cloudbusters at Bentwaters AFB in England.

1999 Günter Hebenstreit presents his study verifying Reich's bioelectrical experiments on human emotion and sexual excitation to the University of Vienna. End of a 5-Year cloudbusting project led by James DeMeo and sponsored by the State of Eritrea, Africa, indicating major increases in regional rains, with new lakes appearing in the Sahara Desert.

Wilhelm Reich and the Cold War

Bibliography

[Note: Reich's books can be obtained directly from the Wilhelm Reich Museum Bookstore, P.O. Box 687, Rangeley, Maine 04970]

Books by Wilhelm Reich
——. *Passion of Youth; An Autobiography, 1897-1922.* Edited by Mary Boyd Higgins. New York: Farrar, Straus and Giroux, 1st edition, 1988.
——. *Early Writings: Volume One.* Translated by Philip Schmitz. New York: Farrar, Straus and Giroux, 1st printing, 1975.
——. *The Impulsive Character and Other Writings.* Translated by Barbara G. Koopman. New York: New American Library, 1st printing, October 1974.
——. *Genitality: In The Theory and Therapy of Neurosis; Early Writings Volume Two.* Translated by Philip Schmitz. Edited by Mary Higgins and Chester M. Raphael. New York: Farrar, Straus and Giroux, 1st printing, 1980.
——. *The Invasion of Compulsory Sex-Morality.* New York: Noonday, 1st printing, 1971.
——. *The Sexual Revolution:* New York, Farrar, Straus & Giroux, 1964.
——. *The Mass Psychology of Fascism.* 1946. Reprint: New York: Farrar, Straus and Giroux, 5th printing, 1973.
——. *Sex-Pol Essays, 1929-1934.* Edited by Lee Baxandall. New York: Random House, 1st edition, 1972.
—— and Karl Teschitz (Karl Motezcisky). *Selected Sex-Pol Essays 1934-37.* London: Socialist Reproduction, 1973.
Reich, Wilhelm. *The Sexual Revolution.* 1945. Reprint: New York: Farrar, Straus and Giroux, 1st paperback edition, 1974.
——. *The Bioelectrical Investigation of Sexuality and Anxiety.* Translated by Marion Farber with Derek and Inge Jordan. Edited by Mary Higgins and Chester M. Raphael, 1st printing, 1982.
——. *The Bion Experiments: On The Origin of Life.* 1938. Translated by Derek and Inge Jordan. Edited by Mary Higgins and Chester M. Raphael. New York: Octagon, 1979.

———. *Beyond Psychology, Letters and Journals 1934-1939*. Edited by Mary Boyd Higgins. New York: Farrar, Straus and Giroux, 1st edition, 1994.

———. *The Function of the Orgasm: Sex-Economic Problems of Biological Energy*. 1948. Translated by Vincent R. Carfagno. New York: Farrar, Straus and Giroux, 1973.

———. *The Cancer Biopathy*. 1948. Translated by Andrew White with Mary Higgins and Chester M. Raphael. New York: Farrar, Straus and Giroux, 1st printing, 1973.

———. *Listen, Little Man!* 1948. Translated by Ralph Manheim. Illustrations by William Steig. New York: Farrar, Straus and Giroux, 1978.

———. *American Odyssey, Letters and Journals 1940-1947*. Edited by Mary Boyd Higgins. New York: Farrar, Straus and Giroux, 1st edition, 1999.

———. *Ether, God and Devil* and *Cosmic Superimposition*. With five chapters translated by Therese Pol. New York: Farrar, Straus and Giroux, 1st printing, 1973.

———. *Children of the Future: On The Prevention of Sexual Pathology*. New York: Farrar, Straus and Giroux, 1st printing, 1983.

———. *The Murder of Christ: Volume One of the Emotional Plague of Mankind*. New York: Touchstone, 1953.

———. *People In Trouble*. Vol. 2 of The Emotional Plague of Mankind. Translated by Philip Schmitz. New York: Farrar, Straus and Giroux, 1st printing, 1976.

———. *Contact With Space: Oranur Second Report 1951-1956*. New York: Core Pilot Press, 1957.

———. *Selected Writings*. New York: Farrar, Straus and Cudahy, 1st printing, 1960.

———. *Reich Speaks of Freud*. Edited by Mary Higgins and Chester M. Raphael, M.D. New York: Noonday Press, 1st printing, 1968.

———, and A. S. Neill. *Record of a Friendship: The Correspondence of Wilhelm Reich and A. S. Neill 1936-1957*. Edited by Beverly R. Placzek. New York: Farrar, Straus and Giroux, 1981.

Biographies about, Books referring to Reich

Boadella, David. *Wilhelm Reich; The Evolution of His Work*. Chicago: Henry Regnery Company, 1973.

DeMeo, James. *In Defense of Wilhelm Reich*. Ashland, OR: Natural Energy Works, 2013.

Bibliography

Goodman, Paul. *Nature Heals: The Psychological Essays of Paul Goodman.* Edited by Taylor Stoehr. New York: E. P. Dutton, 1979.
Herskowitz, Morton. Emotional Armoring. New York: Transactions Press, 1998.
Greenfield, Jerome. *Wilhelm Reich vs. The United States.* New York: W. W. Norton & Company, Inc., 1st edition, 1974.
Mann, W. Edward. *Orgone, Reich and Eros: Wilhelm Reich's Theory of Life Energy.* New York: Simon and Schuster, 1973.
Mann, W. Edward, and Edward Hoffman. *The Man Who Dreamed of Tomorrow: A Conceptual Biography of Wilhelm Reich.* Boston: J. P. Tarcher, Inc., 1980.
Raknes, Ola. *Wilhelm Reich and Orgonomy.* New York: St. Martin's Press, 1970.
Reich, Ilse Ollendorff. *Wilhelm Reich: A Personal Biography.* London: Elek Books Limited, 1969.
Reich, Peter, *A Book of Dreams.* New York: Harper & Row, 1st edition, 1973.
Sharaf, Myron. *Fury On Earth: A Biography of Wilhelm Reich.* London: Hutchinson & Company, 1984.

Government Files Obtained through the Freedom of Information Act
Brady, Mildred Edie and Robert Alexander: BUFILE #100-3368
Straight, Michael: BUFILE # 100-61929
Reich, Wilhelm: BUFILE # 100-14601
US Bureau of Prison Files: Wilhelm Reich, prisoner # 23937-NE

Videos
An Interview with Myron Sharaf. Fort Bragg, CA: Flatland, 1995.
In Conversation with Eva Reich, MD: Wilhelm Reich & the UFO Phenomena: Arizona Desert 1954-55. Fort Bragg, CA: Flatland, 1995.
An Interview with Eva Reich, MD: Gentle Bio-energetics: "Prevention of Neurosis From Birth On." Fort Bragg, CA: 1995.

The Science of Orgonomy
Baker, Elsworth. *Man In The Trap.* New York: Collier Books, 1980.
DeMeo, James. *The Orgone Accumulator Handbook.* 3rd Revised Ed.: Natural Energy Works, 2010.

———. *On Wilhelm Reich and Orgonomy* (*Pulse of the Planet* No. 4). Ashland, OR: Orgone Biophysical Research Lab, 1993.

———. *Saharasia*. Ashland, OR: Natural Energy Works, 2nd Revised Edition, 2006.

——— (Editor). *Heretic's Notebook* (*Pulse of the Planet* No. 5). Ashland, OR: Orgone Biophysical Research Lab, 2002.

———. *Preliminary Analysis of Changes in Kansas Weather Coincidental to Experimental Operations with a Reich Cloudbuster.* Republication of a 1979 University of Kansas study. Ashland, OR: Natural Energy Works, 2010.

Eden, Jerome. *Orgone Energy: The Answer To Atomic Suicide.* New York: Exposition, 1st edition, 1972.

———. *Planet In Trouble: The UFO Assault on Earth.* New York: Exposition, 1st edition, 1973.

———. *Animal Magnetism and the Life Energy.* Hicksville, NY: Exposition Press, 1st edition, 1974.

———. *The Emotional Plague versus Orgonomic Ufology.* Careywood, ID: 1981.

———. *Scavengers from Space: the UFO Hostility Hypothesis.* Careywood, ID: 1988.

———. *Earth of Trial: The Reich Court Case.* 2nd edition, Careywood, ID: 1988.

Hebenstreit, Günter. *Der Orgonakkumulator Nach Wilhelm Reich.* Diplomarbeit, Universität Vien. 1995.

———. *Die Elektrophysiologie erogener und nicht-erogener Hautzonen - Eine experimentelle Untersuchung.* Dissertation, Universität Vien. 2010.

Maglione, Roberto. *Methods and Procedures in Biophysical Orgonometry.* Moncrivello, Italy: 2012.

Maglione, Roberto & Mazzocchi, Alberto: *Biofisica e Medicina Orgonica.* Palermo, Italy: Nuova IPSA, 2013.

Müschenich, Stefan & Gebauer, Rainer. *Der Reich Orgonakkumulator.* Frankfurt, Germany: Nexus Verlag, 1987.

Müschenich, Stefan. *Der Gesundheitsbegriff im Werk des Arztes Wilhelm Reich (1897-1957).* Marburg (Germany): Görich & Weiershäuser, 1995.

Bibliography

The Anglo-American Establishment
Bird, Kai. *The Chairman: John J. McCloy and the Making of the American Establishment.* New York: Simon & Schuster,1992.
Josephson, Emanuel Mann. *Rockefeller, "Internationalist;" The Man Who Misrules the World.* New York: Chedney Press, 1952.
Quigley, Carroll. *Tragedy and Hope.* New York: MacMillan, 1966.
———. *The Anglo-American Establishment: From Rhodes to Cliveden.* New York: Books in Focus, 1981.
Reich, Cary. *Rockefeller: The Life of Nelson A. Rockefeller Worlds To Conquer 1908-1958.* New York: Doubleday, 1st edition, November 1996.
Yergin, Daniel. *The Prize: The Epic Quest for Oil, Money and Power.* New York: Simon & Schuster, 1991.

Psychoanalysis
Deutsch, Helene. *Confrontations With Myself.* New York: W. W. Norton, 1973.
Etkind, Alexander. *Eros of the Impossible: The History of Psychoanalysis in Russia.* Boulder, CO: Westview Press, 1997.
Fallend, Karl. *Wilhelm Reich in Wien.* Vienna: Geyer-Edition, 1988.
Fallend, Karl and Johannes Reichmayr (ed.). *Siegfried Bernfeld oder Die Grenzen der Psychoanalyse.* Materialien zu Leben und Werk, Frankfurt: Stroemfeld/Nexus, 1992.
Jacoby, Russell. *The Repression of Psychoanalysis: Otto Fenichel and the Political Freudians.* Chicago, 1983.
Roazan, P. and B. Swerdloff, eds. *Heresy: Sandor Rado and the Psychoanalytic Movement.* Latham, MD: Jason Aronson, Inc., 1977.
Sterba, Richard. *Reminiscences of a Viennese Psychoanalyst.* Wayne State University Press, 1982.

Sexpol
Gruber, Helmut. *Red Vienna: Experiment in Working-Class Culture 1919-1934.* Oxford: 1991.
Anson Rabinbach. *Austrian Socialism. From Red Vienna to Civil War 1927-1934,* The University of Chicago Press, 1983.
Rackelmann:, Marc. "Wilhelm Reich und der Einheitsverband für proletarische Sexualreform und Mutterschutz: Was war die Sexpol?" *Emotion. Beiträge zum Werk von Wilhelm Reich*, No. 11, Berlin, 1994.

Wilhelm Reich and the Cold War

The Cambridge Five

Borovik, Genrikh and Phillip Knightley. *The Philby Files.* New York: Little and Brown, 1994.

Brown, Anthony Cave. *Treason in the Blood: H. St. John Philby, Kim Philby, and the Spy Case of the Century.* New York: Houghton Mifflin, 1994.

Costello, John. *Mask of Treachery.* London: Collins, 1988.

Knightley, Phillip. *The Master Spy: The Story of Kim Philby.* New York: Alfred Knopf, 1989.

Modin, Yuri. *My Cambridge 5 Friends: Burgess, Maclean, Philby, Blunt, and Cairncross by Their KGB Controller.* Translated by Anthony Roberts. New York: Farrar, Straus and Giroux, 1994.

Penrose, Barrie and Simon Freeman. *Conspiracy of Silence: The Secret Life of Anthony Blunt.* New York: Vintage Books, 1988.

Philby, Kim. *My Silent War.* London: Granada Publishing, 1980.

Straight, Michael. *After Long Silence.* Maine: Thorndike Press, 1981. (large print edition)

West, Nigel and Oleg Tsarev. *The Crown Jewels: The British Secrets at the Heart of the KGB Archives.* New Haven: Yale University Press, 1999.

West, William J. *Spymaster: the Betrayal of MI5.* New York: Wynwood Press, 1990.

WWII-era Emigre Politics

Gardiner, Muriel. *Codename "Mary": Memoirs of an American Woman in the Austrian Underground.* New Haven: Yale University Press, 1983.

Hoffman, Peter. *The History of the German Resistance 1933-1945.* Boston: MIT Press, 1977.

The Politics of Science

Brown, Richard E. *Rockefeller Medicine Men: Medicine & Capitalism in America.* Berkeley: University of California Press, 1979.

Infeld, Leopold. trans. Helen Infeld. *Why I Left Canada: Reflections on Science and Politics.* McGill-Queen's University Press, Montreal 1978.

Lynes, Barry. *The Cancer Cure That Worked! Fifty Years of Suppression.* Ontario: Marcus Books, 1987.

Zinsser, Hans. *Rats, Lice and History.* New York: Black Dog & Leventhal Publishers, 1934, 1935, 1963.

Bibliography

The Secret Life of Mildred Brady
Brady, Joan. *The Unmaking of a Dancer.* New York: Harper & Row, 1982. (novel.)
Brady, Joan. *Theory of War.* New York: Knopf, 1993. (novel.)
Brady, Robert A. *The Spirit and Structure of German Fascism.* Forward by Harold J. Laski. New York: Viking Press, 1st edition, 1937.
Chevalier, Haakon Maurice. *Oppenheimer: The Story of a Friendship.* New York: Braziller, 1965.
Goodman, Walter. *The Committee.* New York: Farrar, Straus and Giroux, 1964.
Kuczynski, Jürgen. *Freunde und gute Bekannte. Gespräche mit Thomas Grimm.* Berlin: Schwarzkopf & Schwarzkopf, 1997.
Matthews, J. B. *Odyssey of a Fellow Traveler.* New York: Mount Vernon Pub., 1938.
Wertham, Frederic. *Seduction of the Innocent.* New York: Rinehart and Company, Inc., 1953.

Soviet Espionage
Albright, Joseph and Marcia Kunstel. *Bombshell: The Secret Story of America's Unknown Atomic Spy Conspiracy.* New York: Times Books, 1st edition, 1997.
Andrew, Christopher, and Vasili Mitrokhin. *The Sword and the Shield*: *The Mitrokhin Archive and the Secret History of the KGB.* New York: Basic Books, 1999.
Dziak, John J. *Chekisty: a History of the KGB.* Lexington, 1988.
Haynes, John Earl and Harvey Klehr. *VENONA: Decoding Soviet Espionage in America.* New Haven: Yale University Press, 1999.
Klehr, Harvey and John Earl Haynes, Fridrikh Igorevich Firsov. *The Secret World of American Communism.* New Haven: Yale, 1995.
Klehr, Harvey and Ronald Radosh. *The Amerasia Case: Prelude To McCarthyism.* UNC Press, Chapel Hill, 1996.
Koestler, Arthur. *The Invisible Writing.* Boston: Beacon Press, 1st edition, 1955.
Koestler, Arthur, et al. *The God That Failed.* Edited by Richard Crossman. New York: Harper & Row, 1st edition, 1963.
"Personnel Security Board, Atomic Energy Commission: Hearings of J. Robert Oppenheimer."
Schwartz, Stephen. *From East to West: California and the Making of the American Mind.* New York: The Free Press, 1998.

Sudoplatov, Pavel and Anatoli; with Jerrold L. and Leona Schecter. *Special Tasks: the Memoirs of an Unwanted Witness-a Soviet Spymaster.* New York: Little, Brown, 1994.
Tannenhaus, Sam. *Whittaker Chambers.* New York: Random House, 1997.
Weinstein, Allen and Alexander Vassiliev. *The Haunted Wood: Soviet Espionage in America-The Stalin Era.* New York: Random House, 1st edition, 1999.
Wolf, Markus. *The Man Without a Face,* New York: Public Affairs, 1997.
Willoughby, Charles A. *The Shanghai Conspiracy: The Sorge Spy Ring.* E. P. Dutton, New York, 1952.

UFO
Corso, Philip and William Birnes. *The Day After Roswell.* New York: Pocket Books, 1st printing, 1997.
Friedman, Stanton. *Top Secret Majic.* New York: Marlowe & Co., 1996.
Robbins, Peter. "Wilhelm Reich and UFOs." *Journal of Orgonomy,* 24:2 and 25:1, Princeton: 1990, 1991.
Vallee, Jacques. *Forbidden Science Journals 1957-1969.* Berkeley: North Atlantic, 1992.
Warren, Larry and Peter Robbins. *Left At East Gate.* New York: Marlowe & Co., 1st edition, 1997.

US Intelligence Agencies
McIntosh, Elizabeth. *Sisterhood of Spies: the Women of the OSS.* Annapolis: Naval Institute Press, 1998.
Meyer, Cord. *Facing Reality: From World Federalism to the CIA.* New York: Harper and Row, NY, 1980.
Russell, Dick. *The Man Who Knew Too Much.* New York: Carroll & Graf, 1992.
Smith, R. Harris. *OSS: The Secret History of America's First Central Intelligence Agency.* Berkeley: University of California Press, 1972; Dell reprint 1973.
Thomas, Gordon. *Journey Into Madness: The True Story of Secret CIA Mind Control and Medical Abuse.* New York: Bantam, 1989.
Winks, Robin, *Cloaks and Gown: 1931-1961, Scholars in the Secret War.* New York: Morrow & Co., 1987.

Bibliography

Weather

The Arizona Daily Star 1952-1964

Battan, Louis J. *Harvesting the Clouds: Advances in Weather Modification*. New York: Doubleday, 1969.

Breur, Georg. *Weather Modification: Prospects and Problems*. Translated by Hans Mörth. Cambridge: University Press, 1980.

Fleagle, Robert Guthrie, et al. *Weather Modification in the Public Interest*. Washington: University of Washington Press, with the American Meteorological Society, 1974.

Halcy, D. S. *The Weather Changers*. New York: Harper & Row, 1968.

Humphreys, W. J. *Rain Making and Other Weather Vagaries*. Baltimore: The Williams & Wilkins Company, 1926.

——. *Weather Proverbs and Paradoxes*. Baltimore: The Williams & Wilkins Company, 1934.

MacDonald Commission of the US Academy of Sciences: 1966. (From the report, *Weather and Climate Modification - Problems and Prospects;* "World Weather Authority - a step toward world government?")

Maglione, Roberto. *Wilhelm Reich and the Healing of Atmospheres*. Ashland, OR: Natural Energy Works, 2007.

Mueller, et al. "Preprints of the Fort Lauderdale Conference on Weather Modification," 1974.

Oroville, H.T. *Final Report of the Advisory Committee on Weather Control, Vol 1 & 2*. Washington, Government Printing Office, 1957.

Powers, Edward. *War and the Weather*. Chicago: 1871, 2nd ed. reprint Delavan, WI: 1890.

Shapley, D. "Rainmaking: Rumored Used Over Laos Alarms Arms Experts, Scientists," *Science* 176 (June 16, 1972)

Taubenfeld, Howard Jack, ed. *Controlling the Weather: A Study of Law and Regulatory Procedures*. New York: Dunellen Co., 1970.

Resources for Information about Wilhelm Reich

American College of Orgonomy
Journal of Orgonomy
P.O. Box 490
Princeton, N.J. 08542
On the Web: www.orgonomy.org

Institute for Orgonomic Science
Annals of the Institute for Orgonomic Science
205 Knapp Road
Lansdale, PA 19446
On the Web: www.orgonomicscience.org

Orgone Biophysical Research Laboratory
Pulse of the Planet
Ashland, Oregon, USA
On the Web: www.orgonelab.org

Orgonics
produces & sells orgone accumulators
On the Web: www.orgonics.com

Wilhelm Reich Museum and Bookstore
Orgonomic Functionalism
P.O. Box 687
Rangeley, Maine 04970
On the Web: www.wilhelmreichtrust.org

Index

A

Adler, Alfred 342
Adler, Friedrich 40, 42, 79, 254, 259, 344
Adler, Solomon 162
Adler, Viktor 254
Akhmerov, Iskhak ("Michael Green") 93-94, 119, 165, 185, 339
Altschuler, Lydia 93
Amerasia Case 77, 94, 131, 133-135, 137, 152, 159, 186, 368
American Civil Liberties Union (ACLU) 119, 355, 361, 382
American College of Orgonomy (ACO) 389, 391, 393, 404
American Federation of Labor (AFL) 44, 119
American Friends for German Freedom 44-45, 48, 76
American Friends Service Committee 45
American Medical Association (AMA) 15, 210
American Psychiatric Association (APA) 20, 210, 275-377
Annals, Orgone Institute 370
Annals, Institute of Orgonomic Science 392, 404
Angleton, James Jesus 100, 127, 163, 181, 374-375, 380, 390
Arbejderbladet (Danish newspaper) 354, 356, 362
Ascoli, Max 138, 157
Atomic Energy Commission (AEC) 31, 224, 269, 271, 369, 374, 381, 401
"Atoms for Peace" 297, 380

B

Baldwin, Roger 119, 151, 362
Bauer, Otto 28, 35-36, 40, 76, 360
Bazhanov, Boris 88
Bendix, Robert 40
Bentley, Elizabeth 165, 244-246, 303, 372
Bentley, Eric 75
Beria, Lavrenty 94, 262, 370, 374, 376-377
Bernfeld, Siegfried 19, 22-32, 75-76, 342, 348, 351-354, 399
Bettauer, Hugo 347-348
Blankenship, Joe ix, 201, 298-301, 314
Blunt, Anthony 100, 116-119, 121, 129-131, 145, 151-152, 154, 156, 159, 182-183, 227, 358-359, 362, 365, 373, 388, 391, 400
Boehm, Felix 85, 96, 355
Bornemann, Ernest 79
Brady FBI Files ix, 4, 329-333
Brady, Joan ix, 217, 230-241, 251, 401
Brady, Mildred Edie ix, 4, 15, 20-22, 37, 45, 57, 71, 74, 92. 95, 99, 110, 126, 131, 144, 146-147, 151-153, 158, 162, 167, 184, 188-189, 207-252, 271, 303-304, 319, 329-333, 358, 360, 366, 370-372, 388, 397, 401
Brady, Robert A. 21-22, 45, 71, 74, 92, 144, 146-147, 151-153, 188-189, 207-252, 271, 303, 319, 321, 340, 351-352, 358, 360-361, 363, 366, 375, 388, 397, 401
Braisford, H.N. 41

405

Brandt, Otto Knobel 221, 318
Brandt, Willy 35-36, 44, 49-55, 58-60, 78-79, 343, 355, 357, 359, 361, 363-364, 368, 381, 385, 389-390
Brunauer, Esther Caukin 42
Bry, Gerhard (aka Paul Kemp) 41
Buckley, William F. 19, 350, 374
Bukarin, Nikolai 26, 66, 69, 362
Burgess, Guy 92, 100, 108, 116-117, 121, 124, 127-131, 150-152, 159, 182, 227, 358, 359, 363, 365, 373, 375, 400
Bursler, Norman Chandler 227-229, 236-237, 247, 321
Buttinger, Joseph 35-38, 40, 42, 76

C

Cameron, Ewen 275-276, 340, 367, 376-377, 382
Caples, Anna 41-42, 361
Carnegie Foundation 111, 219, 221, 248, 370
Carsten, Franz 41
Chambers, Whittaker 57, 162, 245, 372, 402
Chevalier, Haakon 223-224, 237, 248, 251, 303, 363, 401
Childs, Richard Storrs 42
Clay, Lucius D. 279
cloudbuster 16, 177, 192, 198, 204, 267, 272-275, 277, 284-289, 291-295, 314, 316, 377-380, 383, 388-389, 391-393, 398
cloudseeding 198-199, 278, 282-285, 291, 293, 373, 380
"Comrade Thomas" (see Rubinstein, Arnold Thomas)
Condon, Edward U., & *Condon Report* 217, 244, 302-304, 314
Congress of Industrial Organizations (CIO)
Consumers' Advisory Board 218
Consumers' League 340
Consumers' Research ix, 151, 190, 212-215, 217-218, 220-221, 230-231, 238, 241, 249-250, 303, 358
Consumers' Union & *Consumers' Reports* magazine 15, 93, 95, 99, 151, 190, 210, 212-213, 215, 224-227, 229-230, 232-233, 236-238, 241-242, 248-250, 358
Cornford, John 117-118, 122, 148-149, 361
Constable, Trevor James 201, 289-290, 293, 314, 390, 392
Corcoran, Tom 122, 134, 368-369
Corso, Philip 294, 314, 402
Cosmic Orgone Engineering (see cloudbuster)
Council on Foreign Relations 110, 280, 327
Cripps, Stafford 36, 150
Currie, Lauchlin 38, 131, 244, 372

D

DeMeo, James ix, xi, 275, 316, 391-393, 396-397
Derzhinsky, Felix 115-116
Deutsch, Arnold 75, 103-108, 118, 145, 159-163, 182-183, 340, 346-348, 350, 352, 358, 364, 399
Deutsch, Helene 47, 78
Dies, Martin & Dies Committee 214-215, 230-231, 241, 265
Dolivet, Louis 138-145, 155-157
Donovon, William "Wild Bill" 36, 47-49
Douglas, Lewis W. ix, 4, 42, 148, 194-196, 278-282, 284, 286, 291, 293, 295-298, 326-327, 343, 356, 376-377, 379-381, 390
Douglas-Moïse Memo 297, 326-327

Index

dowsing 300
Dreikurs, Rudolf 23
Dukas, Helen 219, 254-255, 258, 262
Dulles, Allen 45-46, 77, 111, 134, 187, 280, 344, 376, 381-382, 387
Dulles, John Foster 111, 281-282, 327, 385
Dunbar, Helen Flanders 167
duTeil, Roger 360-361, 363, 390

E

Eden, Jerome 249, 286, 314, 348, 386, 390, 392, 398
Ehrman, Henry 42
Einstein, Albert 14, 42, 79, 122, 168, 219, 253-264, 322, 340-341, 343-346, 353, 362-366, 374, 376, 380
Eisenhower, Dwight D. ix, 4, 191, 193-195, 268, 277-278, 287, 297, 327, 376-377, 379-380, 385-386, 389
Eisler, Gerhart (aka Hans Berger) 23, 26-27, 74, 361
Eisler, Hanns 74
Eitingon, Max 32, 83-97, 182, 250, 345, 353, 367
Eitingon, Malthus 95
Eitingon, Motty 45, 77, 94-95
Eitingon, Naum (Leonid) 83-97, 182, 339, 365, 377, 391
Eitingon, Waldemar (Vladimir) 77, 90, 94-96, 345
Eliasberg, Stefan & Vera 42
Elices, Norbert 41
Elliot, John Lovejoy 42
Erler, Fritz 42, 59

F

Falk, Erling 49-54, 78, 364
Farago, Ladislas 138-143, 145
Feitelberg, Sergei 29-31, 75-76, 173
Fenichel, Otto 19, 23, 25, 27-28, 31-32, 37, 47, 49, 63-64, 71, 75, 81, 85, 163, 224, 315, 339, 345-347, 351-353, 357-359, 399
Fischer, Ruth (Eisler) 23, 26, 38, 60, 75, 367
Flechtheim, Ossip 42
Food and Drug Administration (FDA) viii, 15, 17, 110, 144, 151, 153, 191, 208, 210-211, 225, 229, 232-236, 248, 267, 304-306, 308, 315, 333-336, 337-338, 371-372, 375-377, 380-381, 383, 391
Fox, Mary 42
Frank, Karl Borromäus (aka Willi Müller, Paul Hagen) ix, 19-30, 32-47, 50, 52, 54, 56, 58-60, 62-64, 67-68, 70-71, 73-81, 103, 122, 179, 254, 324-325, 342, 345-348, 350, 353-354, 356-361, 364, 366, 370, 372, 385
Frank, Lew 113-114, 125
Freud, Anna 35, 393
Freud, Sigmund 1, 10-12, 22-24, 28-32, 36, 42, 46, 57-58, 75, 77, 79, 83-87, 89, 96, 166, 222, 273, 339-342, 345-356, 360, 362, 365, 369, 376
Frey, Josef 351
Friedländer, Paul 23
Friedman, Alice "Litzi" 101, 105-107, 163, 180, 297, 357, 360, 374
Friedman, Stanton 297, 301, 402`
Fuchs, Eduard 67
Fuchs, Klaus "Karl" 219, 373

G

Gaasland, Gertrude (Meyer) 44, 50, 343, 355, 364
Gardiner, Muriel 35, 42, 400
Gardner, Martin 375
Gayn, Mark Julius (aka Joseph Kramer) 94, 131-137, 152, 159, 166, 186, 368
Gebauer, Rainer 315, 392, 398
Geiringer, Hilde 23
Gillmor, D.S. 158, 215-217, 249, 302-304, 314
Goddard, James L. 232-233
Goddard, Robert H. 343, 351
Goldberg, Arthur 45
Goodman, Paul 74, 369, 374
Gottlieb, Sidney 377
Grad, Bernard ix, 276
Green, Michael (see Akhmerov, Iskhak)
Greenfield, Jerome ix, 304, 397

H

Harriman, Averill W. 279
Hays, Arthur Garfield ix, 31, 77, 94-96, 188, 225-227, 355, 368
Hart, Edith Tudor (Suschitsky) 105-106, 180, 350
Hazeland, John S. 53
Hebenstreit, Günter 316, 392-393, 398
Hegna, Trond 53
Held, Adolph 45
Henderson, Leon 231, 241
Hertz, Paul 39, 42, 45
Higgins, Mary Boyd 165, 249, 304-306, 386-387, 393
Hillenkoetter, Roscoe 198, 287, 376
Hiss, Alger 26, 111, 119-120, 122, 161, 228-229, 309, 358, 363, 368-370, 372-373

Hitler, Adolf 52, 78, 340, 345, 348, 352-353, 356, 368, 389
Hitler-Stalin Pact (Molotov-Ribbentrop Pact) 26, 115, 146, 149, 162, 216
Hoel, Sigurd 49, 52-54, 352
Hoover, Calvin 42, 50, 366
Hoover, J. Edgar 135-136, 152-154, 241-242, 246, 337
House Committee on Un-American Activities (HUAC) 27, 126, 139, 214, 227, 230-231
Huisken, David ix, 5

I

Infeld, Leopold 260-262
International Journal of Sex-Economy & Orgone Research 366, 369

J

Jacobson, Edith 19, 41, 47, 63-64, 71, 86, 259, 339, 349, 351, 353, 357, 359
Jaffe, Philip 133-134, 368
Janeway, Eliot 122
Jensen, Ruth Österreich 67, 71, 82, 187
Jeremias, Jerry 43
Journal of Orgonomy 389

K

Kahn, Lore 24-25, 345-346
Kallet, Arthur 212, 220, 237, 358
Kellogg, Paul 42
Koestler, Arthur 19, 26, 41-42, 48, 77, 139, 142, 368, 401
Kopp, Victor 90-91
Kramer, Joseph (see Gayn, Mark Julius)
Krivitsky, Walter 81-82, 120, 142, 346, 363, 365

Index

L

Langmuir, Irving 199, 283-284, 343, 370
Larson, Emmanuel 368
Lash, Joe 150, 167
Laski, Harold J. 250
LeMay, Curtis 201, 292-293, 368-369, 392
Lenin 25-26, 28-29, 41, 53, 61-62, 64, 66-67, 95, 115-116, 165, 244, 340-342, 344, 346-347, 352-353
Lerner, Alba 41
Leunbach, Jonathan Hoegh 357, 360
Lindenberg, Elsa 12, 14, 63, 341, 353, 356, 358
Loeb, Jim 41
Löwenthal, Richard (aka Paul Serling) 41, 77

M

MacArthur, Douglas 127, 290, 314, 327, 365, 387
Maclean, Donald 99-100, 108, 127-130, 154, 159, 182, 227, 400
Maironovsky, Grigori Moisevich 94
Malinowski, Bronislaw 79, 350-351, 356, 366
Marcuse, Herbert 40
Marshall, Thurgood 42
Masters, Dexter 212-213, 229-230, 235-237, 239, 250
Mattech, Karl 42
Mattesh, Kurt 42
Matthews, J.B. 189, 214-215, 226-227, 238, 241, 401
Matusow, Harvey "Job" ix, 200, 309, 311, 315, 374

McCarthy, Joseph, era & hearings xi, 77, 114, 134, 137-138, 153-154, 167, 200, 208, 214-216, 237-238, 309, 374, 378
McCloy, John Jay 167, 195, 268, 278-280, 327-328, 343, 377
McCullough, Robert C. ix, 203-204, 288-291, 377, 379
McCullough, Phyllis ix
McDonald, James E. ix, 283-286, 297, 303, 314, 346, 379, 387-390
Mellon, Richard 111, 138
Mendershausen, Horst 42
Menninger, Karl (and Clinic) 15, 49
Mercader, Caridad & Ramón 93, 391
Meyer, Cord, 157-158, 167, 184, 402
Meyer, Gertrude see Gassland, Gertrude
Meyer, Mary Pinchot 157, 167, 184
Miller, Yevgeni Karlovich 87-90
Mills, Peter 15, 207, 226, 308, 337-338, 378, 380, 382
MJ-12 295, 297-298, 314, 378
Mitterand, Francois 139
MKULTRA 274-276, 340, 376
Moïse, Bill 287-288, 291, 295-297, 326-327, 375, 379, 382
Mosfurco (Moscow Fur Company) 94-95
Mot Dag 49-53, 78, 358-359, 364, 368
Motesiczky, Karl 61, 80, 340, 355, 367
Münzenberg, Willi 41, 70, 140, 142, 155
Müschenich, Stefan ix, 316, 392-393, 398

N

Nagell, Richard Case 135-136, 152
The Nation magazine 32, 57, 149
Nasselstein, Peter ix, 5, 19, 78
National Review magazine 19, 56-57, 350, 374
National Security Agency (NSA) 3, 93, 119, 127, 228, 245-247, 294, 319, 329-330, 376
Neill, A.S. 112, 147, 164, 256, 343, 346, 348, 361, 362
New Beginning (see Org)
New Republic magazine 21, 76, 99, 109-115, 121-123, 125-126, 133, 144, 146, 151, 153, 156-158, 165, 167, 184, 210, 213, 215, 225-227, 235, 248, 250, 315, 322, 343, 370-371, 378, 380-381
New School for Social Research 138, 365-366
Niebuhr, Reinhold 32, 42, 45
Nicole, Leon 140
Nixon, Richard 27, 167, 299, 372, 387, 389, 390

O

Office of Price Administration 158, 212, 223, 230, 241
Office of Strategic Services (OSS) 20, 22, 34, 40, 42, 45-48, 50, 52, 60, 75-78, 122, 128, 133, 187, 280, 295, 366, 368-369, 402
Ollendorff, Ilse ix, 14, 22, 43-44, 49, 56-57, 64, 74, 364, 368, 375, 378-379, 389, 397
Ollendorff, Robert 43
Oppenheimer, J. Robert 189, 223-225, 237, 246, 248, 251, 271, 303, 313, 373-374, 378, 389, 401
Oranur Experiment 15, 165, 189, 265, 268-271, 277, 287-288, 291, 374-375, 379-380, 382, 387, 396
Ording, Arne & Aake 53
Org (New Beginning) 19-82, 254, 346, 352-353, 358-359, 364
orgone energy & accumulator 201, 211, 255, 306, 316, 372, 381
Orgone Biophysical Research Lab ix, 276, 391, 393
Orgone Energy Bulletin 289, 373
Orgone Institute 14, 44, 105, 336-337, 366
Orgonomic Infant Research Center 16, 373
Orgonon xii, 14, 176, 211, 230, 272, 277, 300, 302, 304-305, 333, 369, 371-377, 380-381
Little Orgonon xii, 192, 380
Ossietzky, Carl Von 52, 56
Österreich, Ruth (aka Ruth Gebhardt) 67, 71, 82
Øverland, Arnulf 49, 52, 54

P

Palmer, Dewey 230, 241
Petty, Claire Edward 390
Philby, Kim 35, 75, 99-108, 117-118, 127, 129-131, 147, 154, 159, 163, 180-182, 214, 249, 355, 357-358, 360, 363, 365, 367, 373-375, 380, 387, 388, 400
Philipson, Tage 355, 360
Pink, Annie (see Reich, Annie)
Plevitskaya, Nadyezhda 87, 89

Q

Quigley, Carroll 108-111, 113-115, 123-126, 155-157, 164, 166, 185, 280

Index

R

Radek, Karl 25, 66, 69
Rado, Emmy 19, 45-49, 76-77, 128, 187, 151, 366-367
Rado, Peter 45
Rado, Sandor (psychoanalyst) 19, 45-48, 64, 77, 86, 187, 351-352, 366
Rado, Sandor (cartographer) 77
Reich, Annie (Pink) 11-12, 19, 23, 47, 57, 64, 70-72, 105, 163, 165, 173, 340, 346-350, 352, 354, 356, 359
Reich, Cäcilie & Leon 7-9, 339, 341-342
Reich, Eva ix, 4, 17, 71-72, 82, 105, 165, 174, 205, 251, 266, 270, 286-287, 289, 291, 299-300, 302, 306-307, 311, 314, 347, 349, 352, 354, 359, 364, 375, 384, 386, 397
Reich, Lore ix, 7, 11, 64-65, 67, 70-71, 81, 174, 349, 352, 354, 393
Reich, Peter 14, 45, 291, 368, 390
Reich, Wilhelm basic bio: 7-17
and AEC 269, 271
and Communists 56, 60, 103-105, 144, 159-160, 208-209, 229
and Einstein 253-264
and Emmy Rado 19, 46-47
and Freud 10-13, 31
and Karl Frank 19, 21-22, 25, 44
and Kim Philby 108
and Michael Straight 203-205
and the Org 19-82
and Revolutionary Social-Democrats 102, 351
and Siegfried Bernfeld 22, 25, 27, 30-32
and UFOs 267, 287-288, 293-294, 297, 300-301
and US military 268-269, 271, 287-288, 292
destruction of literature 232-233
expulsion from IPA 63, 84-86
Infant Trust Fund 16, 304, 386
legal issues 225-227, 304-307
Oslo associates of 49-52
in prison 307-312
"psychosis" of 1, 21, 28, 47, 159, 230, 250, 304, 307, 368
Reich Museum 176, 258, 306, 404
Schlamm's memories of 57-58
"Strange Case of Wilhelm Reich" 99, 126, 131, 145, 158, 184, 210, 235, 248, 379
Willy Brandt's memories of 55
Reisner, Konrad 43
Robbins, Peter ix, 293-294, 314, 393, 402
Rockefeller, John D. 222
Rockefeller, family, Foundation & Institute 101, 111, 143, 164, 251, 255, 279-281, 307, 356, 361, 377
Rockefeller, Nelson 113, 133, 138, 156, 168, 315, 368, 382, 384, 390
Roosevelt, Eleanor 37, 47, 74, 111-112, 119-121, 149-150, 167, 189, 271, 340, 359, 362, 364, 370, 374-375, 382
Roosevelt, Franklin D. & Administration 14, 32, 44, 49, 112, 119, 122, 134, 160, 230, 232, 241, 254-255, 279, 327-328, 368
Roosevelt, Theodore 148, 157
Roth, Andrew 131
Roth, Sam 309-311
Rothschilds 109, 116-117

411

Roswell, Arizona 294, 298-301, 351, 371, 376, 380, 402
Rubinstein, Thomas Arnold (aka Comrade Thomas) 25-26, 60, 64-67, 69-72, 79, 81, 165, 187, 340, 346, 359
Ruppelt, Edward J. 302-303, 314, 381
Rusk, Dean 111, 281

S

Sandersen, Fred & Elli 43
Schaefer, Vincent 199, 283, 370
Scheidemann, Philipp 344
Schjelderup, Harald 357
Schlamm, William S. "Willi" 19, 56-60, 79, 81-82, 103, 346-347, 350, 367, 374, 389
Schlink, Frederick J. 213, 217-218, 221, 227, 231, 238, 241, 250, 303
Schmidt, Erich 20, 38, 43
Schmidt, Karl 38-40
Schmidt, Fritz 42
Schmidt, "Ted" 42
Schoettle, Otto 41-42
Schwartz, Stephen 86
Scitovsky, Tibor 41
Sellers, William ix, 286
Service, John Stewart 131, 368
sex-economy 12
Sexpol 12, 22, 56, 58, 61-63, 79-80, 101, 104-105, 147, 347, 350, 352-353, 355, 357-360, 399
Seydewitz, Max 28, 50, 77, 354
Sharaf, Myron 3, 20-22, 25, 64, 74, 210, 266, 304, 367, 370, 374-375, 379
Siersted, Ellen 63
Silver, George 42
Silvermaster, Helen 229, 244, 248, 319

Silvermaster, Nathan Gregory 162, 189, 211, 228-229, 244-248, 303, 315, 319-321, 339, 343, 352
Silvert Michael 17, 291-292, 305-306, 333-334, 336-338, 380-385
Skoblin, Nikolai 87, 89
Sloan Foundation or Sloan-Kettering 280-282, 377
Sømme, Axel 53
Soule, George 151, 167
Sperber, Manes 19
Stalin, Josef (Djugashvilli) 26, 28, 83, 88, 92, 94, 106, 118, 120, 140, 152, 210, 226, 262, 327, 341-342, 347, 350, 363, 365, 367-368, 372, 376, 381, 389
Steig, William 382. 396
Straight, Michael Whitney ix, 74, 76, 94, 99-100, 108, 110, 112-131, 134, 137-139, 142-162, 165, 183-186, 214, 227, 249-250, 280, 303, 340, 344, 357-366, 368, 371, 373, 376, 378, 380-381, 387-390, 397
Straight, Willard & Dorothy 74, 111-115, 139, 213, 341, 343-345, 348
Sudoplatov, Pavel 88-89, 91-94, 313, 341

T

Taurer, Bernard 41
Teschitz, Karl 61-62
Thiele, Theo 42
Thjötta, Theodor 361-362
Thomas, Kenn ix, 5, 258, 296, 393
Thomas, Norman 42
Tippett, Tom 241
Tranmäl, Martin 53

Index

Trotsky, Leon 25-26, 52, 61, 64, 66-67, 83, 87-88, 90-93, 95, 166, 318, 344, 346, 348, 350-351, 353, 355-356, 359-361, 365, 377, 391
Truman, Harry 113-115, 122-127, 139, 156-157, 166-168, 369-376
Twining, Nathan 294, 297-298, 378-380

U

UFO 4, 16, 192, 267, 273, 285-290, 292-294, 297-298, 300-303, 307, 309, 314, 366-367, 374-380, 386-390, 402
Ullstein, Leopold 42

V

Vacor (VACuum ORgone) Tubes 372
VENONA 38, 77, 119, 127, 162, 164-165, 228, 245, 249, 319-321, 329, 370
Vogt, Johan 51, 53
Von Knoeringen, Waldemar 39, 41-42
Vonnegut, Bernard 283
Vonnegut, Kurt 282

W

Walcher, Jacob 50
Wälder-Hall, Jennie 121, 165
Wallace, Henry 21, 113-115, 122-127, 138, 143, 147, 153, 156-158, 164-167, 186, 303, 322-323, 356, 364, 368, 370-373, 388
Warren, Larry 293-294
Washington, William 266, 370, 372
Weather Control (see cloudseeding)
Wehner, Herbert 362

Weimar Republic 20, 30, 69, 94, 221, 254, 344
Wertham, Frederic 99, 162, 370
Wilder, John ix, 5
Willoughby, Charles A. 365, 370, 375, 388
Wolfe, Theodore 14, 110, 167, 226, 259, 363, 372, 378
Wood, Charles A. 211, 249, 371
Wyvell, Lois ix, 249

Z

Zetkin, Clara 67
Zinoviev, Grigori 25, 66-67, 69, 345, 348

Wilhelm Reich and the Cold War

New Title, on a Similar Topic:

In Defense of Wilhelm Reich: Opposing the 80-Years' War of Mainstream Defamatory Slander Against One of the 20th Century's Most Brilliant Physicians and Natural Scientists, by James DeMeo, PhD.
269 pages. Illustrated.

Dr. Wilhelm Reich is the man whom nearly everyone loves to hate. No other figure in 20th Century science and medicine could be named who has been so badly maligned in popular media, scientific and medical circles, nor so shabbily mistreated by power-drunk federal agencies and arrogant judges.

Publicly denounced and slandered in both Europe and America by Nazis, Communists and psychoanalysts, placed on both Hitler's and Stalin's death lists but narrowly escaping to the USA, subjected to new public slanders and attacks by American journalists and psychiatrists who deliberately lied and provoked an "investigation" by the US Food and Drug Administration (FDA), imprisoned by American courts which ignored his legal writs and pleas about prosecutorial and FDA fraud, denied appeals all the way up to the US Supreme Court, which rubber-stamped the FDA's demands for the *banning and burning of his scientific books and research journals,* and finally dying alone in prison – who was this man, Wilhelm Reich, and why today, some 50 years after his death, does he continue to stir up such emotional antipathy? It is a literal *80-Years' War* of continuing misrepresentation, slander and defamation.

Who were and are Reich's attackers? Author and Natural Scientist James DeMeo takes on the book-burners, exposing with clarity and documentation their many slanderous fabrications, half-truths and lies of omission. In so doing, he also summarizes the lesser-known facts about Reich's important clinical and life-energetic experimental findings, now verified by scientists and physicians worldwide, and holding great promise for the future.

www.ingramcontent.com/pod-product-compliance
Lightning Source LLC
Chambersburg PA
CBHW032014230426
43671CB00005B/78